METAL IONS IN BIOLOGICAL SYSTEMS

VOLUME 22

ENDOR, EPR, and Electron Spin Echo for Probing Coordination Spheres

METAL IONS IN BIOLOGICAL SYSTEMS

Edited by

Helmut Sigel
Institute of Inorganic Chemistry
University of Basel
CH-4056 Basel, Switzerland

with the assistance of Astrid Sigel

VOLUME 22
ENDOR, EPR, and Electron Spin Echo for Probing
Coordination Spheres

MARCEL DEKKER, INC.   New York and Basel

Library of Congress Catalog Number: 79-640972

COPYRIGHT © 1987 by MARCEL DEKKER, INC.

ALL RIGHTS RESERVED

Neither this book nor any part may be reproduced or transmitted in any form or by any means, electronic or mechanical, including photocopying, microfilming, and recording, or by any information storage and retrieval system, without permission in writing from the publisher.

MARCEL DEKKER, INC.
270 Madison Avenue, New York, New York 10016

ISSN: 0161-5149
ISBN: 0-8247-7641-0

Current printing (last digit):
10 9 8 7 6 5 4 3 2 1

PRINTED IN THE UNITED STATES OF AMERICA

Preface to the Series

Recently, the importance of metal ions to the vital functions of living organisms, hence their health and well-being, has become increasingly apparent. As a result, the long-neglected field of "bioinorganic chemistry" is now developing at a rapid pace. The research centers on the synthesis, stability, formation, structure, and reactivity of biological metal ion-containing compounds of low and high molecular weight. The metabolism and transport of metal ions and their complexes is being studied, and new models for complicated natural structures and processes are being devised and tested. The focal point of our attention is the connection between the chemistry of metal ions and their role for life.

No doubt, we are only at the brink of this process. Thus, it is with the intention of linking coordination chemistry and biochemistry in their widest sense that the series METAL IONS IN BIOLOGICAL SYSTEMS reflects the growing field of "bioinorganic chemistry." We hope, also, that this series will help to break down the barriers between the historically separate spheres of chemistry, biochemistry, biology, medicine, and physics, with the expectation that a good deal of the future outstanding discoveries will be made in the interdisciplinary areas of science.

Should this series prove a stimulus for new activities in this fascinating "field," it would well serve its purpose and would be a satisfactory result for the efforts spent by the authors.

Fall 1973
                                              Helmut Sigel
                              Institute of Inorganic Chemistry
                                          University of Basel
                                    CH-4056 Basel, Switzerland

Preface to Volume 22

The preceding Volume 21 was devoted to the "Applications of Nuclear Magnetic Resonance to Paramagnetic Species." The present volume carries over to related physical methods which are also based on paramagnetic species, namely, the electron nuclear double resonance (ENDOR), electron paramagnetic resonance (EPR), and electron spin echo (ESE) techniques. The past few years have shown that there is also an enormous potential in employing these methods to probe the coordination sphere of metal ions in biological systems. Recent advances are summarized in this volume with the hope of facilitating wider applications of the new developments.

The potential of ENDOR in probing the coordination environment of metal ions is outlined in the first chapter with the focus on metalloproteins. The next three chapters deal with EPR: Chapter 2 is devoted to a newly developed method that takes advantage of the line-broadening effect of ^{17}O introduced into nucleotide phosphate groups on the EPR spectrum of Mn^{2+} (used as an analog for Mg^{2+}) providing regio- and stereospecific information about metal-nucleotide complexes. Chapter 3 deals with recently developed algorithms that allow the analysis of line intensities (spin concentrations) and the determination of stability constants even in complicated systems with species occurring only in minor concentrations. Chapter 4 concerns the saturation of a paramagnetic spin system with incident microwave power, which has been widely applied in EPR studies; the review focuses on the structural analysis of paramagnetic metal ion sites in proteins. The fifth and final chapter deals with the ESE method, a pulsed version of EPR spectroscopy: the analysis of the ESE envelope modulation opens up new possibilities for studying properties and structures of paramagnetic species in biological systems.

Helmut Sigel
Institute of Inorganic Chemistry
University of Basel
CH-4056 Basel, Switzerland

Contents

PREFACE TO THE SERIES	iii
PREFACE TO VOLUME 22	v
CONTRIBUTORS	ix
CONTENTS OF OTHER VOLUMES	xi

Chapter 1

ENDOR: PROBING THE COORDINATION ENVIRONMENT IN METALLOPROTEINS   1

    Jürgen Hüttermann and Reinhard Kappl

1.	Introduction	1
2.	ENDOR Theory	4
3.	ENDOR Instrumentation and Techniques	13
4.	Survey of Results from Model Complexes and Metalloproteins	19
5.	Concluding Remarks	73
	Abbreviations	75
	References	76

Chapter 2

IDENTIFICATION OF OXYGEN LIGANDS IN METAL-NUCLEOTIDE-PROTEIN COMPLEXES BY OBSERVATION OF THE Mn(II)-^{17}O SUPERHYPERFINE COUPLING   81

    Hans Robert Kalbitzer

1.	Introduction	81
2.	Theory	83
3.	Applications	90
4.	Conclusions and Outlook	99
	Abbreviations and Definitions	100
	References	101

Chapter 3

USE OF EPR SPECTROSCOPY FOR STUDYING SOLUTION EQUILIBRIA   105

    Harald Gampp

1. Introduction   105
2. Analysis of EPR Line Intensities   107
3. Examples   112
4. Conclusions   124
   Abbreviations and Definitions   125
   References   126

Chapter 4

APPLICATION OF EPR SATURATION METHODS TO PARAMAGNETIC
METAL IONS IN PROTEINS   129

    Marvin W. Makinen and Gregg B. Wells

1. Introduction   130
2. Relaxation Theory   131
3. Methods for Collecting Electron Spin-Lattice
   Relaxation Data   144
4. Microwave Power Saturation Studies of Paramagnetic
   Metal Ion Centers in Proteins and Small-Molecule
   Complexes   159
   Abbreviations   199
   References   199

Chapter 5

ELECTRON SPIN ECHO: APPLICATIONS TO BIOLOGICAL SYSTEMS   207

    Yuri D. Tsvetkov and Sergei A. Dikanov

1. Introduction   207
2. ESEEM Spectroscopy   213
3. Biological Applications of ESEEM   240
4. Conclusion   257
   Abbreviations   258
   References   259

AUTHOR INDEX   265

SUBJECT INDEX   279

Contributors

Numbers in parentheses indicate the pages on which the authors' contributions begin.

Sergei A. Dikanov Institute of Chemical Kinetics and Combustion, Academy of Sciences, Siberian Branch, Novosibirsk 630090, USSR (207)

Harald Gampp Institut de Chimie, Université de Neuchâtel, Avenue de Bellevaux 51, CH-2000 Neuchâtel, Switzerland (105)

Jürgen Hüttermann Fachrichtung Biophysik und Physikalische Grundlagen der Medizin, Universität des Saarlandes, D-6650 Homburg (Saar), FRG (1)

Hans Robert Kalbitzer Max-Planck-Institute for Medical Research, Department of Molecular Physics, Jahnstrasse 29, D-6900 Heidelberg, FRG (81)

Reinhard Kappl Fachrichtung Biophysik und Physikalische Grundlagen der Medizin, Universität des Saarlandes, D-6650 Homburg (Saar), FRG (1)

Marvin W. Makinen Department of Biochemistry and Molecular Biology, Cummings Life Science Center, The University of Chicago, Chicago, IL 60637, USA (129)

Yuri D. Tsvetkov Institute of Chemical Kinetics and Combustion, Academy of Sciences, Siberian Branch, Novosibirsk 630090, USSR (207)

Gregg B. Wells Department of Biochemistry and Molecular Biology, Cummings Life Science Center, The University of Chicago, Chicago, IL 60637, USA (129)

Contents of Other Volumes

Volume 1. Simple Complexes*

Volume 2. Mixed-Ligand Complexes*

Volume 3. High Molecular Complexes*

Volume 4. Metal Ions as Probes*

Volume 5. Reactivity of Coordination Compounds

1. THE FORMATION OF SCHIFF BASES IN THE COORDINATION SPHERE OF METAL IONS
    Daniel L. Leussing

2. THE ROLE OF DIVALENT METAL IONS IN PHOSPHORYL AND NUCLEOTIDYL TRANSFER
    Barry S. Cooperman

3. METAL ION-CATALYZED DECARBOXYLATIONS OF BIOLOGICAL INTEREST
    R. W. Hay

4. METAL ION-PROMOTED HYDROLYSIS OF AMINO ACID ESTERS AND PEPTIDES
    R. W. Hay and P. J. Morris

5. CATALYTIC ACTIVITY OF POLY-L,α-AMINO ACID-METAL ION COMPLEXES: NEW APPROACHES TO ENZYME MODELS
    Masahiro Hatano and Tsunenori Nozawa

6. REACTIONS OF MOLYBDENUM COORDINATION COMPOUNDS: MODELS FOR BIOLOGICAL SYSTEMS
    Jack T. Spence

7. INTERACTION OF Cu(I) COMPLEXES WITH DIOXYGEN
    Andreas D. Zuberbühler

AUTHOR INDEX-SUBJECT INDEX

---

*Out of print

Volume 6. Biological Action of Metal Ions

1. ZINC AND ITS ROLE IN ENZYMES
    Jan Chlebowski and Joseph E. Coleman

2. VANADIUM IN SELECTED BIOLOGICAL SYSTEMS
    Wilton R. Biggs and James H. Swinehart

3. THE CHEMISTRY OF BIOLOGICAL NITROGEN FIXATION
    Peter W. Schneider

4. THE METAL ION ACCELERATION OF THE ACTIVATION OF TRYPSINOGEN TO TRYPSIN
    Dennis W. Darnall and Edward R. Birnbaum

5. METAL CHELATES IN THE STORAGE AND TRANSPORT OF NEUROTRANSMITTERS
    K. S. Rajan, R. W. Colburn, and J. M. Davis

6. THE ROLE OF DIVALENT METALS IN THE CONTRACTION OF MUSCLE FIBERS
    F. Norman Briggs and R. John Solaro

AUTHOR INDEX-SUBJECT INDEX

Volume 7. Iron in Model and Natural Compounds*

Volume 8. Nucleotides and Derivatives: Their Ligating Ambivalency

1. X-RAY STRUCTURAL STUDIES OF METAL-NUCLEOSIDE AND METAL-NUCLEOTIDE COMPLEXES
    Robert W. Gellert and Robert Bau

2. INTERACTIONS BETWEEN METAL IONS AND NUCLEIC BASES, NUCLEOSIDES, AND NUCLEOTIDES IN SOLUTION
    R. Bruce Martin and Yitbarek H. Mariam

3. THE AMBIVALENT PROPERTIES OF SOME BASE-MODIFIED NUCLEOTIDES
    Helmut Sigel

4. HEAVY METAL LABELING OF NUCLEOTIDES AND POLYNUCLEOTIDES FOR ELECTRON MICROSCOPY STUDIES
    Thomas R. Jack

5. MACROMOLECULES OF BIOLOGICAL INTEREST IN COMPLEX FORMATION
    S. L. Davydova

AUTHOR INDEX-SUBJECT INDEX

---

*Out of print

CONTENTS OF OTHER VOLUMES xiii

Volume 9. Amino Acids and Derivatives as Ambivalent Ligands

1. COMPLEXES OF α-AMINO ACIDS WITH CHELATABLE SIDE CHAIN DONOR ATOMS
   R. Bruce Martin

2. METAL COMPLEXES OF ASPARTIC ACID AND GLUTAMIC ACID
   Christopher A. Evans, Roger Guevremont, and
   Dallas L. Rabenstein

3. THE COORDINATION CHEMISTRY OF L-CYSTEINE AND D-PENICILLAMINE
   Arthur Gergely and Imre Sóvágó

4. GLUTATHIONE AND ITS METAL COMPLEXES
   Dallas L. Rabenstein, Roger Guevremont, and
   Christopher A. Evans

5. COORDINATION CHEMISTRY OF L-DOPA AND RELATED LIGANDS
   Arthur Gergely and Tamás Kiss

6. STEREOSELECTIVITY IN THE METAL COMPLEXES OF AMINO ACIDS AND
   DIPEPTIDES
   Leslie D. Pettit and Robert J. W. Hefford

7. PROTONATION AND COMPLEXATION OF MACROMOLECULAR POLYPEPTIDES:
   Corticotropin Fragments and Basic Trypsin Inhibitor
   (Kunitz Base)
   Kálmán Burger

AUTHOR INDEX-SUBJECT INDEX

Volume 10. Carcinogenicity and Metal Ions

1. THE FUNCTION OF METAL IONS IN GENETIC REGULATION
   Gunther L. Eichhorn

2. A COMPARISON OF CARCINOGENIC METALS
   C. Peter Flessel, Arthur Furst, and Shirley B. Radding

3. THE ROLE OF METALS IN TUMOR DEVELOPMENT AND INHIBITION
   Haleem J. Issaq

4. PARAMAGNETIC METAL IONS IN TISSUE DURING MALIGNANT DEVELOPMENT
   Nicholas J. F. Dodd

5. CERULOPLASMIN AND IRON TRANSFERRIN IN HUMAN MALIGNANT DISEASE
   Margaret A. Foster, Trevor Pocklington, and Audrey A. Dawson

6. HUMAN LEUKEMIA AND TRACE ELEMENTS
   E. L. Andronikashvili and L. M. Mosulishvili

7. ZINC AND TUMOR GROWTH
   Andre M. van Rij and Walter J. Pories

8. CYANOCOBALAMIN AND TUMOR GROWTH
   Sofija Kanopkaitė and Gediminas Braženas

9. THE ROLE OF SELENIUM AS A CANCER-PROTECTING TRACE ELEMENT
   Birger Jansson

10. TUMOR DIAGNOSIS USING RADIOACTIVE METAL IONS AND THEIR COMPLEXES
    Akira Yokoyama and Hideo Saji

AUTHOR INDEX-SUBJECT INDEX

Volume 11. Metal Complexes as Anticancer Agents

1. ANTITUMOR PROPERTIES OF METAL COMPLEXES
   M. J. Cleare and P. C. Hydes

2. AQUEOUS PLATINUM(II) CHEMISTRY; BINDING TO BIOLOGICAL MOLECULES
   Mary E. Howe-Grant and Stephen J. Lippard

3. CLINICAL ASPECTS OF PLATINUM ANTICANCER DRUGS
   Barnett Rosenberg

4. CARCINOSTATIC COPPER COMPLEXES
   David H. Petering

5. ONCOLOGICAL IMPLICATIONS OF THE CHEMISTRY OF RUTHENIUM
   Michael J. Clarke

6. METAL COMPLEXES OF ALKYLATING AGENTS AS POTENTIAL ANTICANCER AGENTS
   Melvin D. Joesten

7. METAL BINDING TO ANTITUMOR ANTIBIOTICS
   James C. Dabrowiak

8. INTERACTIONS OF ANTICANCER DRUGS WITH ENZYMES
   John L. Aull, Harlow H. Daron, Michael E. Friedman, and Paul Melius

AUTHOR INDEX-SUBJECT INDEX

CONTENTS OF OTHER VOLUMES

Volume 12. Properties of Copper

1. THE COORDINATION CHEMISTRY OF COPPER WITH REGARD TO BIOLOGICAL SYSTEMS
   R. F. Jameson

2. COPPER(II) AS PROBE IN SUBSTITUTED METALLOPROTEINS
   Ivano Bertini and A. Scozzafava

3. COPPER(III) COMPLEXES AND THEIR REACTIONS
   Dale W. Margerum and Grover D. Owens

4. COPPER CATALYZED OXIDATION AND OXYGENATION
   Harald Gampp and Andreas D. Zuberbühler

5. COPPER AND THE OXIDATION OF HEMOGLOBINS
   Joseph M. Rifkind

6. TRANSPORT OF COPPER
   Bibudhendra Sarkar

7. THE ROLE OF LOW-MOLECULAR-WEIGHT COPPER COMPLEXES IN THE CONTROL OF RHEUMATOID ARTHRITIS
   Peter M. May and David R. Williams

AUTHOR INDEX-SUBJECT INDEX

Volume 13. Copper Proteins

1. THE EVOLUTION OF COPPER PROTEINS
   Earl Frieden

2. PROPERTIES OF COPPER 'BLUE' PROTEINS
   A. Graham Lappin

3. THE PROPERTIES OF BINUCLEAR COPPER CENTERS IN MODEL AND NATURAL COMPOUNDS
   F. L. Urbach

4. CERULOPLASMIN: A MULTI-FUNCTIONAL METALLOPROTEIN OF VERTEBRATE PLASMA
   Earl Frieden

5. COPPER MONOOXYGENASES: TYROSINASE AND DOPAMINE β-MONOOXYGENASE
   Konrad Lerch

6. CYTOCHROME C OXIDASE: AN OVERVIEW OF RECENT WORK
   M. Brunori, E. Antonini, and M. T. Wilson

7. THE ACTIVE SITES OF MOLLUSCAN AND ANTHROPODAN HEMOCYANINS
   R. Lontie and R. Witters

8. THE COPPER/ZINC SUPEROXIDE DISMUTASE
   James A. Fee

9. THE CHEMISTRY AND BIOLOGY OF COPPER-METALLOTHIONEINS
   Konrad Lerch

10. METAL REPLACEMENT STUDIES OF BLUE COPPER PROTEINS
    Bennett L. Hauenstein, Jr. and David R. McMillin

AUTHOR INDEX-SUBJECT INDEX

Volume 14. Inorganic Drugs in Deficiency and Disease

1. DRUG-METAL ION INTERACTION IN THE GUT
   P. F. D'Arcy and J. C. McElnay

2. ZINC DEFICIENCY AND ITS THERAPY
   Ananda S. Prasad

3. THE PHARMACOLOGICAL USE OF ZINC
   George J. Brewer

4. THE ANTI-INFLAMMATORY ACTIVITIES OF COPPER COMPLEXES
   John R. J. Sorenson

5. IRON-CONTAINING DRUGS
   David A. Brown and M. V. Chidambaram

6. GOLD COMPLEXES AS METALLO-DRUGS
   Kailash C. Dash and Hubert Schmidbaur

7. METAL IONS AND CHELATING AGENTS IN ANTIVIRAL CHEMOTHERAPY
   D. D. Perrin and Hans Stünzi

8. COMPLEXES OF HALLUCINOGENIC DRUGS
   Wolfram Hänsel

9. LITHIUM IN PSYCHIATRY
   Nicholas J. Birch

AUTHOR INDEX-SUBJECT INDEX

Volume 15. Zinc and Its Role in Biology and Nutrition

1. CATEGORIES OF ZINC METALLOENZYMES
   Alphonse Galdes and Bert L. Vallee

CONTENTS OF OTHER VOLUMES   xvii

2. MODELS FOR Zn(II) BINDING SITES IN ENZYMES
   Robert S. Brown, Joan Huguet, and Neville J. Curtis

3. AN INSIGHT ON THE ACTIVE SITE OF ZINC ENZYMES THROUGH
   METAL SUBSTITUTION
   Ivano Bertini and Claudio Luchinat

4. THE ROLE OF ZINC IN DNA AND RNA POLYMERASES
   Felicia Ying-Hsiueh Wu and Cheng-Wen Wu

5. THE ROLE OF ZINC IN SNAKE TOXINS
   Anthony T. Tu

6. SPECTROSCOPIC PROPERTIES OF METALLOTHIONEIN
   Milan Vašák and Jeremias H. R. Kägi

7. INTERACTION OF ZINC WITH ERYTHROCYTES
   Joseph M. Rifkind

8. ZINC ABSORPTION AND EXCRETION IN RELATION TO NUTRITION
   Manfred Kirchgessner and Edgar Weigand

9. NUTRITIONAL INFLUENCE OF ZINC ON THE ACTIVITY OF ENZYMES
   AND HORMONES
   Manfred Kirchgessner and Hans-Peter Roth

10. ZINC DEFICIENCY SYNDROME DURING PARENTERAL NUTRITION
    Karin Ladefoged and Stig Jarnum

AUTHOR INDEX-SUBJECT INDEX

Volume 16. Methods Involving Metal Ions and Complexes in
           Clinical Chemistry

1. SOME ASPECTS OF NUTRITIONAL TRACE ELEMENT RESEARCH
   Clare E. Casey and Marion F. Robinson

2. METALS AND IMMUNITY
   Lucy Treagan

3. THERAPEUTIC CHELATING AGENTS
   Mark M. Jones

4. COMPUTER-DIRECTED CHELATE THERAPY OF RENAL STONE DISEASE
   Martin Rubin and Arthur E. Martell

5. DETERMINATION OF TRACE METALS IN BIOLOGICAL MATERIALS
   BY STABLE ISOTOPE DILUTION
   Claude Veillon and Robert Alvarez

6. TRACE ELEMENTS IN CLINICAL CHEMISTRY DETERMINED BY NEUTRON ACTIVATION ANALYSIS
   Kaj Heydorn

7. DETERMINATION OF LITHIUM, SODIUM AND POTASSIUM IN CLINICAL CHEMISTRY
   Adam Uldall and Arne Jensen

8. DETERMINATION OF MAGNESIUM AND CALCIUM IN SERUM
   Arne Jensen and Erik Riber

9. DETERMINATION OF MANGANESE, IRON, COBALT, NICKEL, COPPER, AND ZINC IN CLINICAL CHEMISTRY
   Arne Jensen, Erik Riber, Poul Persson, and Kaj Heydorn

10. DETERMINATION OF LEAD, CADMIUM, AND MERCURY IN CLINICAL CHEMISTRY
    Arne Jensen, Jytte Molin Christensen, and Poul Persson

11. DETERMINATION OF CHROMIUM IN URINE AND BLOOD
    Ole Jøns, Arne Jensen, and Poul Persson

12. DETERMINATION OF ALUMINUM IN CLINICAL CHEMISTRY
    Arne Jensen, Erik Riber, and Poul Persson

13. DETERMINATION OF GOLD IN CLINICAL CHEMISTRY
    Arne Jensen, Erik Riber, Poul Persson and Kaj Heydorn

14. DETERMINATION OF PHOSPHATES IN CLINICAL CHEMISTRY
    Arne Jensen and Adam Uldall

15. IDENTIFICATION AND QUANTIFICATION OF SOME DRUGS IN BODY FLUIDS BY METAL CHELATE FORMATION
    R. Bourdon, M. Galliot and J. Hoffelt

16. METAL COMPLEXES OF SULFANILAMIDES IN PHARMACEUTICAL ANALYSIS AND THERAPY
    Auke Bult

17. BASIS FOR THE CLINICAL USE OF GALLIUM AND INDIUM RADIONUCLIDES
    Raymond L. Hayes and Karl F. Hübner

18. ASPECTS OF TECHNETIUM CHEMISTRY AS RELATED TO NUCLEAR MEDICINE
    Hans G. Seiler

AUTHOR INDEX-SUBJECT INDEX

Volume 17. Calcium and Its Role in Biology

1. BIOINORGANIC CHEMISTRY OF CALCIUM
   R. Bruce Martin

2. CRYSTAL STRUCTURE STUDIES OF CALCIUM COMPLEXES AND IMPLICATIONS FOR BIOLOGICAL SYSTEMS
   H. Einspahr and C. E. Bugg

3. INTESTINAL AND RENAL ABSORPTION OF CALCIUM
   Piotr Gmaj and Heini Murer

4. CALCIUM TRANSPORT ACROSS BIOLOGICAL MEMBRANES
   Ernesto Carafoli, Giuseppe Inesi, and Barry Rosen

5. PHYSIOLOGICAL ASPECTS OF MITOCHONDRIAL CALCIUM TRANSPORT
   Gary Fiskum

6. MODE OF ACTION OF THE REGULATORY PROTEIN CALMODULIN
   Jos A. Cox, Michelle Comte, Armand Malnoë, Danielle Burger, and Eric A. Stein

7. CALCIUM AND BRAIN PROTEINS
   S. Alemà

8. THE ROLES OF Ca^{2+} IN THE REGULATION AND MECHANISM OF EXOCYTOSIS
   Carl E. Creutz

9. CALCIUM FUNCTION IN BLOOD COAGULATION
   Gary L. Nelsestuen

10. THE ROLE OF CALCIUM IN THE REGULATION OF THE SKELETAL MUSCLE CONTRACTION-RELAXATION CYCLE
    Henry G. Zot and James D. Potter

11. CALCIFICATION OF VERTEBRATE HARD TISSUES
    Roy E. Wuthier

AUTHOR INDEX-SUBJECT INDEX

Volume 18. Circulation of Metals in the Environment

1. INTRODUCTION TO 'CIRCULATION OF METALS IN THE ENVIRONMENT'
   Peter Baccini

2. ANALYTICAL CHEMISTRY APPLIED TO METAL IONS IN THE ENVIRONMENT
   Arne Jensen and Sven Erik Jørgensen

3. PROCESSES OF METAL IONS IN THE ENVIRONMENT
   Sven Erik Jørgensen and Arne Jensen

4. SURFACE COMPLEXATION
   Paul W. Schindler

5. RELATIONSHIPS BETWEEN BIOLOGICAL AVAILABILITY AND CHEMICAL MEASUREMENTS
   David R. Turner

6. NATURAL ORGANIC MATTER AND METAL-ORGANIC INTERACTIONS IN AQUATIC SYSTEMS
   Jacques Buffle

7. EVOLUTIONARY ASPECTS OF METAL ION TRANSPORT THROUGH CELL MEMBRANES
   John M. Wood

8. REGULATION OF TRACE METAL CONCENTRATIONS IN FRESH WATER SYSTEMS
   Peter Baccini

9. CYCLING OF METAL IONS IN THE SOIL ENVIRONMENT
   Garrison Sposito and Albert L. Page

10. MICROBIOLOGICAL STRATEGIES IN RESISTANCE TO METAL ION TOXICITY
    John M. Wood

11. CONCLUSIONS AND OUTLOOK
    Peter Baccini

AUTHOR INDEX-SUBJECT INDEX

Volume 19. Antibiotics and Their Complexes

1. THE DISCOVERY OF IONOPHORES: AN HISTORICAL ACCOUNT
   Berton C. Pressman

2. TETRACYCLINES AND DAUNORUBICIN
   R. Bruce Martin

3. INTERACTION OF METAL IONS WITH STREPTONIGRIN AND BIOLOGICAL PROPERTIES OF THE COMPLEXES
   Joseph Hajdu

4. BLEOMYCIN ANTIBIOTICS: METAL COMPLEXES AND THEIR BIOLOGICAL ACTION
   Yukio Sugiura, Tomohisa Takita, and Hamao Umezawa

5. INTERACTION BETWEEN VALINOMYCIN AND METAL IONS
   K. R. K. Easwaran

6. BEAUVERICIN AND THE OTHER ENNIATINS
   Larry K. Steinrauf

7. COMPLEXING PROPERTIES OF GRAMICIDINS
   James F. Hinton and Roger E. Koeppe II

8. NACTINS: THEIR COMPLEXES AND BIOLOGICAL PROPERTIES
   Yoshiharu Nawata, Kunio Ando, and Yoichi Iitaka

9. CATION COMPLEXES OF THE MONOVALENT AND POLYVALENT CARBOXYLIC IONOPHORES: LASALOCID (X-537A), MONENSIN, A-23187 (CALCIMYCIN) AND RELATED ANTIBIOTICS
   George R. Painter and Berton C. Pressman

10. COMPLEXES OF D-CYCLOSERINE AND RELATED AMINO ACIDS WITH ANTIBIOTIC PROPERTIES
    Paul O'Brien

11. IRON-CONTAINING ANTIBIOTICS
    J. B. Neilands and J. R. Valenta

12. CATION-IONOPHORE INTERACTIONS: QUANTIFICATION OF THE FACTORS UNDERLYING SELECTIVE COMPLEXATION BY MEANS OF THEORETICAL COMPUTATIONS
    Nohad Gresh and Alberte Pullman

AUTHOR INDEX-SUBJECT INDEX

Volume 20. Concepts on Metal Ion Toxicity

1. DISTRIBUTION OF POTENTIALLY HAZARDOUS TRACE METALS
   Garrison Sposito

2. BIOINORGANIC CHEMISTRY OF METAL ION TOXICITY
   R. Bruce Martin

3. THE INTERRELATION BETWEEN ESSENTIALITY AND TOXICITY OF METALS IN THE AQUATIC ECOSYSTEM
   Elie Eichenberger

4. METAL ION SPECIATION AND TOXICITY IN AQUATIC SYSTEMS
   Gordon K. Pagenkopf

5. METAL TOXICITY TO AGRICULTURAL CROPS
   Frank T. Bingham, Frank J. Peryea, and Wesley M. Jarrell

6. METAL ION TOXICITY IN MAN AND ANIMALS
   Paul B. Hammond and Ernest C. Foulkes

7. HUMAN NUTRITION AND METAL ION TOXICITY
   M. R. Spivey Fox and Richard M. Jacobs

8. CHROMOSOME DAMAGE IN INDIVIDUALS EXPOSED TO HEAVY METALS
   Alain Léonard

9. METAL ION CARCINOGENESIS: MECHANISTIC ASPECTS
   Max Costa and J. Daniel Heck

10. METHODS FOR THE IN VITRO ASSESSMENT OF METAL ION TOXICITY
    J. Daniel Heck and Max Costa

11. SOME PROBLEMS ENCOUNTERED IN THE ANALYSIS OF BIOLOGICAL MATERIALS FOR TOXIC TRACE ELEMENTS
    Hans G. Seiler

AUTHOR INDEX-SUBJECT INDEX

Volume 21. Applications of Nuclear Magnetic Resonance to Paramagnetic Species

1. NUCLEAR RELAXATION TIMES AS A SOURCE OF STRUCTURAL INFORMATION
   Gil Navon and Gianni Valensin

2. NUCLEAR RELAXATION IN NMR OF PARAMAGNETIC SYSTEMS
   Ivano Bertini, Claudio Luchinat, and Luigi Messori

3. NMR STUDIES OF MAGNETICALLY COUPLED METALLOPROTEINS
   Lawrence Que, Jr., and Michael J. Maroney

4. PROTON NMR STUDIES OF BIOLOGICAL PROBLEMS INVOLVING PARAMAGNETIC HEME PROTEINS
   James D. Satterlee

5. METAL-PORPHYRIN INDUCED NMR DIPOLAR SHIFTS AND THEIR USE IN CONFORMATIONAL ANALYSIS
   Nigel J. Clayden, Geoffrey R. Moore, and Glyn Williams

6. RELAXOMETRY OF PARAMAGNETIC IONS IN TISSUE
   Seymour H. Koenig and Rodney D. Brown, III

AUTHOR INDEX-SUBJECT INDEX

Volume 23. Nickel and Its Role in Biology (tentative)

1. NICKEL IN THE ENVIRONMENT
   R. W. Boyle and Heather A. Robinson

2. NICKEL IN AQUATIC SYSTEMS
   Pamela M. Stokes

3. NICKEL AND PLANTS
    Margaret E. Farago and Monica Cole

4. NICKEL METABOLISM IN MAN AND ANIMALS
    Evert Nieboer and Franco E. Rossetto

5. NICKEL ION BINDING TO CITRATE AND TO AMINO ACIDS, PEPTIDES AND PROTEINS
    R. Bruce Martin

6. NICKEL IN PROTEINS AND ENZYMES
    Robert K. Andrews, Robert L. Blakeley, and Burt Zerner

7. NICKEL CONTAINING HYDROGENASES
    José J. G. Moura et al.

8. NICKEL ION BINDING TO NUCLEOTIDES
    R. Bruce Martin

9. INTERACTIONS BETWEEN NICKEL AND NUCLEIC ACIDS: CONSIDERATIONS ABOUT THE ROLE OF NICKEL IN CARCINOGENESIS
    E. L. Andronikashvili

10. TOXICOLOGY OF NICKEL
    Evert Nieboer and Franco E. Rossetto

11. ANALYSIS OF NICKEL IN BIOLOGICAL MATERIALS
    Hans G. Seiler

AUTHOR INDEX-SUBJECT INDEX

Other volumes are in preparation.

Comments and suggestions with regard to contents, topics, and the like for future volumes of the series would be greatly welcome.

METAL IONS IN BIOLOGICAL SYSTEMS

VOLUME 22

ENDOR, EPR, and Electron Spin Echo for Probing
Coordination Spheres

1
ENDOR: Probing the Coordination Environment in Metalloproteins

Jürgen Hüttermann and Reinhard Kappl
Fachrichtung Biophysik und Physikalische
Grundlagen der Medizin
Universität des Saarlandes
D-6650 Homburg (Saar), FRG

1.	INTRODUCTION	1
2.	ENDOR THEORY	4
3.	ENDOR INSTRUMENTATION AND TECHNIQUES	13
4.	SURVEY OF RESULTS FROM MODEL COMPLEXES AND METALLOPROTEINS	19
	4.1. Copper(II) Proteins and Cu(II)-Substituted Proteins	19
	4.1.1. Cu(II) Model Complexes: Tetraphenylporphyrin, Bleomycin, Tetraimidazole	22
	4.1.2. Cu(II) Proteins and Cu(II)-Substituted Proteins	28
	4.2. Cobalt(II) Proteins and Related Compounds	40
	4.3. Iron Proteins	42
	4.3.1. Heme Proteins	43
	4.3.2. Nonheme Proteins	72
5.	CONCLUDING REMARKS	73
	ABBREVIATIONS	75
	REFERENCES	76

1. INTRODUCTION

There is an increasing interest in the application of the electron nuclear double resonance (ENDOR) technique in resolving metal ion and ligand hyperfine interaction in metalloproteins over the past

10 years. The enormous potential of ENDOR as a structural probe monitoring the coordination environment of metal ions has been demonstrated first by Feher and coworkers in the early 1970s on heme proteins [1], although earlier efforts with these compounds date back to 1967 [2]. The first transition metal complex to be studied was Cu phtalocyanine in 1963 [3]. Since then a variety of paramagnetic metalloproteins have been subjected to ENDOR analysis. Although in most cases the spectral information obtained has been incomplete in terms of full determination of principal values of metal ion and ligand hyperfine interaction, the results have nevertheless contributed to an enhanced understanding of metal coordination in proteins and have frequently clarified unambiguously the nature of ligands which has remained obscure in electron paramagnetic resonance (EPR) spectroscopy.

The purpose of this chapter is to bring the reader up to date with the ENDOR literature (1985). The scope is confined to metalloproteins including related ENDOR work on model compounds. The relevant EPR results will also be considered. For more details about transition metal complexes with organic ligands in general, the reader is referred to an extensive and excellent review by Schweiger [4] which covers the literature up to 1980. This author also discusses in detail both experimental and theoretical aspects of ENDOR spectroscopy emphasizing the more recent developments in modulation and polarization schemes. Therefore, we shall confine ourselves to a description of those essential experimental and theoretical parameters necessary for the discussion of metalloprotein results.

Due to the increased application of ENDOR spectroscopy in all fields of paramagnetic resonance investigations, textbooks have become available covering a wide variety of double-resonance aspects to which the reader is referred for descriptions of problems of more general interest [5,6]. Also regular coverage of ENDOR results, including metalloproteins, can be found in the electron spin resonance series [7]. Articles on the application of ENDOR can also be found in monographs dealing with structural methodology in biological and chemical research [8,9]. An account of specific aspects of ENDOR on hemes and hemoproteins has been given by Scholes which includes results up to about 1979 [10]. Hemo- and myoglobin were reviewed more recently by Dickinson and Symons [11].

The main impact of ENDOR spectroscopy compared to conventional EPR derives, especially in the field of metalloproteins, from the enhanced spectral resolution obtainable. Quite typically, metalloproteins have to be studied as solids (frozen aqueous solutions, glasses, single crystals) at low temperatures which yield inhomogeneously broadened lines in EPR due to unresolved hyperfine interaction and orientational distribution of paramagnetic centers. ENDOR essentially has the linewidth of the homogeneous spin packets and can reduce orientational broadening. Roughly, the increase in resolution is of the order 10^3 or more. An additional enhancement of resolution comes from a decrease in spectral density by hyperfine coupling of equivalent nuclei which is additive in ENDOR but increases the number of lines multiplicatively in EPR.

Other advantages of ENDOR come from the possibility of direct determination of the interacting nucleus via the nuclear g factor g_n, the first-order contribution of nuclear quadrupole interaction to the spectra which allows for analysis of quadrupolar splittings, and from the possibility of evaluation of relative signs of hyperfine and quadrupolar interaction. Combined with the high spectral resolution, ENDOR thus provides for a detailed analysis of spin density distribution in the transition metal-ligand complex as a whole if the metal ion is ENDOR-active too. Moreover, the information can frequently extend beyond the coordinated ligands by analysis of purely dipolar interactions of nuclei in the immediate environment of the complex. It therefore appears appropriate to employ the term *coordination sphere* around the paramagnetic ion for the site of ENDOR investigations. This sphere can have a radius of roughly 4-6 Å, within which a spatial resolution of about 0.5 Å can be obtained. It is thus apparent that ENDOR spectroscopy has the potential of a very powerful structural probe of metal coordination stereochemistry in metalloproteins.

2. ENDOR THEORY

In metalloproteins, in principle, two different contributions of nuclear interaction with the unpaired spin need to be considered—that of the central metal ion and that of the ligands. The latter nuclei usually are protons (^{1}H, nuclear spin I = 1/2) and nitrogen nuclei (^{14}N, I = 1; ^{15}N, I = 1/2). There are rare cases in which other nuclei in the coordination environment have been monitored such as chlorine (^{35}Cl, I = 3/2; ^{37}Cl, I = 3/2) [10]. Moreover, only scattered information has been obtained so far about central metal ion interaction. We shall therefore limit the scope of this section to ^{1}H and ^{14}N interactions with an unpaired electron as examples. Orbital momentum contributions to the nuclear interaction will be neglected so that doublet ground states as present in low-spin metalloproteins are best represented by the outlined formalism. A more general account, applicable to high-spin metal ion states too, is found in Ref. 4.

In the conventional formalism of the spin Hamiltonian H, the ENDOR transition frequencies for a proton in a four-level scheme due to interaction with one unpaired electron [electron spin ("effective") S = 1/2] can be derived from [12]:

$$h^{-1}H = h^{-1}\beta H_0 \vec{S} \underline{g} \vec{h} - g_n \beta_n h^{-1} H_0 \tilde{\vec{h}} \vec{I} + \vec{S} \underline{A} \vec{I} \tag{1}$$

the first and second term representing the electronic and nuclear Zeeman terms, respectively, and the third term accounting for the hyperfine interaction. h is the Planck constant, H_0 the external magnetic field, $\hat{h}$ and $\vec{h}$ denote a unit vector along the direction of the magnetic field being written either as column $\vec{h}$ or its transpose $\tilde{\vec{h}}$ (row), β and β_n the electronic and nuclear magnetic moments, and g_n the nuclear g factor. The tensors $\underline{g}$ and $\underline{A}$ denote the electronic g and the hyperfine tensor. $\vec{S}$ and $\vec{I}$ represent the electronic and nuclear spin operators, respectively. In the formalism of (1), the hyperfine tensor $\underline{A}$ is in frequency units.

ENDOR: PROBING THE COORDINATION ENVIRONMENT IN METALLOPROTEINS 5

In the first-order approximation in which $\vec{S}$ is assumed to be quantized along $\underline{g}\vec{h}$ and components of $\vec{S}$ perpendicular to that direction are neglected, (1) can be transformed to

$$h^{-1}H = h^{-1}g_h\beta H_0 S_u - h\underline{\tilde{K}}(S_u)\cdot\vec{I} \qquad (2)$$

with $g_h = \sqrt{\vec{\tilde{h}}\underline{g}^2\vec{h}}$

S_u = component of $\vec{S}$ along $\vec{u}$

$$\vec{u} = \frac{\underline{g}\cdot\vec{h}}{|\underline{g}\cdot\vec{h}|} = \frac{\underline{g}\cdot\vec{h}}{g_h}$$

$$\underline{\tilde{K}}(S_u) = \nu_n\underline{E} - g_h^{-1}S_u(\underline{\tilde{g}}\cdot\underline{A})$$

$$\nu_n = g_n\beta_n h^{-1}\cdot H_0$$

$\underline{E}$ = unit tensor

The term $h\underline{\tilde{K}}(S_u)$ gives the effective magnetic field at the nucleus for a given component of $\vec{S}$ along $\vec{u}$, its direction determining the axis of quantization for the nuclear spin.

The expression for the energy levels of the Hamiltonian (2), then, is

$$h^{-1}E(M_S, m_I) = g_h\beta H_0 M_S - K(M_S)\cdot m_I \qquad (3)$$

with $K(M_S) = \sqrt{\vec{\tilde{h}}\underline{K}^2(M_S)\vec{h}}$

M_S, m_I = electronic and nuclear spin quantum numbers ($\pm 1/2$) which, for a given orientation $\vec{h}$ of H_0 results in a level scheme as depicted in Figure 1. EPR transitions (allowed) occur between levels 1-4 and 2-3 ($\Delta M_S = \pm 1$, $\Delta m_I = 0$). ENDOR transitions involve levels 1-2 and 3-4 ($\Delta M_S = 0$, $\Delta m_I = \pm 1$) at frequencies

$$\nu_\pm = K_\pm = \sqrt{\vec{\tilde{h}}\cdot\underline{K}_\pm^2\vec{h}} \qquad (4)$$

with $\underline{K}_\pm = \nu_n\underline{E} \pm g_h^{-1}(\underline{\tilde{g}}\cdot\underline{A})/2$

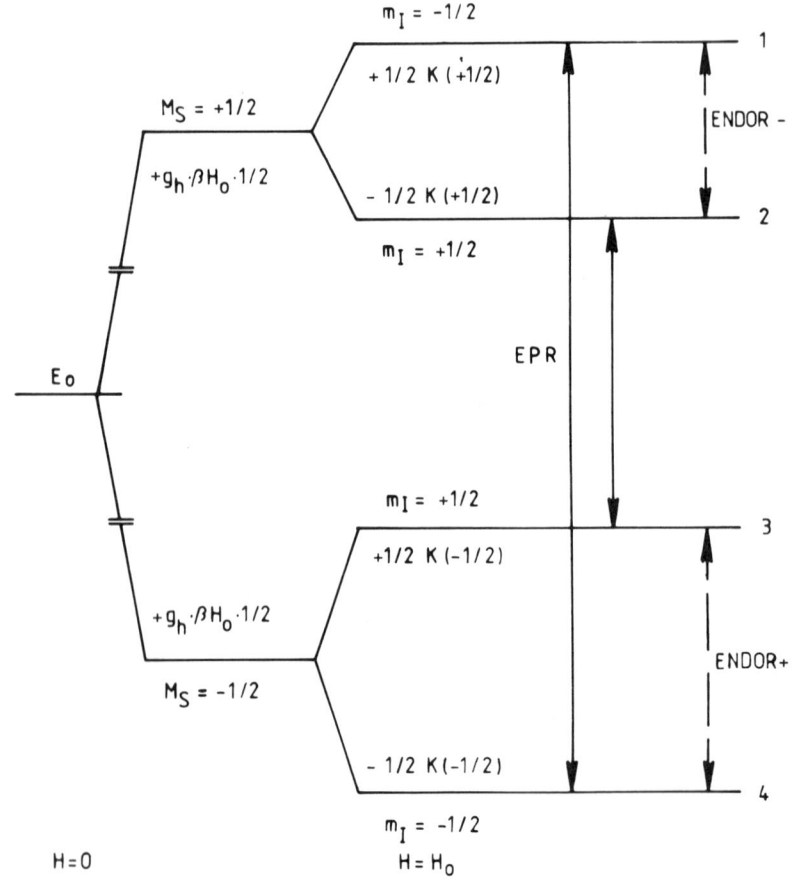

FIG. 1. Energy levels for a two-spin system ($S = 1/2$, $I = 1/2$) coupled by hyperfine interaction. Allowed EPR (arrows) and ENDOR transitions (dashed arrows) in a field $H = H_0$ are given. See text for explanation of symbols.

It becomes apparent from these formulas that a precise determination of the tensors $\underline{\underline{g}}$ and $\underline{\underline{K}}$ requires a single-crystal EPR-ENDOR analysis, EPR for the evaluation of $\underline{\underline{g}}$. If available, the measurement of single crystals in the laboratory frame gives rise to ENDOR resonance lines $\nu_{\pm} = K'_{\pm}$ which follow the usual angular variation

$$K'^{2}_{\pm} = \alpha + \beta \cos 2\theta + \gamma \sin 2\theta \qquad (5)$$

ENDOR: PROBING THE COORDINATION ENVIRONMENT IN METALLOPROTEINS

in a given crystal plane with θ being the angle of rotation of the magnetic field in that plane. Least-squares fit of (5) to the experimental data in three orthogonal planes gives the elements of $\underline{\underline{K}}_{\pm}'^2$ by, e.g., the Schonland procedure [13]. Tensor $\underline{\underline{K}}_{\pm}'^2$ is then brought to the diagonal form $\underline{\underline{K}}_{\pm}$ by a transformation

$$\underline{\underline{K}}_\pm = \underline{\underline{L}} \cdot \underline{\underline{K}}_{\pm}' \underline{\underline{\tilde{L}}} \tag{6}$$

so that

$$K_\pm' = \sqrt{\vec{\tilde{h}} \underline{\underline{L}} \underline{\underline{K}}_{\pm}^2 \underline{\underline{\tilde{L}}} \vec{h}} \tag{7}$$

The above first-order treatment can be simplified if the g anisotropy is small, i.e., $g_h^{-1} \underline{\underline{\tilde{g}}} \simeq \underline{\underline{E}}$. Then it follows from (4) that

$$\underline{\underline{K}}_\pm = \nu_n \cdot \underline{\underline{E}} \pm \underline{\underline{A}}/2 \tag{8}$$

which can be further reduced to yield ENDOR transitions at frequencies

$$\nu_\pm = K_\pm = |\nu_n \pm a/2| \tag{9}$$

if the hyperfine interaction has no anisotropy but only the isotropic value a.

Both latter conditions are usually not fulfilled in metalloproteins. The consequence of hyperfine anisotropy alone is the shift in the "crossover" point, the center of the two ENDOR lines for weak proton interactions ($\nu_n > |\frac{a}{2}|$) which occurs to higher frequencies than ν_n due to off-diagonal elements of $\underline{\underline{A}}$. For strong proton hyperfine interactions (≥ 30 MHz), another correction is necessary due to contributions of terms $S_i I_i$ perpendicular to S_u which are neglected in the first-order treatment [14]. This is a rare situation, however, for proton couplings in transition metal complexes, so that the main task is the evaluation of $\underline{\underline{K}}_{\pm}^2$ from (4) and the subsequent determination of $\underline{\underline{A}}$.

Other corrections to (4) coming, for example, from large metal ion hyperfine interactions and nuclear dipole-dipole couplings are discussed in [4].

For the case of ^{14}N hyperfine interaction (I = 1), the spin Hamiltonian (1) has to be amended by a nuclear quadrupole interaction term $\vec{I}\underline{P}\vec{I}$ in which $\underline{P}$ is the quadrupole tensor (in frequency units) which is symmetric and has a zero trace. The resulting energy correction of hyperfine levels involves nuclear spin states only, for which the basis functions where chosen in (2) to be quantized along $\vec{h}\underline{K}(S_u)$. In this system, the energy correction ΔE can be written as

$$h^{-1}\Delta E_{M_S m_I} = P(S_u)/2 \, [3m_I^2 - I(I+1)] \tag{10}$$

where $P(S_u)$ is a component of $\underline{P}$ with

$$P(S_u) = (\underline{K}(S_u) \cdot \underline{P} \cdot \underline{K}(S_u))/|K(S_u)|^2 \tag{11}$$

For a fixed direction $\vec{h}$ and value H_0 of the external magnetic field, the resulting schematic energy levels for the system S = 1/2, I = 1 are shown in Figure 2. There are three first-order EPR transitions involving levels 1-6, 2-5, and 3-4, respectively, which are not affected by the quadrupolar shift. Depending on which EPR transition is used as ENDOR observer, doublets 1-2, 5-6, and 2-3, 4-5 or a quartet 1-2, 2-3, 4-5, and 5-6 are monitored in ENDOR, their transition frequencies all being dependent on the quadrupolar interaction in first order.

Quite often, hyperfine couplings due to ^{14}N ligands in metalloproteins are small and buried in the EPR linewidth. As a consequence, in ENDOR typically a quartet pattern is observed due to simultaneous excitation of all four nuclear transitions. Assuming $K(S_u) > P(S_u)$ and small g-factor anisotropy, the four lines can be grouped into two pairs of separation $2\nu_n$ (~2 MHz for ^{14}N at X-band frequencies) centered around A/2 and displaced from it by a value 3P with A and P being the effective components of $\underline{A}$ and $\underline{P}$ for the magnetic field orientation considered. A detailed discussion of corrections to this crude approximation has been given in Refs. 4 and 15. Full evaluations of the quadrupolar tensor from (11) are so far rare in the metalloprotein field due to the scarcity of single crystals.

ENDOR: PROBING THE COORDINATION ENVIRONMENT IN METALLOPROTEINS

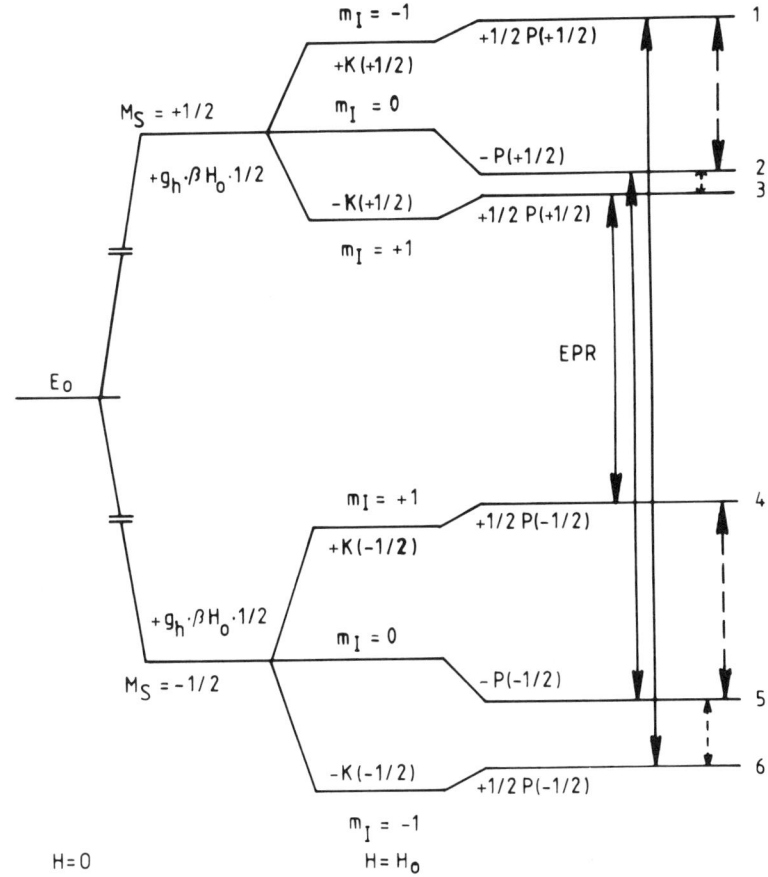

FIG. 2. Energy levels for a two-spin system ($S = 1/2$, $I = 1$) comprising hyperfine and quadrupolar interaction. Allowed EPR (arrows) and ENDOR (dashed arrows) transitions are shown in a field $H = H_0$. For explanation of symbols see text.

The bulk of ENDOR data has been obtained from frozen solutions and glasses of metalloproteins which give rise to powder type EPR and ENDOR spectra. This is not only because of a general difficulty in crystallizing specimen, but also because in frozen solutions the states of the protein in physiologically relevant conformations can more easily be achieved, for example, by adjustment of pH and addition of effectors to the solution prior to freezing. The technique of orientation selection introduced by Rist and Hyde [16] allows the obtaining of useful information on ligand interaction in such samples and even, in specific cases, provides for a full tensor analysis. It involves selection of such portions of the EPR powder spectrum for ENDOR analysis in which the paramagnetic complex has a well defined, narrow range of orientations with respect to the external magnetic field. This is possible when either g or a hyperfine anisotropy dominate the other magnetic interactions. In this case, an extreme part of the EPR spectrum is built up by centers for which the magnetic field is directed along a principal interaction axis. The corresponding ENDOR spectrum is single-crystal-like, usually well resolved, and permits direct extraction of interaction constants. High- and low-spin ferric hemoglobins are typical examples for dominant g anisotropy (axial and rhombic symmetry, respectively). For the former system, at $g \simeq 2.0$ the hemes are oriented with their normal along the magnetic field direction yielding one principal ligand hyperfine tensor element if g and hyperfine tensors are coaxial. At $g \simeq 6.0$, the magnetic field spreads in the heme plane so that the corresponding ENDOR spectrum is two-dimensional covering contributions of all possible in plane interaction values. Methods obtaining ligand hyperfine informations for protons and ^{14}N nuclei for this case have been presented [4].

For EPR powder spectra of rhombic g symmetry like hydroxymethemoglobin, the single-crystal-like ENDOR spectra can be analyzed at the highest and lowest field portions of the EPR spectrum. The intermediate g-turning point selects contributions of molecules from many different orientations for which $g_h = g_{int}$. It gives again essentially a two-dimensional powder ENDOR spectrum but a connection with the heme plane is usually not available. As a consequence, a large number of ENDOR transitions should be found for the intermediate g-turning point providing for strong broadening. Resolvable intensity buildup should then stem mostly from those complexes for which that g factor is a principal element. Observation in the derivative mode suppresses the broad contributions.

ENDOR: PROBING THE COORDINATION ENVIRONMENT IN METALLOPROTEINS

Powder ENDOR spectra often suffer, for metalloproteins, from lack of spectral separation between weakly coupled ^{1}H interactions centered around ν_n and ^{14}N couplings of ligands, when conventional X-band (~9 GHz) EPR-ENDOR spectroscopy is used. This problem is sometimes handled by measuring at two or more frequencies about 0.8 GHz or so apart which is technically simple to achieve. A real advantage in separation due to the large difference in the field-dependent part ν_n for ^{14}N ($\simeq$1 MHz at 9 GHz) and ^{1}H ($\simeq$14 MHz at 9 GHz) can be obtained at Q-band frequencies ($\simeq$35 GHz). This also would enhance the orientation selection preciseness on the EPR side which would be of particular importance for cases of comparable g and hyperfine anisotropy as is frequently encountered in low-spin metalloproteins. Little use of Q-band ENDOR has so far been made [17].

As mentioned in Sec. 1, the ENDOR response in metalloprotein frequently extends beyond the sphere of coordinated ligands and involves nuclei for which the interaction is purely dipolar. In first order, the dipolar coupling a between electronic and nuclear point dipoles can be obtained from the simple formula:

$$a = g_n \beta_n g \beta r^{-3} (3 \cos^2\theta - 1) \cdot \rho \qquad (12)$$

in which r is the distance between dipoles, ρ the electronic spin density, and θ the angle between the $\vec{h}$ and the vector connecting the two dipoles. The maximum dipolar coupling ($\theta = 0^0$), expressed in MHz when r is in Å, can then be calculated as

$$a = 28.1 \cdot r^{-3} \cdot \rho \cdot 5.616 \qquad (13)$$

In this expression the nuclear dipole is a proton. Other dipolar interactions can usually be neglected due to the small values of g_n for nuclei other than ^{1}H. Also, g_h has been taken as g_e = 2.0023, which is only a very crude approximation even for low-spin metalloporphyrins. Nevertheless, the use of (12) or (13) allows to obtain distances from measured dipolar interactions which can be compared to values obtained from x-ray crystallography. This procedure can be helpful in the assignment of weakly coupled proton interactions. It can be refined by taking ρ as a sum of all contributing spin densities and distances of the various parts of the metal ion-ligand complex. Examples of this treatment are given in Sec. 4. An important prerequisite for the application of point-dipole calculations is not only the knowledge of the spin density distribution in the metal ion complex but also the precise bonding geometry. Both parameters are best obtained by single-crystal ENDOR spectroscopy, an enormous task which then obliviates the use of dipolar calculations. In practice, however, one often has to live at best with the combination of spin density and bonding geometry information available from EPR work on single crystals and ENDOR data from ligands in powder spectra. More frequently, even the former set of information is lacking which leaves resorting to model compound studies as the main feasible approach for assigning interactions in proteins to specific ligands.

There are two ENDOR techniques available which, in principle, are helpful amendments in the elucidation of powder-type EPR or ENDOR spectra. One is ENDOR-induced EPR (EI-EPR), which reverses the usual ENDOR conditions of fixed H_0 and variable radiofrequency (rf) sweep to recording an EPR spectrum while a fixed (or, better, corrected for the field dependence of ν_n) frequency is applied. Single-crystal-type EPR spectra are recorded when the frequency corresponds to a principal value of a hyperfine interaction. There are several amendments necessary to this simplified picture (which essentially holds for proton interactions) since coherence between rf frequency and hyperfine transitions critically depends on corrections which may not be linear [4]. Another technique, frequently more feasible, is TRIPLE- (or DOUBLE-) ENDOR, in which the change of ENDOR transitions in intensity is monitored when pumping one transition strongly while a second rf frequency is swept through the whole ENDOR spectrum. Introduced by Cook and Whiffen for the determination of relative signs of hyperfine couplings [18], the technique has also been used to separate overlapped ENDOR spectra from magnetically distinct sites [4], a situation which often occurs in powder ENDOR spectra. While sign determination rests on a reduced or enhanced ENDOR transition intensity corresponding to nuclei in the same or a different spin substate as the one transition pumped additionally, site distinction employs the response only. Thus, if one has a resolved ENDOR line adjacent to an overlapped part in the spectrum, the response of the latter when pumping the resolved line immediately connects lines of one site.

3. ENDOR INSTRUMENTATION AND TECHNIQUES

ENDOR spectroscopy, originally in the hands of solid state physicists who developed highly sophisticated apparatus dedicated to specific scientific problems, has developed to a commercial, more universally applicable level at which studies ranging from solution at room temperature to low-temperature solid state investigations of a fairly broad class of nuclei and coupling frequencies can be performed. The basic instrumentation, in addition to a stable, sensitive EPR spectrometer, involves an rf generator, an rf amplifier, some kind of an impedance matching device, and the ENDOR coil in which the rf field is generated. A typical block diagram of a schematic setup is depicted in Figure 3.

ENDOR spectra can be recorded in three modes. One involves CW radiofrequency (rf) and conventional magnetic field modulation (e.g., 100 kHz) for which H_0 has to be fixed on the maximum or minimum of the EPR first derivative. The result is an absorption-type display. The same type of ENDOR response is gained from AM rf modulation without magnetic field modulation. A derivative ENDOR display, which has the advantage of increased resolution, is obtained by FM modulation of the radiofrequency. For both latter types of recording, H_0 is set to the zero crossing point of the first derivative EPR absorption. Reduction of spurious signal contributions can be achieved by double coding, e.g., magnetic field and FM modulation at two different frequencies.

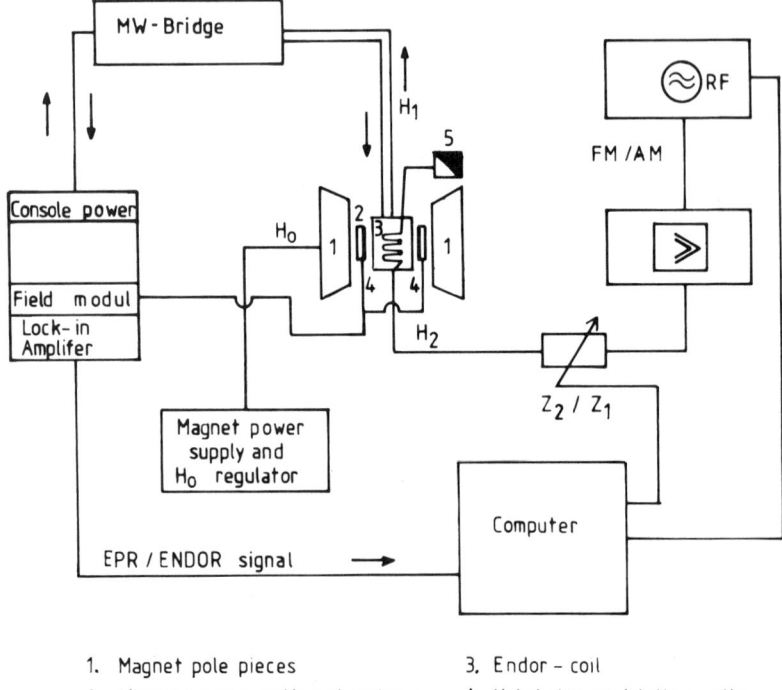

1. Magnet pole pieces
2. Microwave resonating structure
3. Endor - coil
4. Helmholtz modulation coils
5. Dummy load

FIG. 3. Block diagram of an EPR-ENDOR spectrometer.

The critical point in the various experimental assemblies which are discussed in detail in [19] is the insertion of the ENDOR coil into the microwave resonating structure (usually some kind of a cavity) without distortion of its Q together with a broad-band impedance matching of the coil to the output of the amplifier. If the ENDOR coil is broad-banded, i.e., has a resonance frequency far below or beyond the frequency range of interest, impedance matching can be performed by a fixed transformer or even without one if the coil is terminated by a dummy load. Other configurations involve the use of a variable, frequency-dependent transformation which, for example, is controlled by computer and is synchronized to the frequency range of the rf oscillator. Still other setups contain

ENDOR: PROBING THE COORDINATION ENVIRONMENT IN METALLOPROTEINS

an ENDOR coil which is part of a resonance LC circuit in which the operative H_2 field is enhanced (not shown in Fig. 3). Such a configuration, which of course has a limited frequency range, was the first one to be applied in steady-state ENDOR of color centers [20].

The ENDOR response, detected as a change in the EPR signal intensity of a (partially) saturated portion of a spectrum when nuclear transitions are induced by the rf field, critically depends on electronic and nuclear relaxation times. The two time-dependent magnetic fields H_1 (microwave) and H_2 (rf) must be of the size to fulfill both conditions

$$\gamma_e^2 H_1^2 T_{1e} T_{2e} \geq 1$$

and

$$\gamma_n^2 H_2^2 T_{1n} T_{2n} \geq 1 \qquad (14)$$

i.e., both the electronic and the nuclear transitions must be saturated. T_1 and T_2 are the electronic (index e) and nuclear (n) spin-lattice and spin-spin relaxation times, and γ is the gyromagnetic ratio. Both sets of parameters are strongly temperature-dependent. Typically, for metalloproteins, due to the very short times T_{1e} and T_{1n}, one has to operate at temperatures not above roughly 40 K, the best operating conditions sometimes being as low as 2 K.

As a consequence, in the design of an ENDOR cavity one must not only consider the optimized conditions for microwave and rf fields but also has to include the cryogenic requirements. There are several constructions reported in which the whole cavity is immersed into the cooling fluid (liquid He) [21-23]. Among others, an easy way of achieving sub-4.2 K temperature by pumping on the cryostat is afforded by this. On the other hand, sample changement and tuning become tedious in such systems. The other class of configurations involves cooling of the specimen only in an evaporator-type cryostat which can operate down to 3.8 K. A specific version (Oxford Instruments, England) allows to achieve even lower temperatures. The commercial ENDOR version of the Bruker Company

(Karlsruhe, Germany), which operates at X-band frequencies, employs a TM 110 cavity together with an Oxford Cryostat. A schematic drawing is shown in Figure 4. The ENDOR coil is a helix wound on the outside of the quartz end of the cryostat. Depending on the number of turns, the usable frequency range is 1-30 MHz (for 16-20 turns) with no degradation in power. The H_2-field strength achieved depends on the output of the rf amplifier. With 100 W, H_2 fields of ~1 mT in the rotating frame can be applied at 10.5 MHz. A slight drawback in metalloprotein studies for this design is that relatively small field modulation values (~0.4 mT) can be attained which limits the EPR sensitivity somewhat.

We have developed a fairly simple system based on a standard TE 104 ("dual") cavity (Bruker). The schematic setup is shown in Figure 5. The ENDOR coil is an eight-turn Helmholtz-type coil which is photoprinted from a doubly copper-clad Mylar foil. The coil is

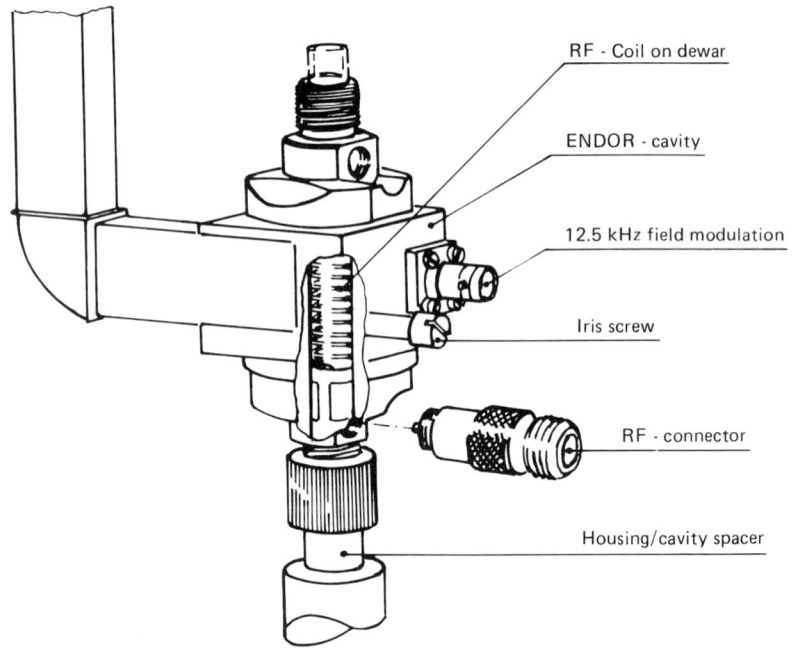

FIG. 4. Cutaway view of the Bruker ENDOR cavity. (Courtesy Bruker Analytische Meßtechnik, Karlsruhe, FRG.)

ENDOR: PROBING THE COORDINATION ENVIRONMENT IN METALLOPROTEINS

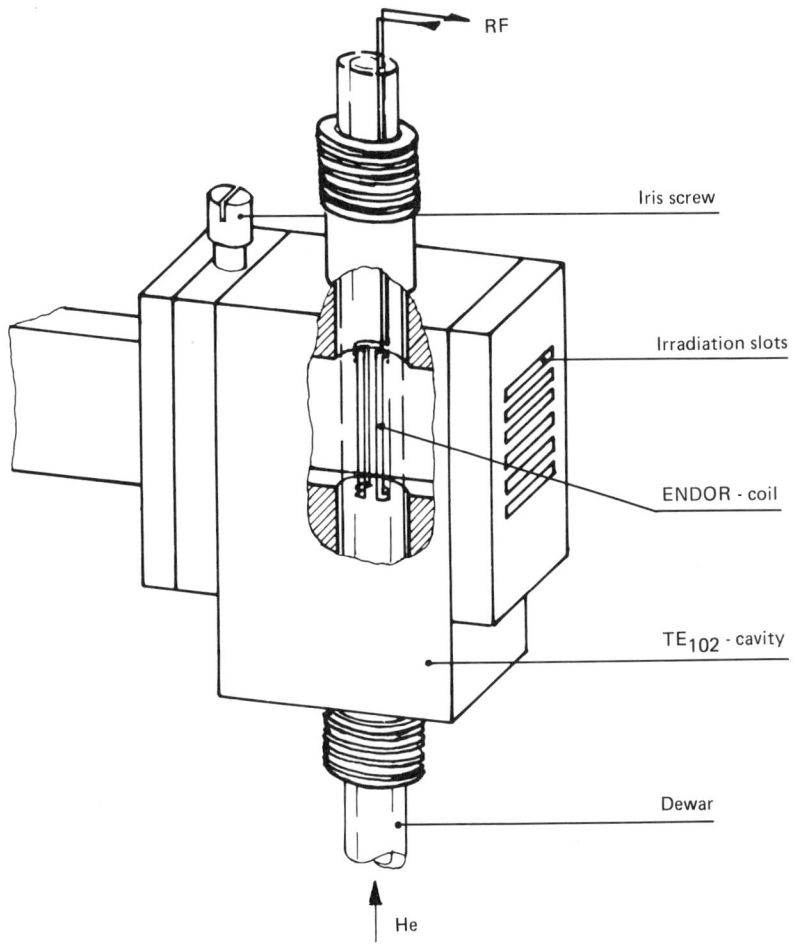

FIG. 5. Cutaway view of the ENDOR cavity used in the authors' laboratory.

inserted into the quartz end of a flow cryostat (Thor Cryogenics, England), so that $\vec{H}_0 \perp \vec{H}_1 \perp \vec{H}_2$. The coil can be used either as terminating load to the output of the amplifier (ENI A 150, for metalloprotein work, ENI 325 LA for organic radicals) or it is terminated by a dummy load. For coupling constants up to 20 MHz, an H_2 field of about 0.1-0.2 mT is achieved, dropping at higher frequencies.

Above 50 MHz, another coil with two turns has to be used. Sample interchange in this system and cavity tuning for different samples is as facile as in conventional EPR at 300 K. Special attention has to be given to the preciseness of the initial insertion of the ENDOR coil into the cavity. A computer (Dietz 621/8) controls the rf frequency sweep of a Wavetek 3000 synthesizer and provides for data acquisition.

For ENDOR-induced EPR, the amplitude of a single ENDOR transition is monitored upon variation of the magnetic field H_0. A prerequisite for this procedure is a strong, nonoverlapped ENDOR transition which has to be established first in conventional ENDOR. Subsequently, this transition is pumped while sweeping the magnetic field through the EPR spectrum, either with or without magnetic field modulation. In the latter case, a singly coded spectrum is obtained when modulating the rf frequency. Experimentally, there are no additional provisions necessary other than for a correction of the ENDOR frequency for the variation of $K_\pm$ with the value of H_0, which can easily be performed by computer. For protons, the rough approximation of correcting only for ν_n is usually sufficient even if $\underline{g}$ and $\underline{A}$ are anisotropic. The potentially enormous advantage of ENDOR-induced EPR is not only in site separation of magnetically inequivalent orientations, e.g., in powder EPR or ENDOR spectra, but moreover in displaying hyperfine interaction of nuclei for which direct ENDOR detection would be difficult due to, say, very small nuclear magnetic moments. This is to say that in ENDOR-induced EPR pumping of proton transitions can reveal ^{14}N and metal ion hyperfine interaction, for example, which could be difficult to obtain directly with ENDOR.

TRIPLE (or DOUBLE) ENDOR in the "low-power" regime ($H_2 \approx 0.1$ mT) also requires little experimental amendment to conventional ENDOR apparatus. An additional rf generator plus amplifier and a power coupler which combines the output of the two rf sources and feeds them into the ENDOR coil are basically sufficient, if the ENDOR spectrometer is of the broad-band type shown in Fig. 3. The limitations in the H_2-field strength obtainable in such a configuration may favor

ENDOR: PROBING THE COORDINATION ENVIRONMENT IN METALLOPROTEINS

the construction of two independent ENDOR coils for the pumping and observing frequency range which can be tuned independently [24]. Such high-power apparatus then also allows for other DOUBLE ENDOR techniques like nuclear spin decoupling and multiple nuclear quantum transition induction. In metalloproteins at low temperatures, these latter, advanced techniques have so far not been applied and probably are of little practical use since ENDOR linewidths set a definitive limit, at least for powder-type ENDOR spectra.

A technique with potential applicability to metalloprotein studies in frozen solutions or glasses is ENDOR with circularly polarized rf fields (CP-ENDOR) introduced by Schweiger and Günthard recently [25]. Basically, like in TRIPLE, the nuclear transitions are separated into those belonging to the same or a different m_S state, one of the groups being suppressed, ideally to zero, in their response. Both site selection of magnetically inequivalent orientations as well as simplification of multiline ENDOR spectra are thus achieved. Experimentally, the output of one rf frequency source is split into two parts, phase-controlled, which feed two ENDOR coils oriented so that their H_2 fields are perpendicular to each other in one plane. Depending on the phase difference between them, a right- or left-hand rotating circularly polarized rf field is produced at the sample. The theory predicts that for small proton couplings ($\gtrsim$15 MHz) as typical for metalloproteins the corrections for anisotropic $\underline{g}$ and hyperfine tensors are small so that reasonable reduction or enhancement ratios of the ENDOR response for a right- or left-hand polarization should be observed [4].

4. SURVEY OF RESULTS FROM MODEL COMPLEXES AND METALLOPROTEINS

4.1. Copper(II) Proteins and Cu(II)-Substituted Proteins

Copper-containing proteins perform a wide variety of functions in animal and plant physiology ranging from oxygen transport in mollusks and arthropods (hemocyanin) to oxidases (cytochrome c oxidase). For some of the proteins, their physiological function is not yet or not fully known (e.g., azurin, stellacyanin). EPR investigations of copper proteins have constituted an area of interest for more than three decades. For a recent review the reader is referred to [26]. Based on EPR findings, the proteins are grouped in three types, 1, 2, and 3. For type 1, the copper can be in a mononuclear Cu(II) site, as in azurin from microorganisms or in stellacyanin from the lac tree, or it can form multinuclear sites like in laccase which has four copper ions or in ceruloplasmin which has seven or eight. The main characteristic of the type 1 EPR spectra is an unusually small value of the Cu hyperfine interaction together with g tensor elements similar to those of model copper(II) complexes which, however, have about twice the value of the hyperfine interaction. Type 1 Cu(II) centers have only been found in proteins and are also characterized by an intense 600-nm absorption ("blue proteins"). In multinuclear copper proteins, one of the sites can be of this type.

Type 2 copper displays EPR features similar to those of model complexes. It is observed in "nonblue" proteins (e.g., superoxide dismutase) and in association with "blue" sites (ceruloplasmin, laccase). Finally, type 3 copper proteins are EPR-silent and inferred, e.g., on the basis of an optical absorption at 300 nm, to be present in ceruloplasmin and laccase in conjunction with types 1 and 2.

From this survey it becomes clear that the question of copper coordination, e.g., in type 1 and 2 spectra, is of central importance. Also, for type 3 copper, the question of Cu(I) ions or exchange-coupled Cu(II) pairs being the source of the apparent diamagnetism is a matter of controversial debate. Both problems can in principle be attacked with ENDOR spectroscopy.

There is another class of proteins which contain copper in association with other metal ions. An important example is cytochrome c oxidase, an oligomeric transmembrane protein which is the terminal enzyme of the eukaryotic electron transport system. It contains two iron ions incorporated into type a heme systems and two copper ions which may be in one of the forms type 1, 2, or 3. One of the copper ions is EPR-silent (type 3) although it is thought to be Cu(II) instead of Cu(I). The apparent diamagnetism is considered to be due to an antiferromagnetic coupling between Cu(II) and Fe(III) on account of magnetic susceptibility measurements, although coupling of Cu(I) and Fe(IV) has also been proposed [27,28]. For the other copper center, the so-called intrinsic copper, the exact coordination of the ion which is now thought to be Cu(II) is a topic of debate. Again, a coupling to Fe(III) arising from a dipole-dipole interaction is advanced in order to explain some of its unusual linewidth effects in EPR, a view which is not unequivocally accepted [26]. In this survey, we shall treat the copper "part" of cytochrome c oxidase in the present section and the heme part in Sec. 4.3.1.3.

Another example of proteins containing copper and other metal ions is bovine superoxide dismutase in which there are two Cu(II) ions and two zinc ions. The protein has two identical polypeptide subunits each of which contains one Cu and one Zn atom. The EPR spectra of the native protein can be classified as being due to type 2 copper(II). X-ray crystallography suggests that four imidazoles coordinate the Cu(II) ion, one of the histidines forming an imidazolate bridge to the zinc atom [29]. A fifth coordination site probably binds water in the native protein [30]. The imidazolate bridge perhaps is the source of a strong magnetic coupling which is observed when the Zn(II) is substituted for Co(II) or Cu(II). The exact nature of this coupling and the imidazolate function are not clear presently [26]. The dismutase activity of the enzyme can be inhibited by cyanide (CN^-) and azide (N_3^-) which bind to Cu(II). Thus, it is interesting to test the change in coordination of Cu(II) in the native and the inhibited protein in order to reveal the aspects of coordination essential to the catalase activity.

4.1.1. Cu(II) Model Complexes: Tetraphenylporphyrin, Bleomycin, Tetraimidazole

Considerable ENDOR work has been performed on planar Cu(II) complexes in which ^{14}N and ^{16}O constitute the atoms of the first coordination shell. The results are discussed extensively in [4] and will be used here only for reference, if applicable. More relevant to the situation in copper proteins are complexes in which all ligands comprise nitrogens.

We start with a description of Cu(II)-tetraphenylporphyrin (CuTPP), structure (I), for which an ENDOR study in $(H_2O)ZnTPP$ as

I

diamagnetic host single crystals has been performed [31]. This compound can also, at least partially, serve as comparison for the coordination of iron in TPP which is a model for the heme complex (see below). For CuTPP, a complete tensor set of metal ion and ^{14}N ligand hyperfine interaction has been obtained together with hyperfine interaction from the porphyrin protons. The data are listed in Table 1. The Cu(II) hyperfine tensor is axially symmetric, its largest principal element coinciding with the maximum g-tensor element, both being aligned parallel to the normal of the Cu(II)-porphyrin plane as is characteristic for planar Cu(II) complexes. The ^{14}N hyperfine tensor also exhibits mostly axial symmetry but with some rhombic distortion. The maximum principal element is directed along the Cu(II)-^{14}N bond. The intermediate element is parallel to the complex normal. It should be mentioned that all four nitrogens are equivalent which necessitates the use of second-order expressions for analysis of hyperfine coupling in the complex plane and numerical spin Hamiltonian diagonalization for the direction parallel to the complex normal [4,31]. Both symmetries, in-plane inversion and perpendicular D_{4h}, can probably be analyzed only in single crystals but should contribute to line broadening in powder-type ENDOR.

The nuclear quadrupole interaction tensor is axially symmetric with the direction of the maximum principal element perpendicular to the Cu(II)-^{14}N bond directions. Such a behavior has been found for Cu(II)-bis(salicylaldoxine) [32] and contrasts with the situation in related complexes of Cu(II) with ^{14}N and ^{16}O like Cu(II)-bis(oxychinolate) in which the maximum quadrupolar interaction coincides with the Cu(II)-^{14}N bond direction [33]. The hyperfine interaction of the pyrrole protons is, for all of them, nearly axially symmetric and exhibits about identical coupling constants. Thus, the origin of interaction is mainly dipolar with little s-spin density contribution. The distances of the proton-metal ion connection calculated, assuming the electron to be confined completely at the Cu(II) site, are in good agreement with x-ray structural data [34]. The high degree of accuracy of coupling constants obtained in this study has allowed for a detailed MO analysis of the spin-density distribution in the complex [35], which, however, will not be discussed here since there is little if any possibility of relating these precise data to available results of other model complexes because the latter have been studied in frozen solutions or glasses and are of comparatively poor informational content.

Nevertheless, the other studies of Cu(II)-^{14}N coordination are relevant with respect to the copper protein results for which only

TABLE 1

ENDOR Parameters (in MHz) of Cu(II) Model Complexes and Proteins

Complex condition	Cu(II) coordination	Hyperfine interaction				Nuclear quadruple interaction		Ref.
		^{63}Cu	^{14}N		^{1}H	^{14}N	^{63}Cu	
Tetraphenylporphyrin (in host crystal)	Square-planar	−615[a] −102.7 −102.7	(4)[c]	54.2 42.8 44.1	(8) 2.52[b] 0.74 0.80	−0.62 0.93 −0.31		31
Bleomycin (frozen sol.)	Square-pyramid		(3) ~30	(1) ~41				37
Tetraimidazole (frozen sol.)	Square-planar		(4) (4)	41.6 39.8 42.3 40.3				38
Stellacyanine (frozen sol.)	Tetrahedral (distorted)	96 87 167	(1) 44 32	(1) 32 44			10 4 6	39 41
	Type 1	96 87 167	(1) 32	(1) 16	(1) 21			43
Azurin (frozen sol.)	Type 1	161 66	(1) 44[d]	(1) 20	(1) 24			43
Plastocyanin (frozen sol.) bean	Type 1	185 60	(1) 49	(1) 20	(1) 27		5	43

Protein	Site					
spinach	Type 1		(2) 55 51.4			48
Laccase (frozen sol.) tree	Type 1	129	(1) 20		7-8	43
			(1) 36.6 (1) 53.2	(1) 4.5 exch.[e]		49
Laccase (frozen sol.)	Type 3		(1) 50/40 (1) 45/36 (1) 36/30	(1) 0.30 / (1) 0.81 / (1) 1.64 / (1) 2.46 exch.		42
Cytochrome c oxidase (frozen sol.)	Cu_A		(1) ~17	(1) 12 / (1) 19		53
		68, 98, 90	(1) ~17	(1) 12 / (1) 19.1		54
	Cu_B		(1) ~16	(1) 12.2 / (1) 16.2		56
			(1) 40/35/28			42
Superoxide dismutase (frozen sol.)	CN^- complex		(1) 47.8 / (1) 37		1.6	58
Human transferrin (frozen sol.)			~31	(1) 14.2 exch. / (1) 17.6 exch.		60

[a] From EPR spectra.
[b] Average value.
[c] Numbers in parentheses indicate manifold of interacting nuclei.
[d] Tentative assignment.
[e] exch.: Proton is exchangeable in D_2O.

powder-type ENDOR data are available, too. One example of model compounds is cupric bleomycin, an antitumor agent, which is thought to have a square pyramidal configuration with Cu(II) coordinated to at least three in-plane nitrogens, one of them being an imidazole nitrogen. The nature of the fourth in-plane ligand, nitrogen or oxygen, is not clear presently but x-ray crystallography of a bleomycin fragment indicates coordination with four in-plane nitrogens [36]. The fifth ligand, in the fragment, is a terminal amine [structure (II)].

II

In their ENDOR study, Antholine and coworkers [37] observed two ^{14}N couplings of about 30 and 41 MHz strength together with several proton resonances which were not analyzed in detail. However, the authors substantiated that one of the protons was easily exchangeable for deuterium and that water probably had access to the cupric ion within about 6 Å. These results were combined with a high-resolution, low-frequency (S-band, 3-GHz) EPR analysis of superhyperfine structure on one of the copper hyperfine lines (m_I = -1/2) to conclude that there is a total of four nitrogen donors to Cu(II), three of which are approximately equivalent (~28 MHz). The four nitrogens are supposed to be in a square-planar configuration, the whole Cu(II) complex perhaps being square-pyramidal as suggested from x-ray data.

It is tempting to speculate that the one large nitrogen coupling (~41 MHz) in Cu(II)-bleomycin is due to the in-plane imidazole coordination. Values of this magnitude have been reported first for Cu(II)-tetraimidazole [structure (III)] by Van Camp and coworkers [38] and were subsequently reproduced [39]. In the latter

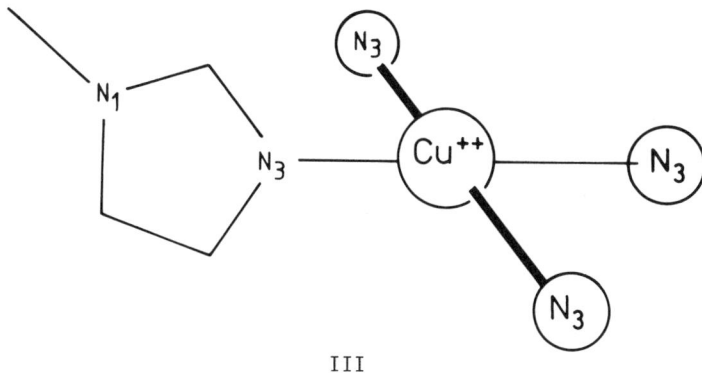

III

study, Yokoi compared imidazole ligation to Cu(II) with other nitrogen donors (ammonia, ethylenediamine, glycine, and pyridine) and found that for imidazole the ^{14}N interaction not only is the largest (~40 MHz compared with 36-39 MHz for the parallel component of the other ligands) but also the least anisotropic. The author suggested that this latter property, resulting in a value of $\tilde{<}1.1$ in the ratio of parallel to perpendicular hyperfine interaction of ^{14}N, should be used as a general indicator of imidazole ligation in cupric complexes in proteins. For the other nitrogen donors, values of 1.2-1.43 were found.

The large linewidth of ~3 MHz and more for the ^{14}N resonances in the Cu(II)-^{14}N complexes in frozen solution has been attributed by Van Camp et al. [38] to a distribution of angles of the imidazole planes in the complex. This linewidth prevented any analysis of quadrupolar coupling since, essentially, the spectrum showed no resolution. The same finding was reported by Yokoi [39]. The former authors also analyzed the remote ^{14}N-ENDOR resonances for which a hyperfine interaction of 1.8 MHz and a quadrupolar coupling of about 0.36 MHz was found, the latter value being the same as found in free imidazole.

The data from bleomycin and tetraimidazole are also collected in Table 1.

4.1.2. Cu(II) Proteins and Cu(II)-Substituted Proteins

4.1.2.1. Stellacyanin, azurin, plastocyanin. The EPR spectrum of stellacyanin, the mononuclear "blue" Cu(II) protein from *Rhus vernicifera*, is characterized by approximately axial $\underline{g}$ and Cu(II) hyperfine tensors $\underline{\underline{A}}$ with some rhombic distortion (g_x = 2.025, g_y = 2.077, g_z = 2.287; A_x^{Cu} = 171 MHz, A_y^{Cu} = 87 MHz, A_z^{Cu} = 105 MHz) [26]. These values for the hyperfine couplings were derived from simulation. Only one component, the maximum value, can be extracted directly from EPR at Q-band frequencies. This coupling is not colinear with the largest principal g axis which is in contrast with the usual findings for square-planar Cu(II) complexes.

^{1}H- and ^{14}N-ENDOR resonances were reported initially by Rist et al. [40] and were later confirmed by Roberts, Hoffmann, and co-workers who, in addition, presented results on the Cu(II) hyperfine interaction [41]. For the protons, several weak interactions were detected together with one strongly coupled (~20 MHz), nearly isotropic nucleus, for which so far no assignment has been attempted. ^{14}N lines were found to occur at all g turning points comprising two broad lines with no nuclear Zeeman or quadrupolar resolution at g_x and g_z whereas only one such line was apparent at g_h = g_y. These findings were interpreted in terms of two axially symmetric, equivalent ^{14}N couplings (parallel component ~44 MHz, perpendicular component ~32 MHz for each) aligned at right angles with respect to their principal axes. This assignment was recently challenged by Yokoi, who attributed the observed lines to two isotropic ^{14}N interactions of ~44 and ~32 MHz [39]. This would require two peaks occurring also at g_h = g_y. In later publications, Hoffmann supports

Yokoi's idea so that stellacyanin should be considered to display two inequivalent isotropic ^{14}N couplings [42,43].

Very broad (linewidth ~20 MHz) copper ENDOR lines were observed at frequencies beyond the ^{1}H and ^{14}N regions. They displayed peculiar spin dynamics and were best detected at fast sweep rates ($\geq$10 MHz/sec) of the rf frequency. Moreover, the nuclear Zeeman splitting could not be resolved and only two out of the expected three lines (nuclear spin I = 3/2 for ^{65}Cu and ^{63}Cu) were observed. For H_0 along g_x, the two lines were associated with a coupling of 167 MHz with a quadrupolar coupling of 6 MHz; for g_z a hyperfine interaction of 96 MHz and a quadrupolar term of 10 MHz were extracted. The usual difficulties for H_0 directed along $g_h = g_y$ in powder spectra led to an estimate of 80-100 MHz for the hyperfine coupling. Refined values were obtained by simulation of the EPR spectrum which yielded 87 MHz for the hyperfine and 4 MHz for the quadrupolar interaction.

With these data, the metal coordination is discussed in terms of a flattened tetrahedral geometry which is brought about mostly by D_{2d} distortions to allow for $4p_z$-orbital mixing with $3d_{x^2-y^2}$ which reduces the isotropic coupling of Cu [41]. Roberts et al. claim that this type of geometry alone is decisive in explaining the unusually small values of Cu hyperfine interaction without any need for a specific nature of the ligand(s). Specifically, the sulfur-to-metal charge transfer, which is thought to give rise to the "blue" property of stellacyanin, is proposed to be unconnected to the magnetic properties. Consequently, approximately the same EPR parameters should be found in systems without nitrogen and sulfur as ligands but with comparable geometry as is the case for Cu(II)-doped Cs_2 $[ZnCl_4]_2$ [44].

For the other two mononuclear blue proteins, plastocyanin and azurin, the copper coordination is known from single-crystal x-ray data [45,46]. There are two imidazole nitrogens and a cysteine sulfur in a trigonal arrangement, the copper being slightly above their plane. The fourth donor is a methionine sulfur. Stellacyanin lacks methionine and the origin and the number of the coordinating nitrogens is not presently clear from x-ray results. The ENDOR data discussed above clearly show coordination with at least two nitrogens. These may be due to imidazole groups as is suggested, for example, by EXAFS studies [47]. The nature of the fourth ligand, replacing methionine, is not known. Roberts et al. argue that the ^{14}N-ENDOR results would allow for a third nitrogen donor since two inversion-related nitrogens as are possible in a trigonal arrangement would not produce different ENDOR spectra from those measured [41]. In view of the invariable coordination of two imidazoles and one thiolate sulfur in a trigonal plane on account of the rather weak bond to methionine in the two other blue proteins, one would, however, be led to favor an arrangement of two imidazoles and cysteine in stellacyanin over a trigonal three-nitrogen coordination. The fourth ligand, weakly bound, would be a variable one, its nature being unresolved in stellacyanin.

If correct, the picture emerges for all mononuclear blue copper sites that the trigonal arrangement of two nitrogens and one sulfur is the main basis for the distorted tetrahedral configuration of Cu(II) which, in turn, is the basis for both the blue and the magnetic properties [cf. structure (IV)]. A comparative ENDOR investigation performed recently on six blue copper proteins by Roberts,

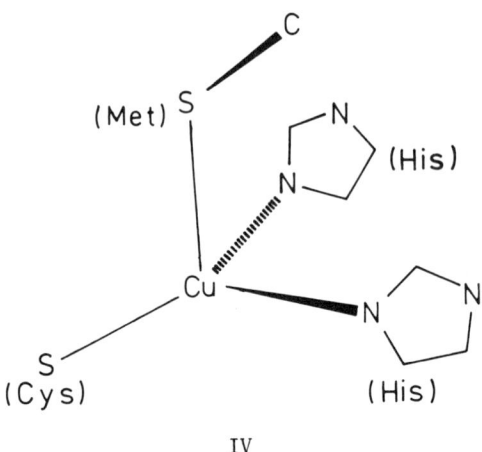

IV

Hoffmann, and coworkers [43] gave results which support this view. The specimens studied were azurin and plastocyanin as mononuclear

ENDOR: PROBING THE COORDINATION ENVIRONMENT IN METALLOPROTEINS

type 1 copper proteins together with tree and fungal laccase. Plastocyanin and azurin display EPR spectra which are nearly identical. Unlike stellacyanin, the symmetry of the g and hyperfine tensors is nearly axial (g_x = 2.042, g_y = 2.059, g_z = 2.226 for plastocyanin) with resolvable Cu hyperfine interaction only along $g_\parallel$. On the other hand, the ^{1}H- and ^{14}N-ENDOR results of all three samples appear to be rather comparable. Roberts et al. assign one strongly coupled proton interaction (27 MHz for plastocyanin, 24 MHz for azurin, and 21 MHz for stellacyanin) which is not exchangeable by deuterium to the methylene protons of cysteine. We mention that this assignment need not be unambiguous. Yokoi and coworkers studied spinach plastocyanin [48]. Although they devoted their attention mainly to nitrogen couplings, their spectra do not reveal the peaks corresponding to the proton coupling in question. Also, for laccase (see below), Roberts et al. [43] find two strongly coupled protons in the type 1 center which were not observed in later work by Desideri et al. [49]. It appears that the technical differences in the various experimental designs used to study the ENDOR response is perhaps responsible for the different results. The majority of results points at least to the existence of one such proton interaction.

For the ^{14}N resonances, Roberts et al. [43] find, for all mononuclear blue sites, two inequivalent nitrogens, which for azurin and plastocyanin have to be ascribed to the imidazoles. One nitrogen is strongly coupled, its interaction ranging from 49 MHz (plastocyanin) to 32 MHz (stellacyanin), the other nitrogen being associated with a coupling around 20 MHz. Yokoi et al. [48], for spinach plastocyanin, report only one ^{14}N response at frequencies of 24-28 MHz for various field positions of the EPR spectrum. They conclude that the two histidine ligands are equivalent in coupling (48-56 MHz), a finding which somewhat contrasts with x-ray diffraction results on the corresponding Cu-N bond lengths of the imidazoles involved.

The copper hyperfine and nuclear quadrupolar coupling for azurin and plastocyanin were measured at the parallel and perpendicular directions [43]. The values are comparable ($\parallel$: 185 MHz, $\perp$: 60 MHz for plastocyanin; $\parallel$: 161 MHz, $\perp$: 66 MHz for azurin hyperfine interaction). An effective axial symmetry of hyperfine coupling was also observed for stellacyanin (see above) but the EPR spectra differences would seem to suggest an enhanced rhombic distortion compared with azurin and plastocyanin. The nuclear quadrupolar interaction of stellacyanin reflects this symmetry but comparison is unfortunately not possible since in azurin and plastocyanin the tensor symmetry could not be clarified.

On the whole, the main features of the blue sites in mononuclear copper proteins as derived from ENDOR agree reasonably well with x-ray data, if available, or calculations to imply a distorted tetrahedral coordination with two imidazoles and one cysteine ligand in a trigonal arrangement. The fourth ligand, methionine in azurin and plastocyanin, seems to be responsible for the spectral differences to stellacyanin in which that ligand is unknown. For plastocyanin, recent calculations and EPR measurements have shown that p_z mixing with $3d_{x^2-y^2}$ of Cu(II) should be negligible and that the main source of the small Cu(II) hyperfine interaction in blue sites should be delocalization, especially onto the cysteine sulfur [50].

4.1.2.2. Laccase. The multinuclear Cu(II)-containing protein laccase is a blue oxidase and has four copper ions of which two are EPR-active in the resting state of the enzyme and display type 1 (blue) and type 2 (nonblue) Cu(II) spectra. Most intensely studied is laccase from *Rhus vernicifera* but fungal sources have also been employed. There appear to be differences between laccases from different origins regarding the distribution of electron acceptor sites during the catalytic cycle in which dioxygen is reduced to water employing four electrons (for details about reactivities, see, for example, [51]).

There are two independent ENDOR studies on the type 1 (blue) copper site in which the type 2 Cu(II) was either reduced [43] or in which its contribution was abolished by depletion of the specimen [49]. In the former work, laccase both from tree and from fungal sources (*Polyporus versicolor*) were studied whereas the other investigation employed specimen from the lac tree only. The EPR spectra of both sources display axial symmetry of the g tensor ($g_{\parallel}$ = 2.289, $g_{\perp}$ = 2.047 [26]). Comparing the two tree laccases, the results appear to be controversial. Desideri et al. [49] present rather well-defined ^{14}N resonances indicating, in the $g_{\parallel}$ direction, two inequivalent nitrogens with couplings corresponding to 36.6 and 53.6 MHz. Such resonances, although less resolved, are observed in the other study too but are assigned, on account of a shift of lines at different microwave frequencies, to protons of 20 and 25 MHz interactions. Only one more weakly coupled nitrogen (~20 MHz) is reported to occur, and another one supposed to be possibly present but obscured by the proton resonances [43]. Desideri et al., however, can find no evidence for such proton couplings. The authors have shown well-resolved ^{1}H resonances in the "weakly" coupled region (8-18 MHz) and found one of the couplings to be exchangeable by D_2O treatment (4.5-MHz coupling in the $g_{\parallel}$ direction). This proves that the solvent is accessible to the copper type 1 site, thus explaining the unusual relaxometry measured by NMR [52].

Roberts and coworkers [43] present results on the Cu(II) hyperfine interaction on both tree and fungal laccase for the $g_{\parallel}$ direction which is found to differ strongly (270 MHz for the fungal and 129 MHz for the tree laccase, respectively). For the fungal laccase, ^{14}N resonances were interpreted as being due to two nitrogens, one coupled weakly (~24 MHz at $g_{\parallel}$), the other one more strongly (~38 MHz at $g_{\parallel}$) [43].

The EPR-silent type 3 copper has attracted some attention recently both in EPR and ENDOR spectroscopy [42]. The copper of the type 3 was activated either during reoxidation of the reduced, native enzyme and/or by depletion of the type 2 copper. For the ENDOR study, which is on tree laccase (*R. vernicifera*), the type 3 copper has been made detectable in EPR by reduction of type 2 copper-depleted specimen. The EPR spectrum is anomalous and resembles neither type 1 nor type 2 copper spectra since it is of strong rhombic symmetry in the g tensor (g_x = 2.03, g_y = 2.15, g_z = 2.277).

ENDOR has yielded hyperfine information on three different nitrogens as near-neighbors. All of the couplings could be attributed to histidine ligands but at least for one of them a histidine can be invoked on account of an observed "remote" nitrogen resonance. All nitrogen couplings show appreciable anisotropy which is atypical for normal histidine coordination.

The copper hyperfine interaction was also resolved to some extent. One value can be obtained along g_z from EPR spectra simulation. Along g_x, ENDOR gives about 218 MHz with a quadrupolar contribution of 10 MHz. This latter interaction for g_z was estimated to be about 7 MHz, leaving both the hyperfine and the quadrupolar interaction with values close to zero for g_y. All values are atypical for normal copper coordination in their degree of anisotropy.

The proton ENDOR region at g_x revealed the presence of several weak interactions (<1 MHz) together with a strong coupling (~2.46 MHz) which was found to exchange against deuterium in D_2O samples. Therefore, one exchangeable proton site should be in close vicinity of the copper.

The coordination of type 3 copper in laccase is therefore considered to be of the type $Cu(N_3, X^-)$ with X offering an exchangeable proton which may be related to the bridge between the two Cu(II) ions of the normal type 3 copper. Alternatively, X could replace the bridge in the uncoupled moiety.

4.1.2.3. Cytochrome c oxidase. This enzyme certainly is one of the most intensely studied in the group of copper proteins employing a wide variety of techniques but the results still allow for no unified, detailed structural picture. The molecule contains two heme irons, a and a_3, and two copper ions. In the fully oxidized enzyme, two paramagnetic features are visible, one from a low-spin ferric iron (a) and one very atypical "copper" signal termed "intrinsic" copper or Cu_A, although there have been proposals that copper should not be involved in the paramagnetic species. Since the total integrated intensity of the EPR spectrum accounts for only half the paramagnetic centers, it was suggested that one iron (a_3) and one copper (Cu_B) should exist as antiferromagnetically coupled binuclear pair.

There are three ENDOR studies available to date which deal with the nature of Cu_A in cytochrome c oxidase either from beef heart

ENDOR: PROBING THE COORDINATION ENVIRONMENT IN METALLOPROTEINS 35

or yeast which, judged from EPR data, can be treated as being rather identical. These studies must be seen against a large background of results from other techniques like EXAFS, low-frequency (S-band) EPR, x-ray absorption edge and of theoretical calculations.

The first ENDOR investigation on Cu_A was on beef heart cytochrome c oxidase by Van Camp and coworkers in 1978 [53]. These authors detected a number of weakly coupled protons, none of which was exchangeable against D_2O. Also, two fairly large (12- and 19-MHz) and isotropic proton couplings were observed which were tentatively assigned to protons from CH or CH_2 groups one covalent bond away from the site of unpaired spin density (usually termed β protons). It was speculated that CH_2 protons near a ligated sulfur should be probable candidates.

For ^{14}N, only two of the expected four lines were observed, roughly split by the nuclear Zeeman frequency, the other pair supposedly being buried under the "free" proton region. The assignment was made that an average coupling of 17 MHz should result, sufficient to consider nitrogen as a near ligand to Cu_A.

Hoffmann and coworkers [54] confirmed the β-proton couplings and the general features of the ^{14}N resonances. They claim that an additional, large proton coupling ($\approx$70 MHz), which was meanwhile suggested to exist by Froncisz et al. [55] on the basis of S-band EPR spectra analysis, should not be considered further on account of lack of ENDOR resonance in the appropriate region. In addition to ^{1}H and ^{14}N interactions, ENDOR resonances from copper were observed which unambiguously confirmed the presence of this metal in the redox center responsible for the EPR signal. The coupling constants found, however, for Cu_A were unusually small and fairly isotropic ranging from ~70 to 90 MHz. Again, these values were smaller than those suggested from S-band EPR analysis [55].

Hoffmann et al. [54] discussed these data in terms of the two alternative models for Cu_A coordination which existed in the literature. One involved a coordination of ("diamagnetic") Cu(I) with a thyil radical RS⋅ from a coordinated cysteinyl in which the copper receives spin density from the radical thus explaining the small

Cu_A hyperfine values and the g tensor elements of the EPR spectrum which rather resemble those of thyil radicals than Cu(II). The other model is a Cu(II) ion in a strongly distorted tetrahedral environment, like in the blue proteins but with more admixture of 4p and 3s orbitals into the 3d manifold. The authors conclude that the data are not consistent with the $Cu_A(I)$-$\dot{S}$-R picture but rather favor a Cu(II) ion in a unique environment, coordinated weakly to one nitrogen (at least) and one RS⁻ fragment with a high degree of covalency.

This view has been challenged more recently by Chan and coworkers [56], who obtained ENDOR signals from yeast cytochrome c oxidase in which histidine nitrogens were isotopically substituted by ^{15}N and the cysteine CH_2 group protons by deuterons. Comparison with the native protein led to the conclusion that the aforementioned proton couplings of ~19 and 12 MHz, which are actually found to be 16.2 and 12.2 MHz in the yeast sample, had to be assigned to the cysteine CH_2 groups. An unsolved ambiguity remains as to whether one or two cysteine residues are involved. Likewise, the ^{15}N resonances which correspond to the ^{14}N coupling detected by Van Camp et al. [53] could be unequivocally assigned to the presence of at least one histidine as ligand to Cu_A. In addition, the authors were able to show in the isotopically substituted EPR spectra that the ~70-MHz coupling of a nucleus with formal I = 1/2 invoked by Froncisz et al. [55] can indeed be detected in X-band spectra, too, when the resolution is enhanced by the isotopic substitution. However, the previous proposal of a dipolar coupling between Cu_A and the heme iron (S_{eff} = 1/2) in cytochrome a was ruled out as source of the interaction on account of its temperature-independent magnitude. Thus, the exact nature of this signal still remains elusive.

In their discussion, Chan and coworkers [56] rule out the distorted tetrahedral coordination of Cu(II) as a model for the Cu_A site since it should require too much 4p mixing into $3d_{x^2-y^2}$ and $3d_{xy}$ orbitals to be reasonable. The authors feel that rather extensive delocalization onto sulfur from copper that should be considered as Cu(I) represents the appropriate description for the oxidized enzyme [cf. structure (V)]. They draw support from EXAFS studies on

```
        HN  ───╲                    ⁻S ─ CH₂ ─ R
            ╲   N ╲                ╱
         (HIS)    ╲              ╱   (CYS)
        R          ╲           ╱
                    Cu (I)
        HN  ───╲   ╱           ╲
            ╲   N                ╲ Ṡ ─ CH₂ ─ R
         (HIS)                      (CYS)
        R
```

V

cytochrome c oxidase which, in comparison with plastocyanin, give much less of a difference in Cu-S (cysteine) bond lengths between the oxidized and reduced states of the enzyme [57]. Thus, oxidation and reduction should not involve the copper but rather the cysteine residue which requires copper to remain in the cuprous state.

The other copper (Cu_B) in cytochrome c oxidase, EPR-silent (type 3) probably due to its coupling with the a_3 heme iron, has been made EPR-detectable recently by trapping an intermediate under turnover conditions [42]. The EPR spectrum of Cu_B strongly resembles that of laccase type 3 copper as do the ENDOR spectra, for which ^1H and ^{14}N resonances were detected. Hoffman and coauthors conclude from this similarity that the coordination sites of both type 3 coppers are identical or nearly so, i.e., involve Cu(II) in an environment of at least three nitrogens of which at least one is an imidazole.

4.1.2.4. *Superoxide dismutase.* ENDOR spectra from the native bovine superoxide dismutase were obtained recently by Van Camp and coworkers together with spectra from the azido (N_3^-) and cyano (CN^-) derivative in an attempt to probe the Cu(II) coordination differences of the native and the inhibited enzyme [58]. Earlier x-ray and EPR studies had established that the Cu(II) ion is coordinated by four imidazole nitrogens (histidines 44, 46, 61, and 118) and that histidine 61 constitutes an imidazolate bridge to the Zn(II) ion. In addition, it was argued that H_2O is bound to the metal as the fifth ligand and that CN^- perhaps replaces it [29]. The EPR spectrum of the native protein differs between human and bovine red blood cells as sources. In the former case, the g tensor is axial ($g_\perp$ = 2.06, $g_\parallel$ = 2.265) whereas in bovine superoxide dismutase there is rhombic symmetry (g_x = 2.029, g_y = 2.108, g_z = 2.265). When CN^- binds to the latter specimen, the spectrum attains axial symmetry ($g_\parallel$ ~ 2.208, $g_\perp$ ~ 2).

In ENDOR, the native dismutase gave highly complex spectra which were only tentatively assigned in terms of possible inequivalent coupling of Cu(II) with the four histidine nitrogens. One well-resolved nitrogen interaction was observed for the azido derivative at the $g_\parallel$ extreme (lowest field Cu hyperfine line, g ~ 2.47). Using the corresponding hyperfine value of 1.21 mT at g = 2.24 ($g_\parallel$), a simulation of the EPR Cu hyperfine line was found to agree with four equivalent ^{14}N couplings assumed.

More satisfactory results were obtained from the cyano complex at the $g_\parallel$ extreme which gave one well-resolved set of ^{14}N hyperfine lines corresponding to a coupling of 47.8 MHz and one less resolved ^{14}N nucleus with 37-MHz coupling. Attention was given also to the low-frequency part of the ENDOR spectrum (1-10 MHz) in which remote nitrogens are expected to show lines. Using isotopic substitution (^{15}N against ^{14}N) in the cyanide and a simplified powder ENDOR simulation together with EPR spectra simulations of the $g_\parallel$ extreme Cu hyperfine line, the authors arrived at a coordination model which involves a square-planar arrangement in which Cu(II) is ligated to three histidine nitrogens and the cyanide which coordinates via the carbon. Within this plane, two imidazoles are thought to be equivalent (47.8 MHz coupling) while the trans-histidine is inequivalent and shows the weaker interaction (37 MHz). We would prefer the assignment to be inversed since the strong interaction is very well resolved which is usually not the case for equivalent nitrogens.

4.1.2.5. *Cu(II)-substituted human transferrin*. Transferrin from the human blood serum is an iron-containing protein with two metal-binding sites. The apoprotein, however, can accommodate a variety of spectroscopically active metal ions. For strong iron binding, equimolar amounts of bicarbonate ions, HCO_3^-, are required. EPR spectra of the Cu(II) transferrin are well resolved and show nearly axially symmetric g (g_x = 2.042, g_y = 2.059, g_z = 2.312) and Cu hyperfine tensors (parallel component ~500 MHz) with some resolution of ligand superhyperfine structure. The latter has been assigned, by spectra simulation, to four nitrogens interacting in the bicarbonate-free and to one nitrogen in the bicarbonate-transferrin complex [59].

In their recent ENDOR study of Cu(II) transferrin, Roberts and coworkers [60] addressed themselves to the problem of Cu(II) coordination in the bicarbonate complex and to the question of whether the two metal-binding sites differ in the copper environment. Samples containing isotopically enriched ^{65}Cu and $^{13}CO_3^{2-}$ and specimen with D_2O exchanged against H_2O were prepared and the possible combinations compared.

^1H ENDOR showed two exchangeable proton resonances with coupling constants of 14.2 and 17.6 MHz at the $g_\perp$ direction which could either be due to two different, isotropic protons or one ^1H nucleus with strong in-plane anisotropy. The interaction was tentatively ascribed to a tyrosinate ligation to Cu(II). Several weakly coupled proton resonances (≤3.5 MHz) were also detected, none of which appeared to be exchangeable.

The ^{14}N resonances in the $g_\perp$ direction were found to compare favorably with the value derived from EPR spectra simulation. There was only one ^{14}N nucleus with a fairly isotropic coupling of ~31 MHz.

Copper ENDOR was obtained in the $g_\perp$ direction for which EPR fails to resolve the hyperfine structure. The resonance found was a single, broad peak at about 45 MHz which was ascribed to a coupling of ~88 MHz.

No indication of direct interaction of the bicarbonate with the Cu(II) was found in $^{13}CO_3$-substituted samples. Likewise, there was no detectable difference in the ENDOR spectra which would be related to a different Cu(II) coordination in the two binding sites.

The ENDOR data from all copper proteins are compiled in Table 1 together with those from the model complexes.

4.2. Cobalt(II) Proteins and Related Compounds

There are only few examples for the natural occurrence of cobalt coordinated in the functional group of biomolecules. One is cobalamin or vitamin B_{12} in which Co(II) is coordinated to four in-plane nitrogens in the corrin framework, which resembles the porphyrins of the hemoproteins, and to a fifth, axial nitrogen from benzimidazole. With respect to metalloproteins, the significance of Co(II) coordination derives mainly from the fact that it can be used to replace other, spectroscopically inactive ions in the prosthetic group. An important example is replacement of iron(II) in hemoproteins, e.g., myo- and hemoglobin, for which the Co(II)-substituted derivatives allow to study, with EPR, structural features of the reversible dioxygen binding to the heme. This aspect, which also relates to dioxygen binding to vitamin B_{12}, has prompted several EPR investigations of Co(II) model complexes, of vitamin B_{12} in the reduced and oxygenated form and of Co(II)-substituted hemo- and myoglobin in both the oxy and deoxy states [61-65]. The specimens were not only studied in frozen solution; a fairly large amount of single-crystal EPR information is available which would seem to provide for a good background for ENDOR studies.

The latter, however, have remained scarce. There is some ENDOR work on four- and five coordinated Co(II) model complexes with nitrogens and oxygens as ligands. The results, reviewed in [4], must remain of little interest in the present context since the few ENDOR studies available on biomolecules to date present an insufficient basis for relation. There is one preliminary investigation on the reduced cobalamin in host single crystals in which well-resolved weakly coupled protons were observed for one magnetic field orientation together with less resolved ^{14}N resonances. Additional powder ENDOR yielded approximately axially symmetric ^{14}N-coupling constants of 45 and 53 MHz and quadrupolar interactions of 0.8, 0.8, and -1.7 MHz. These parameters were assigned to the axial nitrogen of the benzimidazole [61]. No results appear to have been reported on the oxygenated vitamin B_{12} complex.

For Co(II)-substituted hemoproteins, one ENDOR study has been published so far in which attention was given to the differences in the weakly coupled proton region between deoxy-Co(II)- and oxy-Co(II)-hemoglobin. For the former compound, the observed small interactions (0.4-1.1 MHz) were tentatively assigned to the proximal histidine (F8) protons. Oxy-Co(II)-hemoglobin displayed larger proton interactions. By preparation of specimen in D_2O it was shown that one of these (~5.6-5.9 MHz) could be assigned to a proton of $N_{\varepsilon 2}$ of the distal side histidine (E7) which forms a hydrogen bond with the oxygen [66]. The ^{14}N ENDOR lines in both oxy- and deoxyhemoglobin are rather complex in the frozen solution system applied and have not yet been analyzed successfully.

The large amount of EPR results available for reduced and oxygenated vitamin B_{12} and for myo- and hemoglobin allows for a comparison between the two systems. Compared to the hemoproteins, the corrin backbone lacks one methine bridge of the porphyrin and the pyrrole side groups are different, but the general, fivefold coordination with nitrogens as near-neighbors is strongly related. Correspondingly, the main EPR features, g, Co(II) hyperfine, and ^{17}O hyperfine tensors are fairly similar. In the reduced state, the Co(II) hyperfine interaction is nearly identical for myo- and hemoglobin and approximately the same as in reduced vitamin B_{12}. Likewise, oxygenation of vitamin B_{12r} to $B_{12r}O_2$ brings about the same drastic decrease in Co(II) hyperfine coupling as in the hemoproteins, coupled with a ^{17}O hyperfine interaction which corresponds to a spin density of over 70% on the dioxygen. Also, the directional interrelation of g and hyperfine tensors are comparable in both systems. In the oxygenated species, the maximum g-tensor element coincides with the O-O bond direction whereas the cobaltous hyperfine axis system is coupled with the corrin or porphyrin plane [61,62].

Differences between the oxygenated forms of myoglobin and cobalamin become apparent, on the other hand, in the influence of temperature on the Co-O-O configuration. Vitamin $B_{12r}O_2$ displays only one low-temperature configuration (T < -160°C) whereas two distinct sites are found in myoglobin. The latter coalesce into one room-temperature configuration whereas in cobalamin there is considerable disorder in the Co-O-O complex at higher temperatures, perhaps due to the lack of a "distal" endogenous ligand stabilizing the configuration [61-63].

Although the above data suggest that the main features of Co(II) coordination (at low temperature) are independent of variations in the second coordination shell (e.g., corrin vs. porphyrin), there is recent evidence that this holds only as a first approximation since differences in the Co-O-O bond configuration have been found between oxyproto- and oxymesocobalt porphyrin-substituted myoglobin [64]. It would be worthwhile probing such differences with ENDOR in order to achieve a more detailed knowledge of Co(II) coordination stereochemistry.

4.3. Iron Proteins

Iron is one of the most abundant transition metal ions occurring in cells and body fluids. Nearly all of it is bound to protein either chelated into heme or not. The hemoproteins serve as oxygen transport and storage proteins (hemoglobin, myoglobin), as electron transport systems (cytochromes), or as catalases and peroxidases. Nonheme iron proteins can be iron storage or transport proteins or can transport electrons. There has been a formidable amount of spectroscopic work performed on iron proteins, notably NMR, EPR, and Mössbauer spectroscopy together with optical spectroscopic techniques like Raman and resonance Raman spectroscopy (for reviews, see, for example, [67] and [68]).

4.3.1. Heme Proteins

In this survey, we shall obey to the classification into heme and nonheme proteins and start with a description of heme proteins first. The most important and interesting representatives of this group are myo- and hemoglobin. Together with studies on related model complexes, the bulk of ENDOR work on metalloproteins has been performed with these moieties, which constitute the oxygen storage and transport system in the cells and the blood of vertebrates and some invertebrates. Myoglobin consists of a single polypeptide chain of 153 amino acids and one prosthetic heme group with Fe(II) as central ion. The heme group and its adjacent globin amino acid residues are shown in structure (VI) as derived from x-ray diffraction [69].

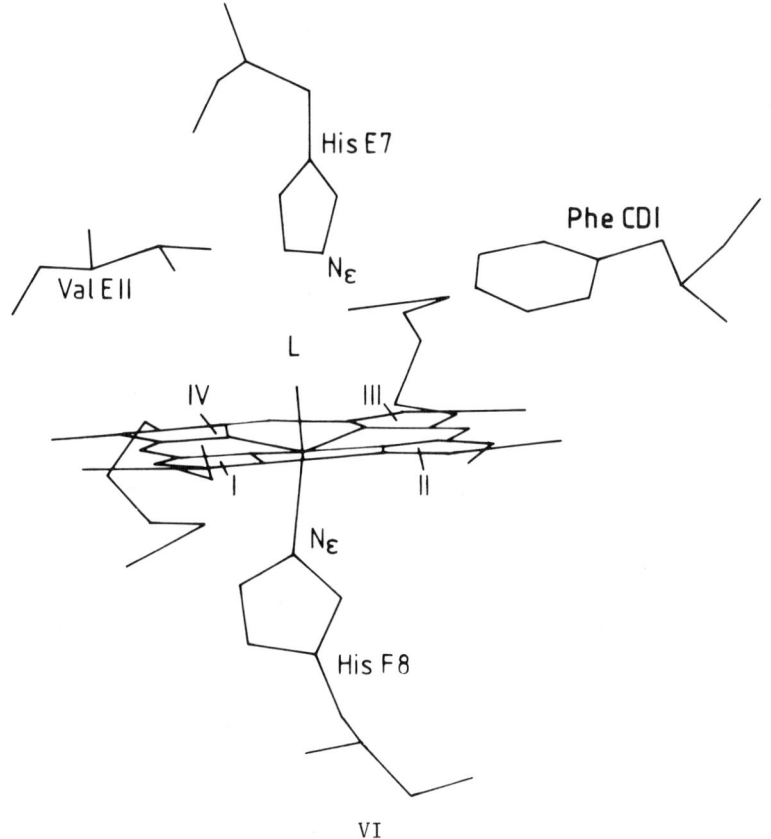

VI

Hemoglobin is a tetrameric protein of two pairs of myoglobinlike subunits, α and β chains. In contrast to myoglobin, it exhibits a cooperative oxygen binding which is associated with conformational changes of the heme group and the globin moieties. The current stereochemical model developed by Perutz [70] describes in detail the induced changes of the subunit tertiary structures and the change in quaternary structure of the whole molecule when the transition from the low-oxygen affinity form T to the high-affinity form R is made upon oxygenation of deoxyhemoglobin.

For EPR and ENDOR spectroscopy, both physiologically important states oxy- and deoxyhemoglobin (HbO_2 and Hb) are inaccessible due to the Fe(II) spin states $S = 2$ in Hb and $S = 0$ in HbO_2. Most of the EPR knowledge about Hb and myoglobin (Mb) which has been accumulated for more than three decades has been derived from Fe(III)-substituted specimen which, depending on the ligand, can be in either a high-spin state ($S = 5/2$), like aquomet Hb and -Mb (HbH_2O, MbH_2O), or in a low-spin state ($S = 1/2$), as in hydroxymet Hb and -Mb ($HbOH^-$, $MbOH^-$) or, say, cyanide-ligated $HbCN^-$ or $MbCN^-$. For this group of hemo- and myoglobins and related model complexes like Fe(III)-protohemin, Fe(III)-tetraphenylporphyrin (TPP), likewise the majority of ENDOR data has been obtained since the work of Feher and coworkers in the early 1970s [1]. Since there are two excellent reviews available which describe the up-to-date aspects of the Fe(III) results [4,10], we shall not go into much detail in their presentation in this chapter. The main attention will be given to the physiologically relevant Fe(II)-heme complexes which can be obtained by specific ligation.

The most prominent example is nitrosyl-hemo- (and -myo-)globin in which the NO radical when ligated to Fe(II) renders the protein to become EPR-active. HbNO can be considered as a good model for Hb since it exhibits reversible binding, cooperativity, and Bohr effect similar to the native Hb [71,72]. Consequently, there have been quite a few EPR investigations of NO-ligated Hb and Mb in both frozen solutions and single crystals. The EPR spectra which have a rhombic g-tensor symmetry were separated, in HbNO, into α- and β-chain contributions and spectral changes upon addition of allosteric effectors and/or pH changes could be attributed to conformational transitions (R ↔ T) of Hb [73-75]. Studies on hybrid hemoglobins yielded evidence for the inequivalence of α and β chains in their mutual influence [76,77] and mutant hemoglobins have shown a definite influence of tertiary structural changes on the EPR spectrum. Investigations of MbNO in single crystals allowed for the determination of the Fe-N-O bonding configuration and revealed that more than one species is stabilized at low temperatures, comparable to the case of $CoMbO_2$ (see above) and that at higher temperatures (approximately -20°C) a different configuration is attained [78,79]. For HbNO single crystals, the principal axes of the $\underline{g}$ and $\underline{A}$(NO) tensors were established for both subunits resulting in a more distorted bonding configuration for the β chains compared with the α chains. Also, the cooperative exchange of O_2 against NO was studied which revealed an enhanced exchange rate of the α chains [80,81].

Another Fe(II)-Hb model system with, however, much reduced possibilities of variation of physiologically relevant parameters is HbO_2^- (or MbO_2^-), which is formed upon exposure of oxygenated samples to ionizing radiation at low temperatures (77 K and below) [82]. Addition of radiation-generated electrons to the Fe(II)-O_2 complex results in the formation of an EPR- and ENDOR-active [Fe(II)-O_2]$^-$ center. The species at 77 K are primary ones for which rhombic g-tensor symmetry has been found. Two different initial centers are present in HbO_2 which can be assigned to α- and β-chain contribution. Likewise, two different primary [Fe(II)-O_2]$^-$ complexes are stabilized in MbO_2 at low temperatures, the majority species displaying g-tensor values comparable to that in the α chains of HbO_2 [83]. Single crystals of MbO_2 yielded, together with isotopic substitution (^{17}O, ^{57}Fe), the bonding geometry and the spin density distribution for the majority species [84]. Upon thermal annealing, the primary species in both HbO_2 and MbO_2 were found to transform into aquomet-Hb and Mb via a surprisingly large number of intermediate species of rhombic $\underline{g}$ symmetry [83].

Another Fe(II) hemoprotein to be considered is cytochrome c oxidase (cf. Sec. 4.1.2.3) for which the heme a_3 is of interest. The O_2 adduct of the reduced enzyme is in a ferrous state of the iron and not accessible to EPR investigations. Replacement of O_2 by NO, however, produces a ferrous heme-NO complex which displays, as for hemo- and myoglobin, an EPR spectrum of rhombic g-tensor symmetry [85]. The other heme iron, heme a in cytochrome c oxidase, is EPR-active in the resting state of the enzyme. A low-spin iron EPR signal is observed which has been ascribed to a sixfold coordination of iron with nitrogens, four in-plane pyrrole and two axial imidazole nitrogens on the basis, for example, of comparison of g tensors for a variety of low-spin heme model compounds [86].

4.3.1.1. NO-ligated Fe(II)-porphyrin compounds. Six-coordinated NO-ligated Fe(II)-porphyrin derivatives with a nitrogenous base as the sixth ligand have been used as models for the heme proteins. An example is tetraphenylporphyrin (TPP) [cf. structure (I)] with Fe(II) as central ion and NO and imidazole as axial ligands (NO-FeTPP-Im). It exhibits, in EPR spectra, a rhombic g-tensor symmetry (g_1 = 2.080, g_2 = 2.003, g_3 = 1.976) and resembles that of MbNO in the overall pattern as can be inferred from the spectra for both compounds shown in Figure 6. Other model compounds studied involve octaethylporphyrin (OEP) which, in comparison to TPP, contains the methine bridge protons which are also present in Mb and Hb. As sixth ligand, pyridine can be exchanged against imidazole. ENDOR studies on this group of compounds have been performed in low-temperature glasses using toluene or CCl_4 as glass-forming solvents. ^{14}N hyperfine interaction was resolved for both the NO and the nitrogenous base ligands. Proton resonances have revealed the nature of the weakly coupled protons [85,87, this work]. Table 2 lists representative values for the NO-FeTPP-Im system. The $^{14}N(NO)$ hyperfine interaction is roughly axially symmetric with values of $A_\perp \sim$ 32-35 MHz and of $A_\parallel \sim$ 64 MHz, the parallel direction being associated with the intermediate g-tensor element at which position the hyperfine coupling can also be derived from EPR spectra. The imidazole ^{14}N

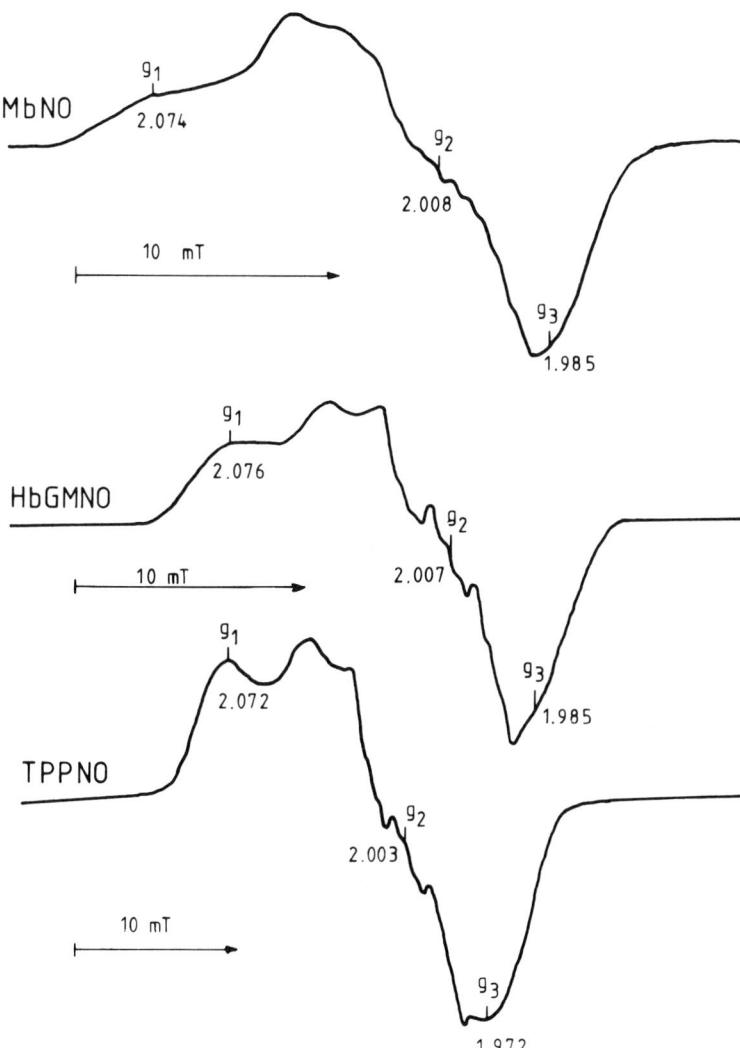

FIG. 6. EPR spectra (first derivative) of NO-ligated specimen of myoglobin (MbNO, top), monomeric *Glycera* hemoglobin (HbGMNO, middle), and Fe(II)-tetraphenylporphyrin-imidazole (TPPNO, bottom) showing comparability of rhombic g tensors indicated.

TABLE 2

^{14}N and ^1H Hyperfine Couplings (in MHz) from Frozen Glasses of NO-Ligated Fe(II)-TPP-Imidazole [85, this work]

g Factor	^{14}N (NO) interaction		^{14}N Imidazole interaction		^1H interaction and assignment	
g_1: 2.080	A_1	: 31.7 ± 0.3[a]	A_1	: 16.65 ± 0.2	0.3[b]	
			P_1	: 1.0 ± 0.2	0.6[b]	
					1.8[b]	Methine protons (in OEP)[d]
g_2: 2.003	A_{max}	: 64.4 ± 1.5[c]	A_{max}	: 19.6 ± 1.5[c]	0.36[b]	
	A_2	: 34.8 ± 1.0[a]	A_2	: 17.9 ± 0.2	0.70[b]	
					1.70[b]	Methine protons (in OEP)[d]
					3.60	Carbon-bound His (F8) protons
g_3: 1.976	A_3	: 32.7 ± 0.4[a]	A_3	: 16.5 ± 0.3	0.4	
			P_3	: 0.9 ± 0.2	0.70	
					1.70	Methine protons
					3.20	Carbon-bound His (F8) protons

[a]From Ref. 85.
[b]Not well-expressed shoulders.
[c]From respective EPR spectra.
[d]OEP = octaethylporphyrin.

hyperfine coupling gives rise to ENDOR resonances in the 6- to 11-MHz region which translates to interactions of 16-20 MHz, the largest coupling again being associated with g_2. Within experimental limits, there is no quadrupolar interaction detectable at g_2 whereas along g_1 and g_3 values of about 1 MHz are found.

For the ^{1}H interactions, the variation in lines between TPP and OEP allows for assigning a coupling of 1.7-1.8 MHz to the methine protons. Likewise, a line pair corresponding to an interaction of 3.2-3.6 MHz can be assigned to carbon-bound protons of the axial imidazole ligand by employing N-methyl-substituted and fully deuterated imidazole ligands. The additional, smaller interactions shown to occur in Table 2 can be due to the other "tensor" elements of the above-mentioned interactions or to the peripheral porphyrin protons.

NO-ligated dimethylhemin in CCl_4 was studied as a model for proton interactions in five-coordinated systems. It shows a strongly reduced range of proton couplings (0.3-0.4 MHz) [87].

4.3.1.2. NO-ligated myo- and hemoglobin. NO-ligated human Hb (HbA) and its isolated α and β subunits together with sperm whale Mb were studied in frozen solutions with ENDOR under a wide range of conditions including the influence of allosteric effectors (inositol hexaphosphate, IHP) and of pH variations between 6.4 and 8.4 [87]. Emphasis was given to the proton interactions in the "weakly" coupled region as well as to imidazole ^{14}N interaction of the proximal histidine [$N_{\varepsilon 2}$ of His (F8)]. More recently, we have obtained results from MbNO single crystals and from frozen solutions of the monomeric fraction of Hb from *Glycera dibranchiata* (HbGM) which has the distal histidine (E7) replaced by a leucine residue. The EPR spectrum of the latter specimen is also included in Figure 6.

Typical ENDOR responses for the three frozen solution specimens MbNO, HbNO, and HbGMNO (NO-ligated monomeric *Glycera* Hb) in the frequency range 1-11 MHz which comprises the pyrrole and His (F8) ^{14}N interactions are shown, taken along g_1, in Figure 7. Although there are discernible differences in the pyrrole range, the signals assigned to the imidazole interaction (6- to 11-MHz range) apparently

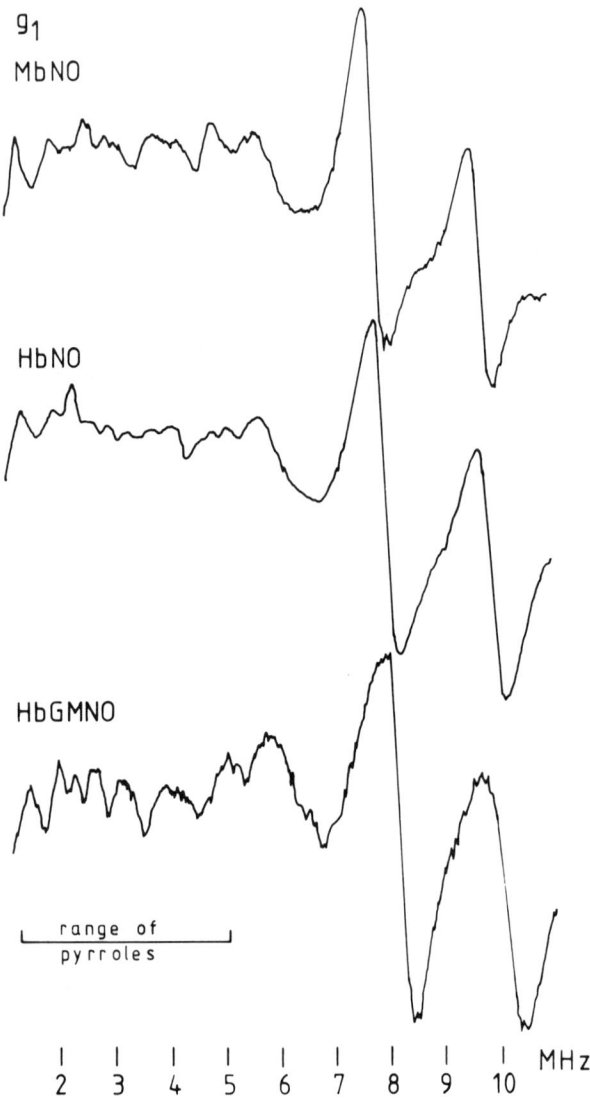

FIG. 7. ENDOR spectra (FM modulation) from NO-ligated myo- (MbNO, top), hemoglobin (HbNO, middle), and monomeric *Glycera* hemoglobin (HbGMNO, bottom) in the frequency range 1-11 MHz taken at field position corresponding to g_1 (cf. Fig. 6). Range of ^{14}N interactions for pyrrole nitrogens is indicated. The stronger lines at 6.5, 8.5, and 10.5 MHz stem from ^{14}N couplings of $N_{\epsilon 2}$ [His (F8)].

are rather similar in the three samples. They consist of a strong
doublet at about 8 and 10 MHz split approximately by the twice
nitrogen nuclear Zeeman term (~2 MHz) and a weaker low frequency
line at about 6 MHz. We have previously assigned only the doublet
feature to the histidine $N_{\varepsilon 2}$-interaction [87] but now tend, in the
light of single-crystal measurements discussed below, to include the
weaker line into the set which then is a triplet resulting from a
quadrupolar interaction incidentally having just half the value of
the nuclear Zeeman interaction. The resulting ^{14}N hyperfine interaction along g_1 then is about 1 MHz smaller than previously reported
[87].

Table 3 lists the ^{14}N interactions for the monomeric species
MbNO and HbGMNO in frozen solutions for both N(NO) and $N_{\varepsilon 2}$ (imidazole).
In order to obtain means for comparison between imidazole interactions
of many different samples, we have previously treated the couplings
obtained at $g_h = g_2$ as being due to the principal tensor element to
a rough, first approximation [87]. For MbNO, the values obtained for
the $N_{\varepsilon 2}$ interaction from single crystals which are listed in Table 4
can be used for comparison to document that the approximation underestimates the maximum coupling value which occurs nearly along g_2 and
thus should, if resolvable, rather be extracted from EPR data. For
the quadrupolar interaction, all values from powder ENDOR spectra are
smaller than those from the single crystals, an observation reflecting
the differences in ENDOR linewidths between single crystal and powder.

Although it is difficult to relate the direction cosines for
the $\underline{A}$ and $\underline{P}$ tensors obtained from single crystals directly to powder
spectra without a full simulation, we can infer from the behavior of
the orientational variation of the single-crystal spectra the assignment of the powder ENDOR pattern obtained along g_1. Upon rotation
in the abc*-axis system, the single-crystal g factor closely approaches
the value for g_1 ($\Delta g = 0.004$) at $H_0 \parallel c^*$. In this orientation, clearly
a quadrupole split nitrogen coupling is observed in the single crystals,
its low-frequency line varying from about 6.2 to 5.8 MHz within a window of orientations of ±20° around c*. We thus have to conclude that

TABLE 3

^{14}N Hyperfine (A) and Nuclear Quadrupole Couplings (P) (in MHz) from Frozen Solutions of NO-Ligated Myoglobin (MbNO) and in the Monomeric Fraction of *Glycera dibranchiata* Hemoglobin (HbGMNO) [85, this work]

	MbNO		HbGMNO	
g Factor	^{14}N (NO)	^{14}N ($N_{\epsilon 2}$ (His (F8)))	^{14}N (NO)	^{14}N ($N_{\epsilon 2}$ (His (F8)))
g_1	A_1: 27.7 ± 0.3[a]	A_1 : 15.8 ± 0.1 P_1 : 1.0 ± 0.1	A_1 : 38.1 ± 0.2 P_1 : 0.56 ± 0.2	A_1 : 16.20 ± 0.1 P_1 : 1.0 ± 0.1
g_2	A_2: 56.0 ± 1.5[a]	A_{max}: 18.5 ± 1.5[b] A_2 : 16.7 ± 0.1 P_2 : 0.4 ± 0.1	A_{max}: 64.1 ± 1.5 A_2 : 39.7 ± 1.0[c]	A_{max}: 19.1 ± 1.5[b] A_2 : 17.3 ± 0.1 P_2 : 0.44 ± 0.1
g_3		A_3 : 16.75 ± 0.1 P_3 : 0.42 ± 0.1	A_3 : 38.4 ± 1.0[c]	A_3 : 17.4 ± 0.1 P_3 : 0.43 ± 0.1

[a]From Ref. 85.
[b]From respective EPR spectra.
[c]Broad lines, unresolved quadrupole interaction.

TABLE 4

^{14}N Hyperfine ($\underline{\underline{A}}$) and Nuclear Quadrupole Tensor ($\underline{\underline{P}}$) in MHz for N$\varepsilon$2 [His (F8)] Interaction in NO-Ligated Single Crystals of Myoglobin

$\underline{\underline{A}}$	Direction cosines[a]			$\underline{\underline{P}}$	Direction cosines[a]		
	a	b	c*		a	b	c*
20.4	0.8364	±0.5388	0.1004	1.67	0.8919	±0.4401	0.1042
15.2	±0.3604	0.6787	±0.6399	0.86	±0.4058	0.8804	±0.2455
16.8	−0.4130	±0.4991	0.7618	0.98[b]	−0.1997	±0.1767	0.9638

a_{iso} = 17.46

[a] Upper signs refer to molecular site A, lower to B.
[b] Sp (P) = 0.17 is within error margins.

the corresponding line at g_1 in the powder spectra belongs to the $N_{\epsilon 2}$ interaction pattern.

The $\underline{\underline{A}}$ tensor ($N_{\epsilon 2}$) derived from single crystals is not very anisotropic. A rough estimate of the spin density yields about 4% on N ($N_{\epsilon 2}$) with 1.2% 2 s-character [88]. The maximum hyperfine interaction of 20.4 MHz agrees well with the value obtained from EPR spectra along g_2. The reason for the failure of observing this coupling in powder ENDOR is not fully clear presently. A similar finding was made for NO-ligated cytochrome c oxidase heme a_3 [85].

The single-crystal results presented have been derived from a first-order analysis neglecting, for example, the g-tensor anisotropy. We expect that minor changes have to be introduced in the final analysis which presently is in preparation. Nevertheless, the directional interrelations between $\underline{\underline{g}}$, $\underline{\underline{A}}$ ($N_{\epsilon 2}$), and $\underline{\underline{P}}$ ($N_{\epsilon 2}$) can be used for presenting an approximate description of imidazole configuration on the proximal side in relation to the Fe-N-O geometry. The maximum $\underline{\underline{A}}$ and $\underline{\underline{P}}$ elements include an angle of about 6° and are co-linear within 10-15° with the intermediate g-tensor element g_2 which in turn exhibits the least deviation from the heme normal (30-40°) and has been correlated with the Fe-N (NO) bond [78]. We therefore reason that A_{max} and P_{max} represent the direction of the Fe-$N_{\epsilon 2}$ (His F8) bond which then forms an angle of 40-50° with the heme normal. A further information can be gained for the direction of the imidazole plane with respect to the Fe-N-O plane when associating the minimum $\underline{\underline{P}}$-element direction with the imidazole plane normal. This direction deviates by about 20° from the Fe-N-O plane normal supporting the view that both planes are approximately aligned. A view of the arrangement of Fe-N-O and the His (F8) imidazole with respect to a plane comprising the pyrrole nitrogens N IV and N II together with the heme normal is shown in Figure 8. One should note, however, that the Fe-N-O plane is tilted toward the back (direction of N III) and the imidazole plane points toward the front (direction of N I), respectively, when viewed against the N IV to N II hemenormal plane.

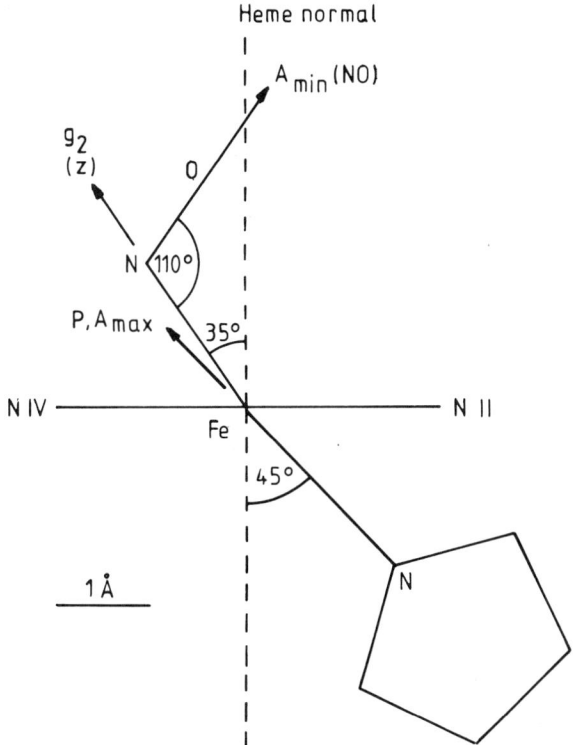

FIG. 8. Bonding arrangement of [His (F8)] $N_{\epsilon 2}$-Fe-NO in MbNO as derived from ENDOR on single crystals projected on a plane containing pyrroles N II and N IV [cf. Struct. (VI)] and the heme normal. Directions of relevant $\underline{\underline{g}}$, $\underline{\underline{A}}$, and $\underline{\underline{P}}$ elements are indicated. See text for details.

If one takes, for the sake of comparability, the average value of the imidazole hyperfine interactions as measured in ENDOR from frozen solution spectra including the "apparent" value for g_2, one finds the trend that the coupling is largest in model compounds (17.0 MHz for NO-FeTPP-Im) and in *Glycera* HbGMNO (16.95 MHz) and decreases to 16.4 MHz in MbNO. For the tetrameric molecule HbNO and its α and β subunits, the weakest imidazole coupling is found in α chains (16.49 MHz) whereas β chains and the T-state tetrameric HbNO + IHP exhibit about 16.95-MHz interaction [87]. This trend reflects the bond strength of Fe and $N_{\epsilon 2}$, which apparently is smaller in the α chains compared with the β subunits. This finding relates to the observation that the α chains lose the $N_{\varepsilon 2}$ interaction completely when a tertiary transition from the relaxed r to the tense t structure is forcibly induced by pH variation, whereas the interaction remains intact in the β chains [87].

The situation is more complex in the tetrameric HbNO. The average in the quaternary T-state $N_{\varepsilon 2}$ coupling is somewhat (0.3 MHz) smaller than for the isolated β chains (in t). Since hybrid studies have confirmed that the α chains lose the $N_{\varepsilon 2}$ interaction in the tetrameric molecule also [89], the difference must be accounted for by a slightly different conformation between isolated β chains and those embedded in the tetrameric Hb. For the R-state tetramer, rather broad ^{14}N ENDOR resonances were observed which were attributed to a superposition of the corresponding subunit spectra but separation was not possible.

We note that the trend discussed above for the average $N_{\varepsilon 2}$ interaction has a relation with the corresponding N (NO) hyperfine interaction. The data listed in Table 3 show that a weak $N_{\varepsilon 2}$ coupling, as for MbNO, correlates with a small value of ^{14}N interaction of N (NO) whereas, for example, in HbGMNO a strong $N_{\varepsilon 2}$ bond corresponds to a larger value of the NO coupling. Since an increase in N (NO) indicates a weaker Fe-N (NO) bond, this finding is at variance with the usual assumptions deriving from the so-called trans effect in which the NO-ligand and the $N_{\varepsilon 2}$ bond strengths are considered to influence each other such as to relate a smaller $N_{\varepsilon 2}$ coupling to a larger N (NO) interaction [90].

The ENDOR spectra from frozen solutions of all NO-ligated specimens, HbNO, HbGMNO, and MbNO proved to be information-rich in the weakly coupled proton region (10-20 MHz) [87, this work]. Up to nine line pairs centered around the proton ν_n (~14 MHz at X-band frequencies) could be distinguished. The corresponding coupling values for some of them are listed in Table 5 for the three g-turning points of observation.

In MbNO, there is a unique exchangeable proton with couplings of 5.4 MHz at g_1 and 2.0 MHz at g_3. Its probable origin is the

TABLE 5

^{1}H Hyperfine Couplings (in MHz) for Various Protons in Frozen Solutions of NO-Ligated Myo- (MbNO), Hemoglobin (HbNO) in the R State and of Monomeric *Glycera dibranchiata* Hemoglobin (HbGMNO) [87,95, this work]

g Factor	MbNO	HbNO	HbGMNO	Assignment based on calculations for MbNO
g_1	0.35	0.35	0.40	
	0.60	0.60	0.65	
	1.40	1.40	1.35	Methine protons M_1 and M_3 (tentative)
	2.35	2.20	2.15	Methine protons M_2 and M_4
	3.10[a]	3.20[a]	3.00[a]	
	4.10		4.30[a]	Carbon-bound His (F8) protons
	5.40[b]	5.00[b]		$N_{\epsilon 2}$ [His (E7)] hydrogen bond
g_2	0.35	0.35	0.35	
	0.70	0.60	0.75	
	1.40	1.30	1.30	
	1.95[a]	1.95[a]	1.95[a]	
	2.20[a]	2.50[a]	2.65[a]	
	2.95[a]			
	3.70	3.80	3.85	
	4.10			
	5.50[a]	5.10[a]	5.20[a]	
g_3	0.35	0.40	0.40	
	0.70	0.60	0.65	Methine protons M_2 and M_4
	1.40[a]	1.35[a]	1.25[a]	Methine protons M_1 and M_3 (tentative)
	2.00[b]	2.00[b]	2.00[b]	$N_{\epsilon 2}$ [His (E7)] hydrogen bond (MbNO, HbNO)
		3.10		Aspartic acid (tentative) in HbGMNO
	3.60	3.80[a]	3.70	Carbon-bound His (F8) protons
	4.80[a]	4.90[a]	5.40[a]	
	5.80[a]	6.00[a]	6.50[a]	

[a]Broad shoulders, weakly expressed.
[b]Deuterium-exchangeable.

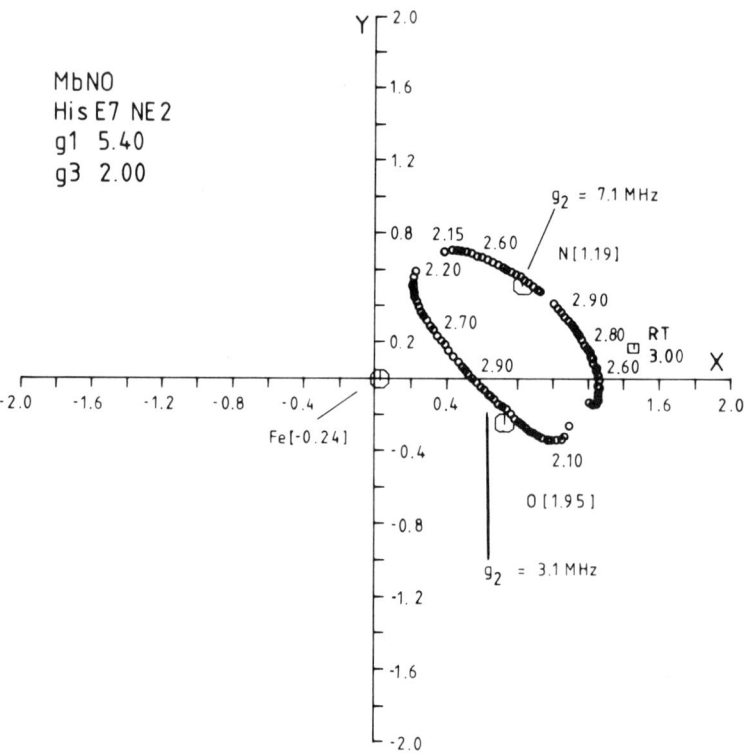

FIG. 9. Heme plane projection of calculated positions (○) of the $N_{\epsilon 2}$ [His (E7)]-bound proton in MbNO for two assumed couplings along g_2 (3.1 and 7.1 MHz) with fixed, experimental couplings at g_1 (5.4 MHz) and g_3 (2 MHz). The Fe, N, and O positions are shown (◯) in the xy frame, their z coordinates indicated in brackets. The x-ray-derived position of the proton at 300 K (RT) is given as □. The numbers at the calculated positions give the z-coordinates (in Å). Scales for x and y axes are in Å.

proton at the nitrogen $N_{\epsilon 2}$ of the distal histidine (E7) which is supposed to form a hydrogen bond to the ligand [91]. Using dipolar calculations in which a spin density distribution of 70% on the iron, 25% on NO, and 5% on $N_{\epsilon 2}$ [His (F8)] was employed, we find "allowed" proton positions for the $N_{\epsilon 2}$ [His (E7)] proton shown in Figure 9 when allowing the coupling at g_2 to vary between 3.1 and 7.1 MHz, values which are, however, not observed in the powder spectra for reasons probably related to the usual problems at this g turning point. Comparison with the room temperature proton position shows that it has shifted only by about 0.5 Å. Other spin density distributions yield unreasonably large shifts.

An exchangeable coupling of similar size is also found in HbNO which also should be assigned to a hydrogen bonded $N_{\epsilon 2}$ (E7) proton. As discussed above for the $N_{\epsilon 2}$ [His (F8)] interactions, the α chains lose this coupling upon a tertiary transition from r to t [87].

Incidentally, HbGMNO also displays an exchangeable proton interaction along g_3 with a coupling of 2 MHz but has no corresponding resonances at g_1 or g_2. Since the Glycera Hb lacks the distal histidine we have to look for another source. It appears possible that aspartic acid, which neighbors the E7 leucine in HbGM [92], is causing the observed interaction. Another explanation would be that the interaction at g_3 is common to all specimen. This would require a proximal side interaction for which dipole calculations have yielded no reasonable source.

For myoglobin (MbNO), another set of proton couplings, 1.4 MHz at g_1 and 0.7 MHz at g_3, can best be assigned to the methine protons M2 and M4 on account of dipole couplings which translate into possible proton positions shown in Figure 10. The difference to crystal structure-derived room temperature coordinates is seen to be very small as is expected for these protons since they are inflexible and should stay in position upon cooling. The other pair of methine protons is inequivalent and probably connected to a coupling of 0.6 MHz and 0.7 MHz observed at g_1 and g_3, respectively. Other possible combinations of experimental couplings do not yield reasonable methine proton locations.

A group of couplings of 4.1 MHz at g_1 and 3.6 MHz at g_3 can be attributed to carbon-bound proton(s) of the proximal histidine imidazole ring on account of results from NO-ligated heme models (see above). Additional evidence comes from calculations of proton locations taking into account the low-temperature position of the

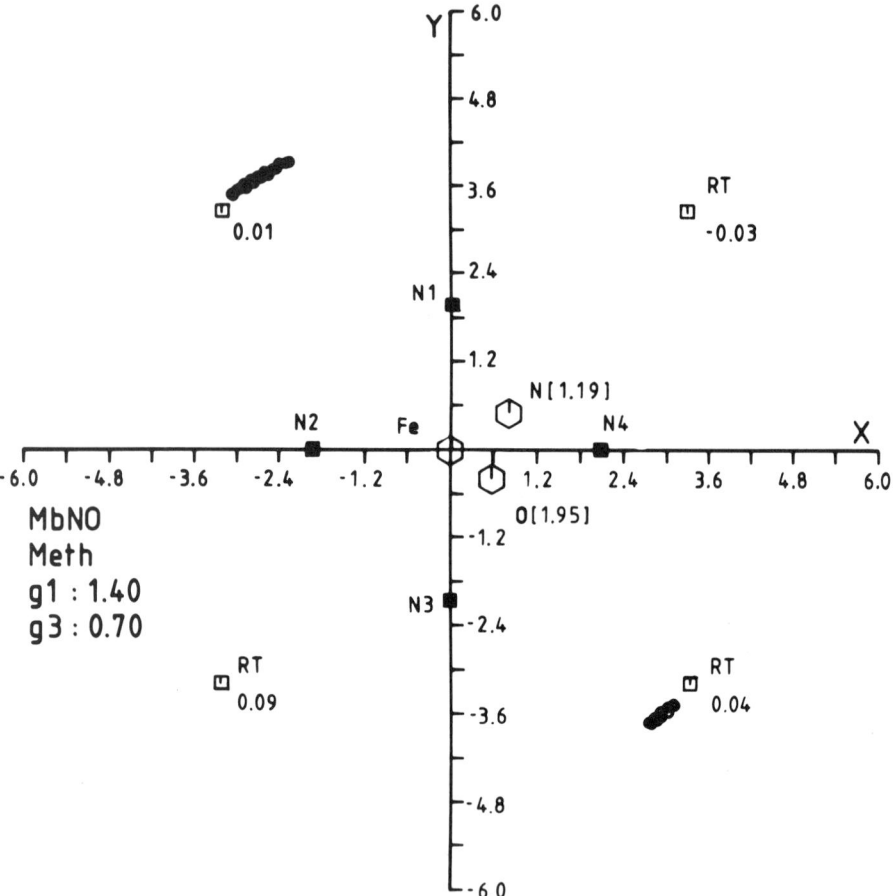

FIG. 10. Heme plane projection of calculated proton positions (o) from experimental ENDOR couplings at g_1 (1.4 MHz) and g_3 (0.7 MHz) showing Fe-N-O coordinates in x and y (◯) with z values in brackets and positions of pyrrole nitrogens (■) and methine protons (□) as derived from x-ray structural data at 300 K (RT). Calculated positions are seen to agree well with two methine protons.

$N_{\varepsilon 2}$ of His (F8). Allowed proton locations were found 2.3 Å from this nitrogen which is the expected distance for carbons CE1 and CD2.

The remaining proton interactions observed in myoglobin can presently not well be related to actual locations in the heme environment since calculations give ambiguous results. We hope to derive the necessary information shortly from the analysis of the single-crystal data which is presently under way.

For HbNO, the situation is more complex due to contributions of α- and β-chain couplings. In all specimen, the tetramer in both the R and T states and the isolated subunits, an interaction of about 1.4 MHz is found at g_1 which could be grouped with an interaction of ~0.4-0.6 MHz at g_3 to indicate the methine protons as probable source like in MbNO. Likewise, the interaction of 3.2 MHz at g_1 and ~3.8 MHz at g_3 assigned, in MbNO, to carbon-bound protons of the proximal histidine (F8) can be found. The coupling at g_1 was not detected previously [87] but is found, weakly expressed, in more concentrated samples. The large, exchangeable interaction (5.0 MHz) at g_1 also is of more recent origin whereas the probable counterpart coupling of 2.0 MHz at g_3 was observed previously. As noted above, this group of couplings can reasonably well be ascribed to the $N_{\epsilon 2}$-bound proton of the distal histidine (E7). Other couplings have been tentatively discussed in [87] and the work performed since then, e.g., on hybrid hemoglobins [89], has not given further clues for final assignments. Since some of the couplings resemble those found in MbNO, we hope to get more information from its single-crystal analysis presently under way.

4.3.1.3. NO-ligated heme a_3 in cytochrome c oxidase. The ferrous heme a_3-NO complex has been studied recently by Scholes and coworkers [85] using ^{14}NO- and ^{15}NO-ligated specimen in frozen solutions. The EPR spectrum is of rhombic symmetry (g_x = 2.091, g_z = 2.006, g_y = 1.980) with well-resolved hyperfine structure along g_z which can be assigned to the two axial ligands, N (NO) and $N_{\epsilon 2}$ (His (F8)). ^{14}N- and ^{15}N-ENDOR resonances were obtained at all field values across the EPR spectrum but were best resolved along g_x, which gives a single-crystal type ENDOR response rather free from disturbances of other g directions since g_x is sufficiently larger than g_z. The hyperfine interaction found along g_x for the NO ligand stayed approximately constant when moving the magnetic field to g_y via g_z and the EPR-detectable, larger interaction along

g_z was not well resolved in ENDOR but rather gave a broad shoulder at the expected frequency. Likewise, two sets of N_{ϵ_2} (His (F8)) interactions were observed along g_z, only one of which corresponded to the EPR-resolved hyperfine interaction, the other set being closely related to the couplings observed along g_x and g_y. The couplings found at the various field positions both from ENDOR and EPR are listed in Table 6.

From these data, Scholes and coworkers [85] calculate an unpaired spin density of N (NO) employing essentially an axial hyperfine tensor which gives 2.8% 2s-character and 21% 2p-density on NO. Thus, NO still retains much of π^*-character of free NO but ligation to Fe(II) covalently redistributes spin density away from the ligand. For the proximal histidine ^{14}N (N_{ϵ_2})-interaction, 1.2% 2s-character and 1.9% of 2p was calculated indicating binding of N_{ϵ_2} to the iron through an sp^2 orbital directed nearly along g_z. The authors use the nitrogen hyperfine interactions found to simulate the EPR spectra. Satisfactory agreement was found when a noncollinearity between $\underline{\underline{g}}$ and $\underline{\underline{A}}$ (NO) tensors was introduced by rotating the minimal and maximal hyperfine values 15-20° away from the direction of g_y and g_z with g_x as axis of rotation. No rotation of the rather isotropic histidine ^{14}N tensor was found to be necessary.

Proton ENDOR resonances measured across the EPR spectrum gave rise to four to five weakly coupled proton responses which are not visible at all field positions, indicating, at least partially, the dipolar character of their interaction. There was no interaction which exchanged in specimen prepared from D_2O. Comparison with proton couplings in MbNO and in NO model compounds gave evidence for possible assignments. Table 6 lists the values of couplings found together with their assignment. One interaction of 3.5-4 MHz at intermediate g factors is somewhat tentatively assigned to CH protons of the proximal histidine for which dipolar calculations give an angle of about 35° between the maximum proton interaction and the Fe-N_{ϵ_2} direction. In these calculations, a spin density distribution leading to 73% on the iron and to 23.5% on the NO

TABLE 6

^{14}N Hyperfine (A) and Quadrupolar Couplings (P) and ^1H Hyperfine Couplings in MHz from Frozen Solutions of NO-Ligated Heme a_3 in Cytochrome c Oxidase [85]

g Factor	^{14}N (NO)	^{15}N (NO)	^{14}N ($N_{\epsilon 2}$ (His (F8)))	^1H
g_x: 2.086	A_x : 30.56 ± 0.1	A_x : 42.36 ± 0.25	A_x : 16.51 ± 0.1	0.82 ± 0.04
	P_x : 1.03 ± 0.05		P_x : 0.45 ± 0.05	1.43 ± 0.04 Methine protons
				2.23 ± 0.02
g_z: 2.000	A_{max}: 59.9 ± 1[a]	A_{max}: 83.4 ± 2[a]	A_{max}: 19.3 ± 1.0[a]	1.99 ± 0.1 CH_2-β proton (tentative)
		A_z : 42.6 ± 0.1	A_z : 16.2 ± 0.1	3.52 ± 0.02 CH proton of His (F8)
			P_z : 0.65 ± 0.05	
g_y: 1.97		A_y : 42.2 ± 0.6	A_y : 16.1 ± 0.2	2.10 ± 0.1
			P_y : 0.67 ± 0.01	1.43 ± 0.1

[a] From EPR spectrum.

ligand and 3.5% on N_{ε_2} was used. The corresponding interaction is observed in MbNO and in NO-FeTPP-Im, too (cf. Sec. 4.3.1.1) but is missing in fully deuterated NO-FeTPP-Im. Another coupling, also common with MbNO, is that of 1.43 MHz along g_x which is attributed to the methine protons. Scholes and coauthors [85] argue that its maximal value should be oriented nearly parallel to g_x. Dipolar calculations yield 24° deviation from the Fe mesoproton vector. Finally, a coupling of about 1.99-2.23 MHz occurring at all g turning points was found to be unique to heme a_3 of cytochrome c oxidase only. The isotropic character led the authors to assume a β proton, perhaps on a CH_2 group of a cysteine which was proposed to form the bridge between Fe and Cu of the a_3 center, as responsible for the interaction.

4.3.1.4. *O_2^--ligated myo- and hemoglobin.* Oxymyoglobin (MbO_2) when exposed to ionizing or UV radiation at temperatures ≤77 K exhibits EPR spectra which, apart from huge contributions from organic radicals at g ≃ 2.0, are ascribed to two primary $[FeO_2]^-$ species MbIO and MbIIO which both have rhombic g symmetry (MbIO: g_x = 2.221, g_y = 2.119, g_z = 1.976; MbIIO: g_x = 2.248, g_y = 2.139, g_z = not determined due to overlap with g ≃ 2.0 signals) [82-84]. MbIIO is a minority species having detectable intensity only at neutral pH in frozen H_2O solutions. Single crystals of MbO_2 also stabilize both $[FeO_2]^-$ centers at 77 K. For MbIO, ^{17}O and ^{57}Fe isotopic substitution in single crystals have allowed for determination of the spin density (70% on Fe and 30% equally distributed on the two oxygens) as well as of the bonding configuration for MbIO. The Fe-O-O bond was found to be in a bent "end-on" configuration with a bond angle of 118°, the Fe-O bond being tilted away by about 20° from the heme normal [84]. Compared with the x-ray structure of MbO_2 at 300 K, the bond angle is about identical (115°) but the dioxygen projection over the heme plane as well as the Fe-O tilt are different [91].

Analogous exposure of HbO_2 yields two discernible primary $[FeO_2]^-$ species, too, which can be assigned to one center each for the α and β subunits (HbαO: g_x = 2.215, g_y = 2.121, g_z = 1.967; HbβO: g_x = 2.251, g_y = 2.147, g_z = 1.967). The relative contribution of HbαO and HbβO was found to vary, in frozen solutions, with the concentrations of the glass-forming agents added. Upon thermal annealing the primary species transform irreversibly via up to seven different species of rhombic EPR symmetry in HbO_2 and up to three in MbO_2 into ferric aquo- and hydroxymet Hb and Mb above 200 K.

^{14}N- and ^{1}H-ENDOR was reported recently on the variety of different $[FeO_2]^-$ complexes in MbO_2 and HbO_2 [83]. For the dominant species MbIO as well as for HbαO and HbβO, among others, a strong interaction of an exchangeable proton was observed with a maximum coupling of about 10.5 MHz, which is roughly twice the interaction found for the "isoelectronic" $CoHbO_2$. The plausible assignment of this interaction is to the proton of $N_{\epsilon 2}$ of the distal histidine (E7) in a hydrogen bond to the dioxygen, an interpretation supported by neutron diffraction studies on MbO_2 at 300 K [93]. Calculations of possible proton locations responsible for this coupling were conducted for MbIO since both the Fe-O-O bond configuration as well as the spin density distribution are available [84]. Since only along and in the vicinity of g_x the coupling was observable, the calculations were performed with variable adopted values for g_z ranging from 0.5 to 6 MHz. Two regions of proton locations were found reproducing the 10.5-MHz value, shown in Figure 11, one being located 3-4 Å above the projection of the outer oxygen onto the heme plane, the other, more reasonable choice being removed by about 1.2-1.9 Å from the outer oxygen and about 2.4-2.6 Å above the heme plane. The distances to the outer oxygen were found to lie in the range of hydrogen bonds and to agree with the value of 1.98 Å from neutron diffraction [93]. It is worth noting that x and y components of the proton positions calculated were found to vary slightly (~±0.1 Å) with the various coupling values of g_z adopted whereas a more pronounced variation (2.4-3.1 Å) was resulting for the z component. Translating this into the experimental finding that the linewidth of the 10.5-MHz interaction was fairly large (~0.9 MHz), one would imply a slight

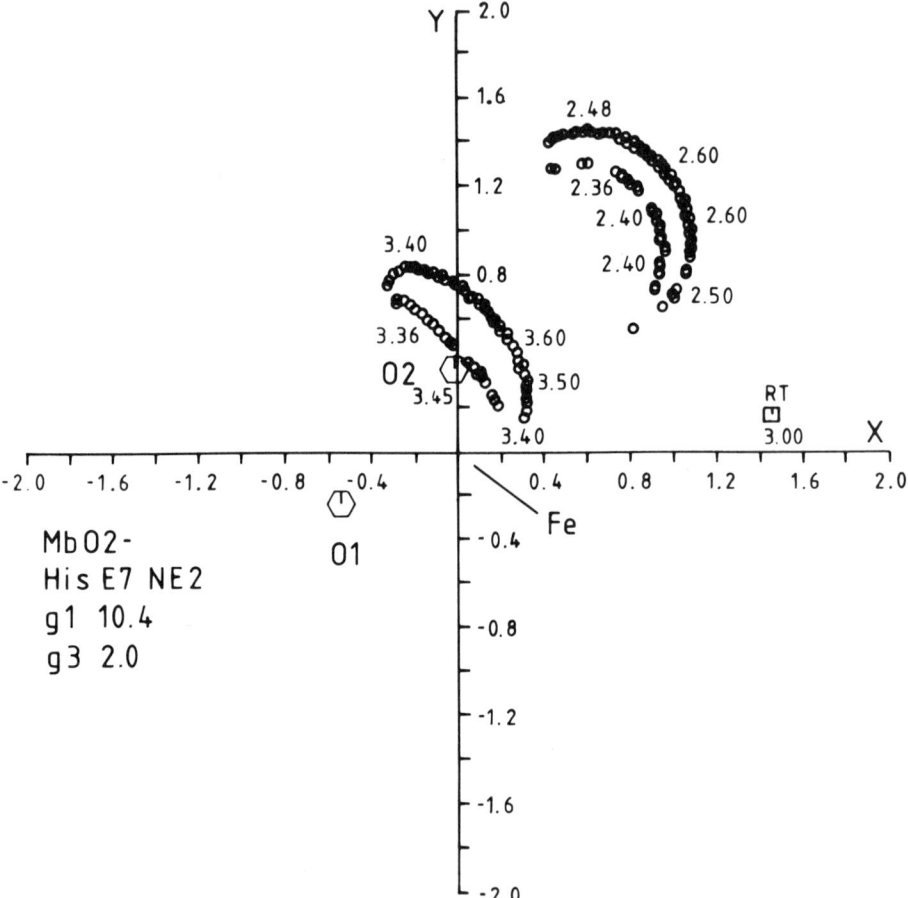

FIG. 11. Heme plane projection of calculated positions (o) of $N_{\varepsilon 2}$ [His (E7)]-bound proton in MbO_2^- using experimental ENDOR coupling at g_1 (10.4 MHz) and assumed, fixed coupling at g_3 (2.0 MHz) leaving coupling at g_3 variable. Positions of $Fe-O_1-O_2$ are indicated (⌬), together with position of proton at 300 K (RT, □). Numbers at calculated positions are z coordinates.

molecular disorder of the distal histidine. When comparing the position of the distal histidine $N_{\varepsilon 2}$ at 77 K with the x-ray-deduced coordinates at 300 K [91], one finds a shift between 0.4 and 1.4 Å, similar to the MbNO observations.

In frozen solutions of HbO_2, an exchangeable coupling was detected for both subunit species $Hb\alpha O$ and $Hb\beta O$ for which the maximum interaction was found to be 11 MHz. Employing isolated chains, the directions for the maximum couplings were found to differ. For the α chains, g_z was the direction for maximum coupling whereas in the β chains that value was found for magnetic fields in between g_x and $g_h = g_y$. This implies that in all three hemoproteins the geometry of $[FeO_2]^-$ is somewhat different, as is also reflected from differences in other proton ENDOR resonances in the three specimens.

Additional support for the assignment of the exchangeable proton to the distal histidine $N_{\varepsilon 2}$ position was gained very recently from ENDOR spectra of the $[FeO_2]^-$ complex in the monomeric fraction of the annelid *Glycera dibranchiata* (HbGM) for which three slightly differing primary species were found to be formed at 77 K [94,95]. For none of these, an exchangeable proton interaction was found reflecting the distal histidine residue being exchanged by leucine in HbGM which, on account of its terminal CH_3 group, is unable to form hydrogen bonds [92].

Besides the unique, exchangeable proton in the primary $[FeO_2]^-$ complex of MbO_2 or HbO_2, up to six weak ^1H interactions were found in addition [83]. Again, possible proton positions were calculated for species MbIO to allow for at least tentative assignments. A reasonable coincidence was found for the proton at carbon CZ of phenylalanine (CD1) (cf. Fig. 12). Taking experimentally observed interactions of 0.6 MHz at g_x and 1.5 MHz at g_z the calculations give an expected coupling along g_y of 2.1 MHz which correlates well with the observed value of 2.15 MHz. Also, the difference calculated for the proton position from this set of couplings is only 0.2-0.6 Å off the 300-K x-ray data [91]. Other proton assignments are not as unambiguous since experimental couplings can be reproduced by more than one proton. A reasonable range of couplings for the methine protons can be defined of about 0.4 MHz at g_x, 0.8 MHz at g_z, and 1.6 MHz at g_y. More reliable assignments of the other interactions observed would require a single-crystal analysis, which we failed so far to have success with.

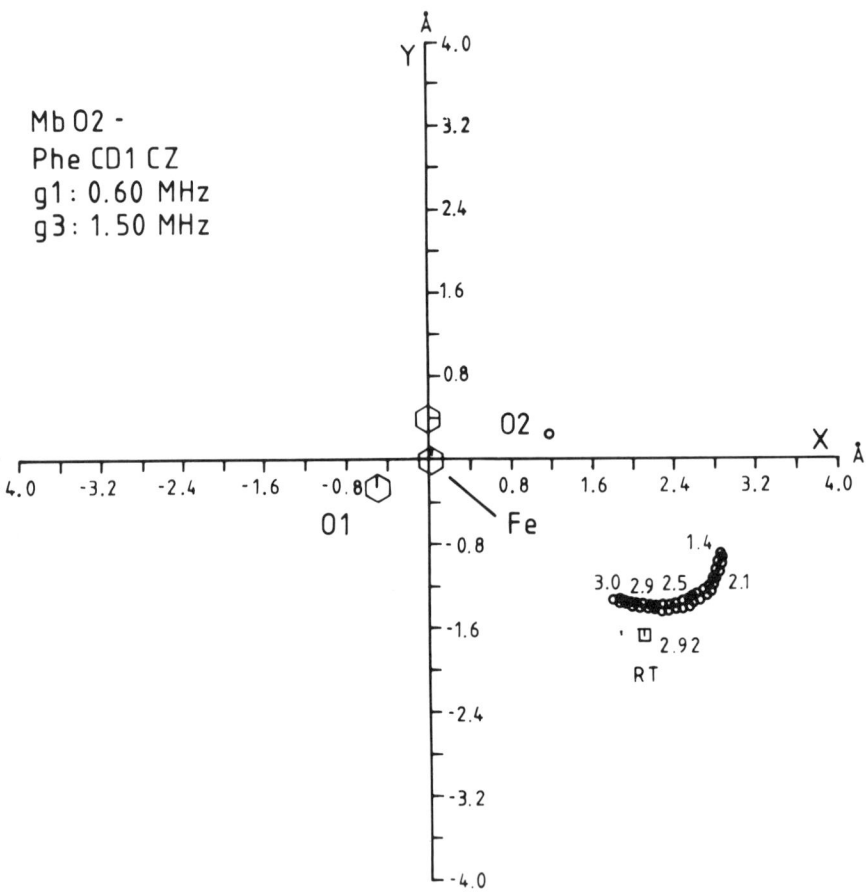

FIG. 12. Heme plane projection of calculated positions (o) of carbon CZ-bound proton of phenylalanine residue (CD1). Other details as in Fig. 11.

Proton resonances for the intermediate, rhombic EPR species which are formed upon annealing have been observed and described [83]. The main information gained so far is that in the first transition of MbIO, HbαO, and HbβO to another species the unique, exchangeable proton is lost. The mechanism probably is proton transfer to the $[FeO_2]^-$ complex to form a hydroperoxide. ^{14}N-ENDOR lines of all primary and intermediate species showed only little resolution both for the monomeric MbO_2 and the tetramer HbO_2. Along g_x of MbO_2, a coupling to the proximal N_{ε_2} of His (F8) can be assigned (A_x = 12.6 MHz, P_x = 0.52 MHz) together with well-resolved pyrrole nitrogens which apparently can be grouped into two inequivalent pairs. Pyrrole nitrogen resolution was also obtained along g_z but complex line patterns at $g_h = g_y$ prevented a full assignment. Similar difficulties were found to hold for the histidine interaction (F8) for which no clear description could be arrived at so far. The situation is even more complex in the tetramer HbO_2 due to the superposition of α- and β-chain-located $[FeO_2]^-$ complexes. In isolated β chains, the line pattern in the frequency range 1-10 MHz showed, along g_x, some relation to that in MbO_2 but differed at other field values.

For the annealed centers of MbO_2 and HbO_2, spectral changes in the pyrrole- and histidine range could be detected but no clear interpretation could possibly be derived from the complex spectra.

Summarizing the properties of the $[FeO_2]^-$ complex, the detailed EPR and ENDOR analysis has shown that the species is not a monitor for the native stereochemistry of oxygenated hemoglobins since its formation involves drastic ligand reorientations compared with the Fe-O-O bond configuration derived from x-ray structural data [91]. On the other hand, the complex is an extremely sensitive monitor of the Fe-O-O bond geometry which is, of course, controlled by the heme environment. It could thus be used as a probe of structural consequences of changes in the heme pocket amino acid residues, e.g., in mutant studies as was shown for the *Glycera* hemoglobin [94,95]. At present, however, the system is still too complex to allow for an interpretation of the whole gamut of spectral details.

4.3.1.5. Fe(III)-heme proteins and model compounds. The electronic configuration ($3d^5$) of Fe(III) allows for three values of the effective spin (S = 5/2, 3/2, 1/2) classified as high-, intermediate-, and low-spin complexes. In Fe(III) metalloproteins, only high- or low-spin states of the iron are found which correspond to binding of weak (H_2O, F^-, HCO_2^-, OCN^-) or strong (OH^-, CN^-, N_3^-, imidazole) ligands. A full account of the EPR results on high- and low-spin myo- and hemoglobins has appeared recently in which some of the ENDOR results have been included [11]. Also, a detailed

description of the ENDOR work which is mainly on high-spin complexes
has been given by Scholes [10] and, more recently, by Schweiger [4].
Both latter reviews, although not of very recent origin, contain the
information available including model compounds nearly completely
since little work on Fe(III)-heme complexes has been performed since
1980. We therefore refer the reader to these reviews for the earlier
literature and outline shortly work which has appeared since.

Pattison and Kim [96] used dilute frozen solutions of high- and
low-spin metmyoglobin. With H_2O as ligand, the authors reproduced
the data by Scholes and coworkers on the ^{14}N resonances [10]. The
main emphasis was then on investigation of details in the "free
proton" (~14 MHz) region for high- and low-spin complexes. The
results, of little structural implication, were discussed in terms
of distant ENDOR mechanisms.

Scholes and coworkers reported on the first single-crystal
analysis of high-spin aquomet-myoglobin MbH_2O [15]. The complete
hyperfine and quadrupolar interaction tensors of the heme and histidine (F8) nitrogens were evaluated including ^{15}N-isotope substitution for the pyrrole nitrogens. Most interestingly, the latter were
found to be pairwise inequivalent, two of them with larger couplings
being assigned to pyrroles with methyl and propionic acid side groups,
the two with smaller interactions to those with methyl and vinyl side
groups. The detailed information obtained was related to spin densities which, on the average, gave 2.5% 2s-, 5.6% p_σ- and 0.9% $2p_\pi$-
density for the heme and 2.92%, 7.6%, and 0.9% respective contributions for the histidine nitrogen $N_{\varepsilon 2}$ [His (F8)]. The inequivalence
of the heme nitrogens was assigned to an electronic inequivalence
explained by a difference of ~0.02 Å in bond lengths between iron
and one heme nitrogen compared with its diagonally opposite partner.

Another heme complex studied recently with ENDOR is cytochrome a
in cytochrome c oxidase. The a site, generally thought to be the primary electron acceptor in cytochrome c oxidase, is known to contain a
low-spin Fe(III) coordinated by four in-plane heme nitrogens, but the
nature of the axial ligands was controversial. Martin and coworkers
[97] used native and [1,3-$^{15}N_2$]histidine substituted yeast cytochrome

ENDOR: PROBING THE COORDINATION ENVIRONMENT IN METALLOPROTEINS

c oxidase for which, by comparison with isotopically substituted ^{15}N-containing model compounds [Fe(III)bisimidazole TPP, Mb-imidazole], convincing evidence could be obtained to show that at least one of the axial ligands in the cytochrome a site is histidine. Since the coupling observed is closely related to that in bisimidazole TPP, the authors suggest that bisimidazole coordination should also prevail for cytochrome a.

4.3.1.6. *Fe(IV)-heme proteins: Cytochrome c peroxidase compound ES, and horseradish peroxidase compound I.* Cytochrome c peroxidase catalyzes the oxidation of ferrocytochrome c to ferricytochrome c by hydroperoxide through the action of an intermediate, denoted compound ES, which is formed by reaction with stoichiometric quantities of H_2O_2. Compound ES is two oxidizing equivalents above the original ferric state. For horseradish peroxidase (and catalase), the corresponding intermediate is compound I. In both systems, one of the two oxidizing equivalents is ascribed to a ferryl heme iron [Fe(IV), S = 1]. The other equivalent is thought to be a cation free radical on the porphyrin ring in compound I of horseradish peroxidase whereas for ES a spatially removed protein free radical is invoked. Thus, the heme in compound I of horseradish peroxidase can be depicted schematically as $[-Fe(IV)-]^+$ whereas the spatially removed radical $\dot{R}$ in compound ES requires the notation $[-Fe(IV)-] + \dot{R}$.

For compound ES, the nature of $\dot{R}$ is controversial, the proposals ranging from a heme-bound peroxy radical to a tryptophanyl radical. In two ENDOR studies, Hoffman and coworkers contradict both these suggestions and instead invoke a dimeric thioether radical cation $(R_2SSR_2)^+$ to be responsible for the EPR signal [98,99]. The basis for this is mainly the definite absence of nitrogen ENDOR resonances together with fairly large proton couplings of ~11-22 MHz all of which are apparently isotropic.

The apparent conflict has been somewhat reduced quite recently by an EPR single-crystal investigation [100] in which clear evidence for two different paramagnetic species in compound ES was obtained.

One is isotropic at g = 2.004 with a hyperfine structure due to three equivalent protons. The other is at very low temperatures (5 K) dominating and axially symmetric in the g tensor and probably presents the species on which Hoffman and coworkers thought to perform the ENDOR measurements. It appears possible that the minority species gave the ENDOR results since the couplings measured in ENDOR were also found to be isotropic and gave values comparable to EPR. If so, the nature of the dominating species, for which pulsed EPR indicated an interaction between the ferryl heme and the free radical site as well as nitrogen superhyperfine structure [101], is still open to debate and perhaps a tryptophanyl residue as is known from x-ray data to lie in the immediate heme vicinity cannot be ruled out to contribute.

For horseradish peroxidase compound I, Roberts and coworkers find ENDOR resonances from ^{14}N nuclei and protons of the β and probably also the α type which are interpreted in terms of a porphyrin π-cation radical spin-coupled with the oxyferryl [Fe(IV)] ion [102]. In a subsequent report, the authors use ^{17}O-ENDOR of isotopically enriched samples to prove that one oxygen atom of the oxidant H_2O_2 remains with the intermediate, which again lends support to the thesis that the other oxidizing equivalent, in addition to the porphyrin cation, must be associated with an oxyferryl center of the type Fe(IV)=O [103].

4.3.2. Nonheme Proteins

Among the nonheme iron proteins, the iron-sulfur proteins form a group with considerable biological relevance. Iron and sulfur form clusters in a variety of proteins, the sulfur coming from cysteinyl residues or being of inorganic origin. The clusters are of variable size ranging from 2Fe-2S (e.g., two-iron ferredoxins) to 12Fe-12S (e.g., hydrogenase from *Clostridium pasteurianum* W5). ENDOR spectroscopy on these systems, for which EPR usually has no hyperfine information, has been performed with considerable success. The contributions have been summarized in two review articles [4,104]

ENDOR: PROBING THE COORDINATION ENVIRONMENT IN METALLOPROTEINS

which cover the literature up to about 1980. Comparable with the case of Fe(III) heme proteins and model compounds, the coverage appears fairly exhaustive since little additional reports have come to our attention.

There is a recent investigation on the oxidized hydrogenase from *C. pasteurianum* W5 in which Mössbauer spectroscopy was coupled with ENDOR including ^{57}Fe substitution [105]. It was shown that the EPR spectrum of apparent rhombic symmetry had to be considered as composite. Eight of the 12 Fe atoms of the oxidized enzyme appeared to be in two ferredoxin-type clusters [4Fe-4S]$^{2+}$ and the remaining iron was assigned to a spin-coupled structure on account of ^{57}Fe-ENDOR which gave evidence for at least two distinct iron sites.

Another nonheme iron protein for which ENDOR has given some information on structural aspects is the molybdenum-iron (MoFe) protein of nitrogenase from *Azetobacter vinelandii*. Hoffmann and coworkers studied the resting state of this enzyme in which the cofactor contains 6-8 Fe and 4-6 S per molybdenum and has an EPR spectrum characteristic of a total spin S = 3/2 [106,107]. ^{1}H-, ^{95}Mo-, and ^{57}Fe-ENDOR was observed. It was concluded that Mo is integrated into the cluster and probably spin-coupled. For the iron, six inequivalent sites were resolved supporting the existing models of the stoichiometry of the cluster.

5. CONCLUDING REMARKS

ENDOR spectroscopy has established itself over the past two decades as an independent, powerful branch of electronic magnetic resonance spectroscopy. Especially in the field of metalloproteins, the theoretical promise of high resolution and, consequently, enhanced insight into structural details in complex paramagnetic systems over conventional EPR spectroscopy has been fulfilled to a very large extent. With this technique, the problem of structure and function of metalloproteins can be attacked for a wide range of transition metal centers at a level revealing the nature and bonding of ligands in the first coordination shell and beyond. As has been discussed in the foregoing, for quite a few metalloproteins, the ENDOR results have given unique and unambiguous information not attainable with other spectroscopic techniques.

Despite this undoubted success, there are a few limitations apparent at this stage which reveal that ENDOR spectroscopy, especially of very complex systems for which no single crystals can be obtained, is to some extent still a technique for which further development is necessary. There is, of course, the problem of sensitivity which requires, in several cases, a limiting amount of sample and facilities for multiscan averaging of signals. This is usually a minor step, however, compared with the problem of directly determining tensor parameters from powder spectra for which there is a pressing demand for metalloproteins. The theoretical foundation of lineshape analysis and powder simulation to date is yet inadequate and most of the definite answers to questions of structure and bonding have come from isotopic replacement of nuclei or well-defined model complexes. The main difficulty to overcome in this context is the strong influence of spin dynamics on transition probabilities and lineshapes.

As is the case for other spectroscopic techniques, ENDOR data become most useful in connection with other structural information, especially from x-ray crystallography, since, among others, it is difficult to obtain the number of equivalent nuclei from the ENDOR spectra. We expect that the recent, more refined ENDOR techniques together with pulsed EPR spectroscopy which has become available over the last few years will form a sound spectroscopic basis for enhanced future applications of electronic magnetic resonance in the field of metalloproteins.

ACKNOWLEDGMENT

Work reported from the authors' laboratory was supported by grants from the Deutsche Forschungsgemeinschaft. The contributions of our coworkers Drs. M. Höhn, W. Leibl, W. Nitschke, and E. Wagner are also gratefully acknowledged, as is the help of Dr. H. Zorn with the preparation of model compounds.

ABBREVIATIONS

AM	amplitude modulation
CP-ENDOR	ENDOR with circularly polarized fields
CW	continuous wave
EI-EPR	ENDOR-induced EPR
ENDOR	electron nuclear double-resonance spectroscopy
EPR	electron paramagnetic resonance spectroscopy
EXAFS	extended x-ray absorption fine structure
FM	frequency modulation
Hb	hemoglobin
HbA	adult hemoglobin
HbGM	monomeric fraction of *Glycera dibranchiata* hemoglobin
HbH_2O, MbH_2O	aquomethemo- and -myoglobin, respectively
$HbOH^-$, $MbOH^-$	hydroxymethemo- and -myoglobin, respectively
HbO_2, HbNO	O_2- and NO-ligated hemoglobin, respectively
His (F8), His (E7)	Histidine residues in polypeptide chain of hemoglobin located at helical positions F8 and E7, respectively
IHP	inositolhexaphosphate
Im	imidazole
LC circuit	resonance circuit employing inductive and capacitive elements
Mb	myoglobin
MbO_2, MbNO	myoglobin O_2- and NO-ligated myoglobin, respectively
MO	molecular-orbital
OEP	octaethylporphyrin
Q	measure of absorbed vs. dissipated energy in a microwave resonator
Q band	microwave frequency band (26-40 GHz); used frequency in EPR: 35 GHz
rf	radiofrequency (MHz range) to induce ENDOR transitions

S band	microwave frequency band; used frequency in EPR: 4 GHz
TPP	tetraphenylporphyrin
TRIPLE (DOUBLE-ENDOR)	ENDOR with two rf photons
X band	frequency band for microwaves (8-12 GHz); used frequency in EPR: 9 GHz

REFERENCES

1. G. Feher, R. A. Isaacson, C. P. Scholes, and R. L. Nagel, Ann. N.Y. Acad. Sci., 222, 86 (1973).
2. P. Eisenberger and P. S. Pershan, J. Chem. Phys., 47, 3327 (1967).
3. R. M. Deal, D. J. E. Ingram, and R. Srinivasan, Proc. Colloq. Ampere., 12, 239 (1963).
4. A. Schweiger, in Structure and Bonding, Vol. 51, Springer-Verlag, Berlin, 1982.
5. L. Kevan and L. D. Kispert, Electron Spin Double Resonance Spectroscopy, John Wiley and Sons, New York, 1976.
6. M. M. Dorio and J. H. Freed (eds.), Multiple Electron Resonance Spectroscopy, Plenum Press, New York, 1979.
7. Electron Spin Resonance (A Specialist Periodical Report), The Royal Society of Chemistry, London.
8. D. J. Kosman, in Structural and Resonance Techniques in Biological Research (D. L. Rousseau, ed.), Academic Press, Orlando, 1984.
9. H. C. Box, Radiation Effects: ESR and ENDOR analysis, Academic Press, New York, 1977.
10. C. P. Scholes, in Multiple Electron Resonance Spectroscopy (M. M. Dorio and J. H. Freed, eds.), Plenum Press, New York, 1979, p. 297.
11. L. C. Dickinson and M. C. R. Symons, Chem. Soc. Rev., 12, 387 (1983).
12. H. Muto and M. Iwasaki, J. Chem. Phys., 59, 4821 (1973).
13. D. S. Schonland, Proc. Roy. Soc., 73, 788 (1959).
14. A. Kwiram, J. Chem. Phys., 51, 2484 (1971).
15. C. P. Scholes, A. Lapidot, R. Mascarenhas, T. Inubushi, R. A. Isaacson, and G. Feher, J. Am. Chem. Soc., 104, 2724 (1982).
16. G. H. Rist and J. S. Hyde, J. Chem. Phys., 52, 4633 (1970).

17. R. J. Cook, *J. Scient. Instrum. (J. Phys. E.)*, *43*, 548 (1966).
18. R. J. Cook and D. H. Whiffen, *Proc. Phys. Soc.*, *84*, 845 (1964).
19. C. P. Poole, Jr., *Electron Spin Resonance* (A Comprehensive Treatise on Experimental Techniques), 2nd ed., John Wiley and Sons, New York, 1983.
20. W. Seidel, *Z. Phys.*, *165*, 239 (1961).
21. H. L. Van Camp, C. P. Scholes, and R. A. Isaacson, *Rev. Sci. Instr.*, *47*, 516 (1976).
22. H. C. Box, H. G. Freund, and K. T. Lilga, *J. Chem. Phys.*, *46*, 2130 (1967).
23. R. A. Isaacson, C. Lulich, S. B. Oseroff, and R. Calvo, *Rev. Sci. Instr.*, *51*, 1409 (1980).
24. J. Forrer, A. Schweiger, and H. H. Günthard, *J. Phys. E.*, *10*, 470 (1977).
25. A. Schweiger and H. H. Günthard, *Mol. Phys.*, *42*, 283 (1981).
26. J. F. Boas, in *Copper Proteins and Copper Enzymes*, Vol. 1 (R. Lontie, ed.), CRC Press, Boca Raton, 1984, p. 5.
27. T. H. Moss, E. Shapiro, T. E. King, H. Beinert, and C. Hartzell, *J. Biol. Chem.*, *253*, 8072 (1978).
28. C. H. A. Seiter and S. G. Angelos, *Proc. Natl. Acad. Sci. USA*, *77*, 1806 (1980).
29. J. S. Richardson, K. A. Thomas, B. H. Rubin, and D. C. Richardson, *Proc. Natl. Acad. Sci. USA*, *72*, 1349 (1975).
30. J. A. Fee and B. P. Gaber, *J. Biol. Chem.*, *247*, 60 (1972).
31. T. G. Brown, J. L. Petersen, G. P. Lozos, J. R. Anderson, and B. M. Hoffman, *Inorg. Chem.*, *16*, 1563 (1977).
32. A. Schweiger and H. H. Günthard, *Chem. Phys.*, *32*, 35 (1978).
33. R. G. Rist and J. S. Hyde, *J. Chem. Phys.*, *50*, 4532 (1969).
34. E. W. Fleischer, C. K. Miller, and L. E. Webb, *J. Am. Chem. Soc.*, *86*, 2342 (1964).
35. T. G. Brown and B. M. Hoffman, *Mol. Phys.*, *39*, 1073 (1980).
36. Y. Iitaka, H. Nakamura, T. Nakatani, Y. Muraoka, A. Fujii, T. Takita, and H. Umezawa, *J. Antibiot. (Tokyo)*, *31*, 1070 (1978).
37. W. E. Antholine, J. S. Hyde, R. C. Sealy, and D. H. Petering, *J. Biol. Chem.*, *259*, 4437 (1984).
38. H. L. Van Camp, R. H. Sands, and J. A. Fee, *J. Chem. Phys.*, *75*, 2098 (1981).
39. H. Yokoi, *Biochem. Biophys. Res. Commun.*, *108*, 1278 (1982).

40. G. H. Rist, J. S. Hyde, and T. Vänngård, Proc. Natl. Acad. Sci. USA, 67, 79 (1970).
41. J. E. Roberts, T. G. Brown, B. M. Hoffman, and J. Peisach, J. Am. Chem. Soc., 102, 825 (1980).
42. J. Cline, B. Reinhammar, P. Jensen, R. Venters, and B. M. Hoffman, J. Biol. Chem., 258, 5124 (1983).
43. J. E. Roberts, J. F. Cline, V. Lum, H. Freeman, H. B. Gray, J. Peisach, B. Reinhammar, and B. M. Hoffman, J. Am. Chem. Soc., 106, 5324 (1984).
44. M. Sharnoff and C. W. Reiman, J. Chem. Phys., 43, 2998 (1965).
45. P. M. Colman, H. C. Freeman, J. M. Guss, M. Morata, V. A. Morris, J. A. M. Ramshaw, and M. P. Venkatappa, Nature, 272, 319 (1978).
46. E. T. Adman, R. E. Stenkamp, L. C. Sieker, and L. H. Jensen, J. Mol. Biol., 123, 35 (1978).
47. J. Peisach, L. Powers, W. E. Blumberg, and B. Chance, Biophys. J., 38, 277 (1982).
48. H. Yokoi, Y. Ohba, and T. Takabe, Biochem. J., 215, 209 (1983).
49. A. Desideri, L. Morpurgo, E. Agostinelli, G. J. Baker, and J. B. Raynor, Biochim. Biophys. Acta, 831, 8 (1985).
50. K. W. Penfield, A. A. Gewirth, and E. I. Solomon, J. Am. Chem. Soc., 107, 4519 (1985).
51. O. Farver and I. Pecht, in Copper Proteins and Copper Enzymes, Vol. 1 (P. Lontie, ed.), CRC Press, Boca Raton, 1984, p. 183.
52. A. Rigo, E. F. Orsega, P. Viglino, L. Morpurgo, M. T. Graziani, and G. Rotilio, Biochem. Biophys. Res. Commun., 111, 824 (1983).
53. H. L. Van Camp, Y. H. Wei, C. P. Scholes, and T. E. King, Biochim. Biophys. Acta, 537, 238 (1978).
54. B. M. Hoffman, J. E. Roberts, M. Swanson, S. H. Speck, and E. Margoliash, Proc. Natl. Acad. Sci. USA, 77, 1452 (1980).
55. W. Froncisz, C. P. Scholes, J. S. Hyde, Y. H. Wei, T. E. King, R. W. Shaw, and H. Beinert, J. Biol. Chem., 254, 7482 (1979).
56. T. H. Stevens, C. T. Martin, H. Wang, G. W. Brudvig, C. P. Scholes, and S. I. Chan, J. Biol. Chem., 257, 12106 (1982).
57. R. A. Scott, S. P. Cramer, R. W. Shaw, H. Beinert, and H. B. Gray, Proc. Natl. Acad. Sci. USA, 78, 664 (1981).
58. H. L. Van Camp, R. H. Sands, and J. A. Fee, Biochim. Biophys. Acta, 704, 75 (1982).
59. T. Vänngård, in Biological Applications of Electron Spin Resonance (H. M. Swartz, J. R. Bolton, and D. C. Borg, eds.), Wiley-Interscience, New York, 1972, p. 411.
60. J. E. Roberts, T. G. Brown, B. M. Hoffman, and P. Aisen, Biochim. Biophys. Acta, 747, 49 (1983).

61. E. Jörin, A. Schweiger, and H. H. Günthard, *J. Am. Chem. Soc.*, *105*, 4277 (1983).
62. J. C. W. Chien and L. C. Dickinson, *Proc. Natl. Acad. Sci. USA*, *69*, 2783 (1972).
63. L. C. Dickinson and J. C. W. Chien, *Proc. Natl. Acad. Sci. USA*, *77*, 1235 (1980).
64. H. Hori, M. Ikeda-Saito, and T. Yonetani, *J. Biol. Chem.*, *257*, 3636 (1982).
65. H. Hori, M. Ikeda-Saito, J. S. Leigh, Jr., and T. Yonetani, *Biochemistry*, *21*, 1431 (1982).
66. M. Höhn and J. Hüttermann, *J. Biol. Chem.*, *257*, 10554 (1982).
67. S. A. Fairhurst and L. H. Sutcliffe, *Prog. Biophys. Mol. Biol.*, *34*, 1 (1978).
68. G. Palmer, in *The Porphyrins*, Vol. 4 (D. Dolphin, ed.), Academic Press, New York, 1979, p. 313.
69. S. E. V. Phillips and P. B. Schoenborn, *Nature*, *292*, 81 (1981).
70. M. F. Perutz, *Proc. Roy. Soc.*, *B208*, 135 (1980).
71. Q. H. Gibson and F. J. W. Roughton, *Proc. Roy. Soc.*, *B147*, 44 (1957).
72. J. S. Griffith, *Proc. Roy. Soc.*, *A235*, 23 (1956).
73. J. E. Bennett, D. J. E. Ingram, P. George, and J. S. Griffith, *Nature*, *176*, 394 (1955).
74. T. Shiga, K. J. Hwang, and I. Tyuma, *Biochemistry*, *8*, 378 (1969).
75. H. Rein, O. Ristau, and W. Scheler, *FEBS Lett.*, *24*, 24 (1972).
76. Y. Henry and R. Banerjee, *J. Mol. Biol.*, *73*, 469 (1973).
77. R. Cassoly, *J. Mol. Biol.*, *98*, 581 (1975).
78. H. Hori, M. Ikeda-Saito, and T. Yonetani, *J. Biol. Chem.*, *256*, 7849 (1981).
79. W. Nitschke, Diploma thesis, Univ. Regensburg (FRG), 1982.
80. S. G. Utterback, D. C. Doetschman, J. Szumowski, and A. K. Rizos, *J. Chem. Phys.*, *78*, 5874 (1983).
81. D. C. Doetschman, A. K. Rizos, and J. Szumowski, *J. Chem. Phys.*, *81*, 1185 (1984).
82. M. C. R. Symons and R. L. Peterson, *Proc. Roy. Soc.*, *B201*, 285 (1978).
83. R. Kappl, M. Höhn-Berlage, J. Hüttermann, N. Bartlett, and M. C. R. Symons, *Biochim. Biophys. Acta*, *827*, 327 (1985).
84. W. Leibl, W. Nitschke, and J. Hüttermann, *Biochim. Biophys. Acta*, *870*, 20 (1986).

85. R. LoBrutto, Y. H. Wei, R. Mascarenhas, C. P. Scholes, and T. E. King, *J. Biol. Chem.*, *258*, 7437 (1983).

86. W. E. Blumberg and J. Peisach, in *Probes of Structure and Function of Macromolecules and Membranes*, Vol. 2 (B. Chance, T. Yonetani, and A. S. Mildvan, eds.), Academic Press, New York, 1971, p. 215.

87. M. Höhn, J. Hüttermann, J. C. W. Chien, and L. C. Dickinson, *J. Am. Chem. Soc.*, *105*, 109 (1983).

88. C. Dousmanis, *Phys. Rev.*, *97*, 967 (1955).

89. M. Höhn, PhD thesis, Univ. Regensburg (FRG), 1983.

90. M. Christahl, H. Twilfer, and K. Gersonde, *Biophys. Struct. Mechanism*, *9*, 61 (1982).

91. H. C. Watson, *Progr. Stereochem.*, *4*, 299 (1969).

92. T. Imamura, T. O. Baldwin, and A. Riggs, *J. Biol. Chem.*, *247*, 2785 (1972).

93. S. E. V. Phillips, *J. Mol. Biol.*, *142*, 531 (1980).

94. N. Bartlett and M. C. R. Symons, *Biochim. Biophys. Acta*, *744*, 110 (1983).

95. E. Wagner, Diploma thesis, Univ. Regensburg (FRG), 1985.

96. M. R. Pattison and Y. W. Kim, *J. Magn. Reson.*, *43*, 193 (1981).

97. C. T. Martin, C. P. Scholes, and S. I. Chan, *J. Biol. Chem.*, *260*, 2857 (1985).

98. B. M. Hoffman, J. E. Roberts, T. G. Brown, C. H. Kang, and E. Margoliash, *Proc. Natl. Acad. Sci. USA*, *76*, 6132 (1979).

99. B. M. Hoffman, J. E. Roberts, C. H. Kang, and E. Margoliash, *J. Biol. Chem.*, *256*, 6556 (1981).

100. H. Hori and T. Yonetani, *J. Biol. Chem.*, *260*, 349 (1985).

101. K. Lerch, W. B. Mims, and J. Peisach, *J. Biol. Chem.*, *256*, 10088 (1981).

102. J. E. Roberts, B. M. Hoffman, R. Rutter, and L. P. Hager, *J. Biol. Chem.*, *256*, 2118 (1981).

103. J. E. Roberts, B. M. Hoffman, R. Rutter, and L. P. Hager, *J. Am. Chem. Soc.*, *103*, 7654 (1981).

104. R. H. Sands, in *Multiple Electron Resonance Spectroscopy* (M. M. Dorio and J. H. Freed, eds.), Plenum Press, New York, 1979, p. 331.

105. G. Wang, M. J. Benecky, B. H. Huynh, J. F. Cline, M. W. Adams, L. E. Mortenson, B. M. Hoffman, and E. Münck, *J. Biol. Chem.*, *259*, 14328 (1984).

106. B. M. Hoffman, R. A. Venters, J. E. Roberts, M. Nelson, and W. H. Orme-Johnson, *J. Am. Chem. Soc.*, *104*, 4711 (1982).

107. B. M. Hoffman, J. E. Roberts, and W. H. Orme-Johnson, *J. Am. Chem. Soc.*, *104*, 860 (1982).

2

Identification of Oxygen Ligands in Metal-Nucleotide-Protein Complexes by Observation of the Mn(II)-^{17}O Superhyperfine Coupling

Hans Robert Kalbitzer
Max-Planck-Institute for Medical Research
Department of Molecular Physics
Jahnstrasse 29, D-6900 Heidelberg, FRG

1.	INTRODUCTION	81
2.	THEORY	83
	2.1. EPR on Manganese-Protein Complexes	83
	2.2. ^{55}Mn(II)-^{17}O Superhyperfine Interaction	87
3.	APPLICATIONS	90
	3.1. Elongation Factor Tu	90
	3.2. Creatine Kinase	93
	3.3. Adenylate Kinase	96
	3.4. Myosin Subfragment 1	98
	3.5. 3-Phosphoglycerate Kinase	98
4.	CONCLUSIONS AND OUTLOOK	99
	ABBREVIATIONS AND DEFINITIONS	100
	REFERENCES	101

1. INTRODUCTION

In most of the known biochemical pathways nucleoside triphosphates play an essential role as sources of metabolic energy or modulators of the enzymatic activity. During these processes they must interact

with the active center or with regulatory sites of the involved
enzymes. Nucleoside triphosphate-utilizing enzymes are usually
activated by Mg^{2+} ions which are assumed to be complexed with the
phosphate groups of the nucleotides at the active center of the
enzyme. Obviously, for a detailed understanding of the catalytic
mechanism the knowledge of the structure of the metal-nucleotide
complex before, after, and, if possible, during the reaction is
mandatory.

There are different levels of structural information to be
obtained: regiospecific information about the type and number of
phosphate groups involved in the binding of the metal ion (α-, β-,
γ-phosphate; mono-, bi-, tridentate complex), stereospecific information about the individual oxygen atoms of the phosphate groups coordinated (diastereomers created by the binding of the metal ion to
the α or β group), and spatial information about the three-dimensional
arrangement of the metal-nucleotide complex.

According to the type of information wanted, different techniques have been developed. Almost all of them are based on the use
of analogs for the nucleotide and/or the metal ion, either replacing
the diamagnetic Mg^{2+} ion by paramagnetic ions for EPR or NMR studies
or using nucleotide-metal analogs with known structures for kinetic
experiments.

An elegant approach, which was introduced by Jaffe and Cohn
[1], relies on the observation that Cd^{2+} has a preference for sulfur
over oxygen coordination in thiophosphate analogs and that Mg^{2+} preferentially coordinates to oxygen. From a reversal of stereospecificity it is possible to conclude which diastereomeric metal-nucleotide
complex is the natural substrate during the rate-limiting step of the
enzymatic reaction (for a recent review, see [2]). However, using
the thiophosphate analogs, no information about the terminal phosphate group can be obtained.

An alternative method employs exchange-inert Cr(III) and
Co(III) complexes as probe for the structure of the bound metal-nucleotide complex [3-5].

IDENTIFICATION OF OXYGEN LIGANDS

Nuclear magnetic resonance offers a number of different possibilities for studying the environment of the metal ion. These methods have already been described in the preceding Vol. 21 of this series and will not be discussed further. Their advantage is that they can provide real spatial information; their main disadvantage is that results are often very difficult to interpret.

The most recently introduced method, which the next sections will cover in detail, takes advantage of the line-broadening effect of ^{17}O introduced into nucleotide phosphate groups on the EPR spectrum of Mn^{2+} used as an analog for Mg^{2+}. It provides regio- and stereospecific information about the metal-nucleotide complex and, in addition, by the use of ^{17}O-labeled H_2O, information about the number of water molecules in the first coordination sphere of the metal ion.

2. THEORY

2.1. EPR on Manganese-Protein Complexes

In virtually all enzymatic reactions the naturally occurring diamagnetic Mg^{2+} ion can be substituted by the paramagnetic Mn^{2+} ion, changing only the magnitude of individual rate constants but not altering the general reaction scheme. This is not surprising since the ions are rather similar with respect to many physicochemical properties such as ionic radii and coordination preferences, though Mn^{2+} has a somewhat higher affinity for N-donor binding sites than Mg^{2+}.

For many years it was claimed that the EPR spectrum of manganese bound to macromolecules cannot be observed in solution due to the extensive line broadening expected for the high rotational correlation time of the complex. In the last decade, however, a relatively large number of well-resolved EPR spectra of manganese bound to proteins have been reported, invalidating the above theoretical expectation (for a recent review, see [6]).

The EPR spectrum is determined by the Hamiltonian H of the system. For high-spin ^{55}Mn(II) (S = 5/2, I = 5/2), the appropriate spin Hamiltonian can be written as the sum of the electronic Zeeman interaction H_{eZ}, the nuclear Zeeman interaction H_{nZ}, the hyperfine interaction H_{hfs} between nuclear spin I and electron spin S, and the zero field splitting term H_{zfs}:

$$H = H_{eZ} + H_{nZ} + H_{hfs} + H_{zfs} \tag{1}$$

In spite of a few known exceptions (e.g., [7,8]), in most cases the g tensor of high-spin Mn(II) can be assumed as nearly isotropic [6,9] and can be replaced by a scalar with a value near the free electron g factor. Then H_{eZ} can be written as

$$H_{eZ} = g\beta B_0 S_z \tag{2}$$

where the static magnetic field B_0 is applied in the z direction of the laboratory frame (β: Bohr magneton).

The nuclear Zeeman interaction H_{nZ} can be described by

$$H_{nA} = -\gamma_n \hbar B_0 I_z \tag{3}$$

where γ_n is the magnetogyric ratio, $\hbar$ the Planck constant h/2π, and I_z the z component of the nuclear spin I. As far as fully allowed EPR transitions ($\Delta M = \pm 1$, $\Delta m = 0$) are concerned H_{nZ} can be omitted, but it leads in higher order perturbation theory to "forbidden" transitions of the type $\Delta M = \pm 1$, $\Delta m = \pm 1$, which can be observed also in some manganese complexes with proteins such as in complexes with glutamine synthetase [10].

The hyperfine tensor describing the interaction of the nuclear and electronic spins of manganese can be considered as nearly isotropic; consequently, the hyperfine interaction H_{hfs} can be written in the form:

$$H_{hfs} = AS \cdot I \tag{4}$$

where A is the hyperfine coupling constant. Typical values found for A are -9.5 mT in octahedrally coordinated and -6.0 to -7.0 mT in tetrahedrally coordinated manganese complexes.

IDENTIFICATION OF OXYGEN LIGANDS

The zero field splitting Hamiltonian H_{zfs} commonly used for the simulations of EPR spectra of manganese-protein complexes has the form:

$$H_{zfs} = D\{S_{z'}^2 - \tfrac{1}{3}S(S+1)\} + E\{S_{x'}^2 - S_{y'}^2\} \qquad (5)$$

with the axial zero field splitting parameter D and the rhombic zero field splitting parameter E. The prime indicates that the Hamiltonian refers to the molecular principal axis system which in general does not coincide with the laboratory frame. Although in principle for a complete description of the system not only the quadratic terms [Eq. (5)] but, depending on the symmetry of the complex, also multipole terms of the fourth degree may contribute to the H_{zfs} [9,11], they can usually be neglected [6] for biological applications.

The observed EPR spectrum is strongly influenced by the rotational correlation time τ_r of the complex, which determines the type of spectrum obtained. According to the Stokes-Einstein relation, one obtains a correlation time τ_r in the order of 3 nsec even for a very small protein with a molecular weight of 10,000. In many manganese-protein complexes an anisotropy of the zero field splitting in the order of 0.3-1 GHz has been found [6,12]; under these conditions the rotational correlation time is not small enough for averaging out the zero field splitting and, accordingly, one observes a powder pattern-like EPR spectrum. The correspondent spectra can be obtained from the above-introduced Hamiltonian by perturbation theory as described by several authors in detail [6,9,12,13]. The central $|½,m\rangle \leftrightarrow |-½,m\rangle$ fine structure transitions are only at higher order dependent on the orientation of the manganese complex and the correspondent six hyperfine lines are usually easily detectable, whereas the other transitions are smeared out over a wide spectral range and are very often too weak to be detected [6,12,13].

As a general rule the central fine structure transitions in EPR spectra of manganese-protein complexes are much better resolved at higher magnetic fields. Figures 1 and 2 demonstrate this feature on the manganese complexes with elongation factor Tu [14,15]; going from X band (9.4 GHz) to the Q band (33.9 GHz) leads to a fourfold decrease

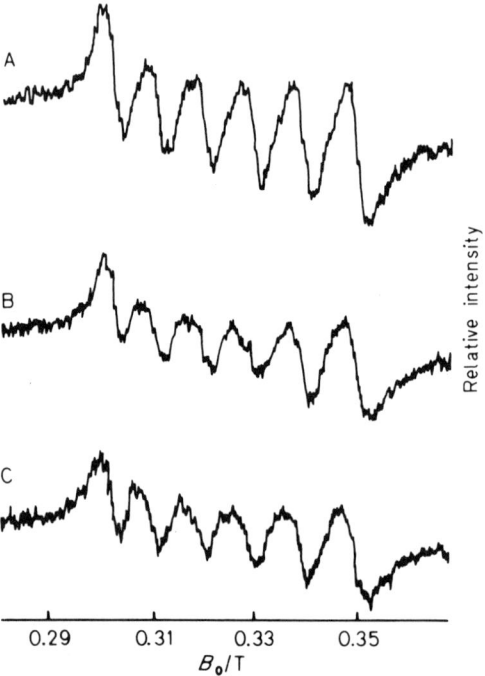

FIG. 1. X-band spectra of manganese complexes with elongation factor Tu (EF-Tu). (A) EF-Tu·Mn, (B) EF-Tu·Mn·GDP, (C) EF-Tu·Mn·GTP. (Reproduced by permission from Ref. 14.)

in linewidth. This result can easily be understood from the theory because the residual angular dependence of higher order terms decreases with the static magnetic field B_0 proportionally to B_0^{-1} and B_0^{-2} [12]. A similar decrease in linewidth is predicted for molecules tumbling rapidly enough for averaging out the zero field splitting interaction but leaving it as the dominant relaxation mechanism. The corresponding theory has been elaborated by several authors [9,16-20]. For $\omega_0 \tau_C \gg 1$, a condition which is usually fulfilled for macromolecular complexes at X- or Q-band frequencies ω_0 (see above), the linewidth of the central $|\frac{1}{2},m\rangle \leftrightarrow |-\frac{1}{2},m\rangle$ fine structure transitions decreases with $1/\omega_0^2 \tau_C$, whereas the remaining components give a broad background signal with a linewidth proportional to τ_C.

IDENTIFICATION OF OXYGEN LIGANDS

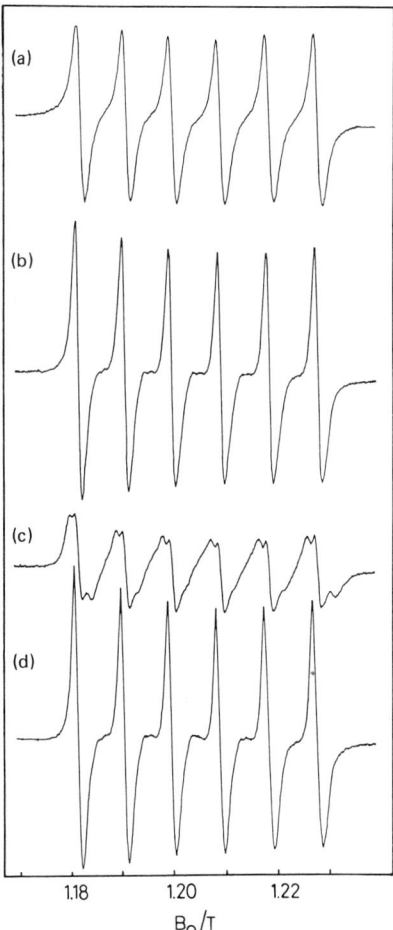

FIG. 2. Q-band spectra of manganese complexes with elongation factor Tu(EF-Tu). (a) EF-Tu·Mn, (b) Ef-Tu·Mn·GDP, (c) EF-Tu·Mn·GTP, (d) EF-Tu·Mn·GDP·P_i. (Reproduced by permission from Ref. 15.)

2.2. ^{55}Mn(II)-^{17}O Superhyperfine Interaction

A powerful tool for identifying the ligands of a paramagnetic metal ion is the observation of the hyperfine splitting in the EPR spectrum induced by the nuclear magnetic moment of the ligand. Although these studies are possible with any nucleus with a spin greater than zero, in our context the interaction with a stable isotope of oxygen, ^{17}O, is especially interesting. ^{17}O has a nuclear spin I = 5/2 and a natural abundance of only 0.037%. In contrast, the predominantly occurring stable isotopes ^{16}O (natural abundance 99.759%) and ^{18}O (natural abundance 0.204%) have no nuclear spin.

For including the interaction with the ^{17}O nuclear moment in our Hamiltonian, at least three additional terms have to be added to Eq. (1). These are, respectively, the superhyperfine interaction H_{shfs} between the electronic spin S of manganese and the nuclear spin I^1 of the ^{17}O ligand, the nuclear Zeeman interaction H_{nZ1} of I^1 with the external magnetic field, and, eventually, the quadrupolar interaction H_{q1}. The first two terms are analogous to the hyperfine term H_{hfs} and H_{nZ} already introduced:

$$H_{shfs} = S \cdot A^1 \cdot I^1 \qquad (6)$$

$$H_{nZ1} = -\gamma_{n1} \hbar B_0 I_z^1 \qquad (7)$$

In contrast to the wealth of experimental data concerned with hyperfine coupling of the manganese nuclear spin, there is almost nothing known about the magnitude of the individual components of the hyperfine tensor A^1 in manganese-^{17}O complexes. Only the trace of A^1, A^1 in the $Mn(H_2^{17}O)_6^{2+}$ complex is known from ^{17}O NMR measurements: A^1 could be estimated as 6.0 MHz (corresponding to 0.21 mT) [21], a value which is much smaller than the linewidth of 1-3 mT usually found in manganese-protein complexes at 33.6 GHz. Under this condition the superhyperfine splitting cannot be resolved but merely causes an inhomogeneous broadening of the hyperfine lines. In addition, pure $H_2^{17}O$ is not commercially available (the maximum enrichment obtainable is approximately 55%), and therefore the observed spectrum is always a superposition of spectra from labeled and unlabeled species. It appears evident that even at the highest frequencies usually available (Q-band) the information content of the EPR spectrum is not sufficient for estimating all parameters from a rigorous theoretical treatment of the complete Hamiltonian.

Because of these difficulties the perturbation introduced by the ^{17}O ligand has been treated only qualitatively in the literature.

IDENTIFICATION OF OXYGEN LIGANDS

The simplest approach introduced by Reed and Leyh [22] assumes that the unresolved scalar hyperfine splitting is the factor dominating the change in the EPR spectrum observed after substitution of ^{16}O by ^{17}O causing a line broadening of the order of $5A^1$, i.e., of the order of 1 mT. Even if the oxygen atom is not coordinated to the manganese ion, the dipolar coupling between the electron spin S and the nuclear spin I^1 is still left. With a distance of 0.223 nm, the manganese-oxygen bond length, between both spins a maximum dipolar splitting of 0.34 mT has been calculated [22]. This value can be taken as the practical upper limit for the effect which a noncoordinated ^{17}O atom could have on the manganese EPR spectrum. Consequently, if a line broadening by a ^{17}O-labeled molecule is observed which is clearly larger than this limit, it follows that the ^{17}O atom is coordinated to the manganese ion [22].

Using regiospecifically labeled substrates (e.g., ATP ^{17}O-labeled at the α-, β-, or γ-phosphate group), one can decide to which group(s) the metal ion is coordinated from the line broadening in the EPR spectrum. Moreover, using stereospecifically labeled substrates it is possible to conclude which of the stereoisomers is the true substrate of a given enzyme. It has to be pointed out that the opposite conclusion, namely, that a missing effect on the spectrum proves the lack of a coordination, is not valid as long as it is not known what magnitudes of hyperfine coupling constants do really occur in such complexes.

It can be a problem to decide simply by inspection of the EPR spectrum whether the line broadening is sufficiently large for implying a direct coordination to the metal ion. It is certainly more appropriate to quantify the effect by simulation of the EPR spectrum. Although the anisotropy of the superhyperfine coupling tensor is not known, the assumption appears reasonable that it is comparable to the dipolar splitting calculated above, i.e., of the order of a few MHz. An anisotropy of this magnitude can be averaged out even by the relatively slow tumbling of the macromolecular complex in solution and, to a first approximation, only the isotropic part, the trace of the tensor A^1, A^1 is left. The time modulation of the anisotropy induces relaxation processes which can be accounted for by an appropriate line broadening. The superhyperfine constants obtained with that method for different manganese-nucleotide-protein complexes agree well with the expected range of a few tenths of a mT [15,23].

3. APPLICATIONS

3.1. Elongation Factor Tu

Elongation factor Tu (EF-Tu) takes part in the synthesis of polypeptides in cells. It forms a complex with Mg^{2+}, GTP, and aminoacyl-tRNA which interacts with the codon-programmed ribosome resulting in the binding of aminoacyl-tRNA to the ribosomal acceptor site. In the next step GTP is hydrolized and EF-Tu·Mg·GDP is released from the ribosome [24-26]. The hydrolizing activity (GTPase activity) of elongation factor Tu is very low in the absence of the physiological activators ribosome and aminoacyl-tRNA, but enhanced very much in the presence of the antibiotic kirromycin. Mn^{2+} ions can substitute Mg^{2+} ions as cofactor in these reactions and have an even higher affinity to the known high-affinity binding site for Mg^{2+} in EF-Tu [27,28].

The structure of the active center has been probed by manganese EPR on EF-Tu from *B. stearothermophilus* [14,15] and from *E. coli* [29, 30]. The EPR spectra of Mn(II) bound to EF-Tu and its nucleotide complexes (Figures 1 and 2) appear to be very similar in the proteins from both microorganisms.

Addition of GDP to a solution containing EF-Tu·Mn leads to a decrease in linewidth in the EPR spectrum, giving a first hint that the metal ion and the nucleotide are bound at the same site of the protein. The use of ^{17}O-labeled GDP proves this suggestion: the EPR spectrum of EF-Tu·Mn complexed to unlabeled GDP and GDP labeled at the α-phosphate group are almost identical, but that of EF-Tu complexed to GDP labeled at the β-phosphate group shows a reproducible line broadening for the proteins of both microorganisms [14,29]. Spectrum simulation considering only the isotropic superhyperfine coupling results in a coupling constant of 0.25 mT [23]. This shows that the metal ion is coordinated to the β-phosphate group in the EF-Tu·Mn·GDP complex. The lack of a broadening effect suggests that the metal ion is not coordinated to the α-phosphate group.

The EPR resonance lines of the EF-Tu·Mn·GTP complex are relatively broad even at Q-band frequencies [Fig. 2(c)]. At room temperature the signal slowly changes its shape: the broad signal disappears and concomitantly a new resonance with much smaller linewidth appears dominating the spectrum after 3 hr [Fig. 2(d)]. The linewidth is significantly smaller than that of EF-Tu·Mn·GDP, which should be the final product of the hydrolysis of GTP. Subsequently, biochemical experiments with radioactively labeled GTP revealed that the spectrum could be assigned to the hitherto not observed intermediate state complex EF-Tu·Mn·GDP·P_i [15]. Experiments with GTP labeled at the α-, β-, or γ-phosphate group permitted the conclusion that in the EF-Tu·Mn·GDP·P_i complex the metal ion is again only complexed to the β-phosphate group, but most probably not to the α-phosphate group and to P_i. The calculated superhyperfine coupling constant of 0.16 mT was significantly smaller than that in the EF-Tu·Mn·GDP complex.

When the metal ion binds to the protein it can retain a part of the water molecules of its hydration shell. Replacing $H_2^{16}O$ by $H_2^{17}O$ leads to a remarkable broadening of the EPR lines of all manganese-protein complexes (Fig. 3). Analyzing the spectra with the method described in Sec. 2.2 with a superhyperfine coupling constant of 0.22 mT, i.e., a constant similar to that in $Mn(H_2^{17}O)_6^{2+}$ [21], gives the best fit for four water molecules in EF-Tu·Mn, three water molecules in EF-Tu·Mn·GDP, and five water molecules in EF-Tu·Mn·GDP·P_i. In Figure 4 the structures of the various EF-Tu-metal-nucleotide complexes are depicted as they can be proposed from EPR measurements [14,15,29] and the studies using thio analogs of GDP and GTP [14,27].

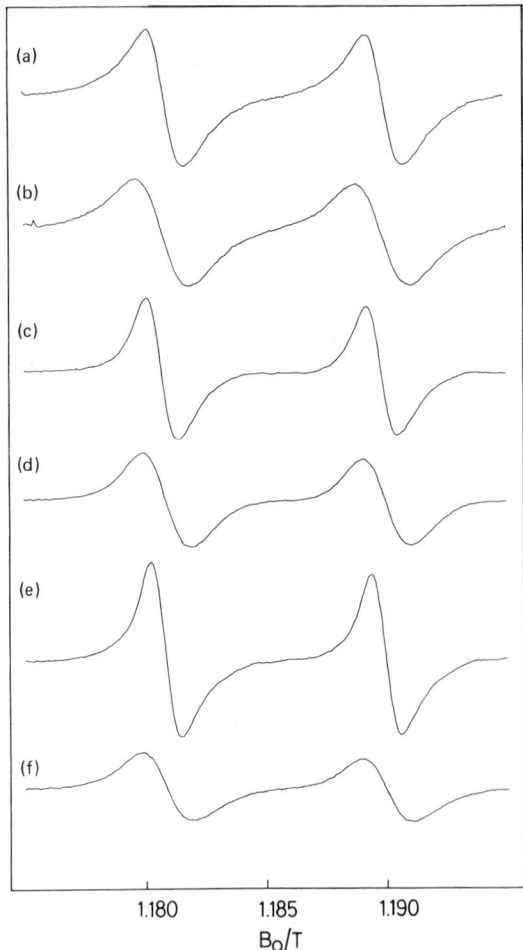

FIG. 3. Effect of ^{17}O-enriched water on the EPR spectra of manganese complexes with EF-Tu. (a) EF-Tu·Mn in normal water, (b) in $H_2^{17}O$-enriched water, (c) EF-Tu·Mn·GDP in normal water, (d) in $H_2^{17}O$-enriched water, (e) Ef-Tu·Mn·GDP·P_i in normal water, (f) in $H_2^{17}O$-enriched water. ^{17}O enrichment 52.4%, only the two lowest field hyperfine components are shown. (Reproduced by permission from Ref. 15.)

IDENTIFICATION OF OXYGEN LIGANDS

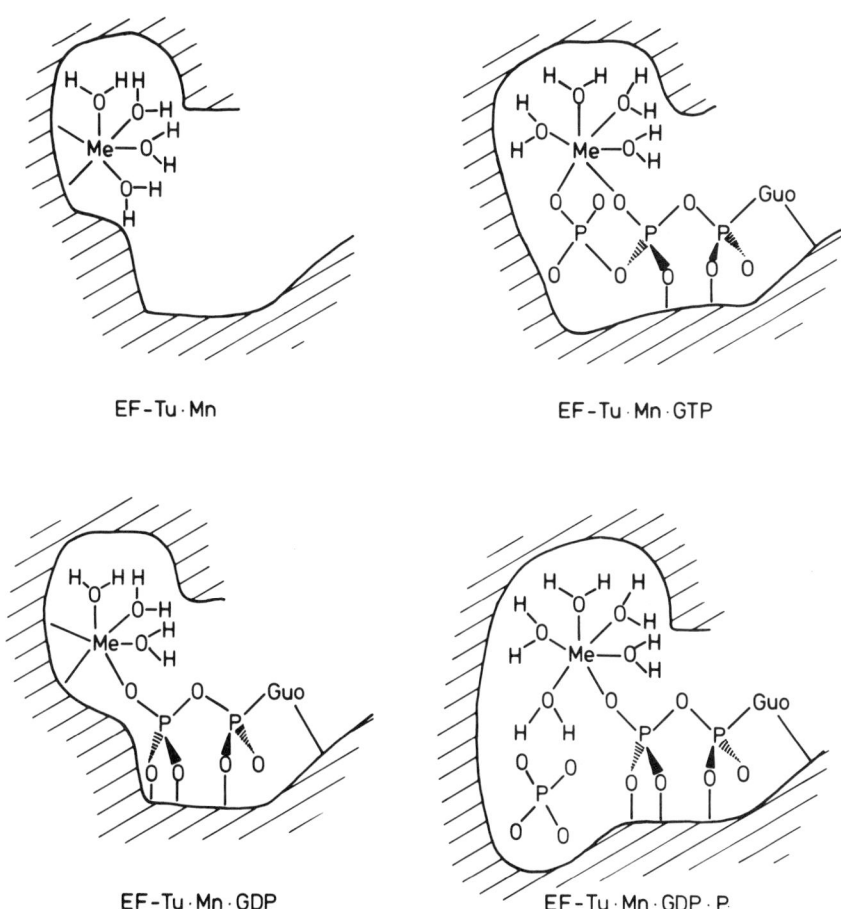

FIG. 4. Proposed coordination scheme of the divalent ion in various EF-Tu-nucleotide complexes. (Reproduced by permission from Ref. 15.)

3.2. Creatine Kinase

Creatine kinase (CK, EC 2.7.3.2) catalyzes the reaction:

$$\text{Phosphocreatine} + \text{ADP} \xrightleftharpoons{\text{Mg}^{2+}} \text{creatine} + \text{ATP}$$

The rabbit enzyme is a dimer composed of two identical subunits of molecular weight ~41,300; its biochemical and structural properties have recently been reviewed [31,32]. The enzyme has been studied extensively by spectroscopic methods including ^{31}P NMR (e.g., [33]). ^{1}H NMR (e.g., [34]), and EPR spectroscopy [22,35-37]. It appears to be especially suited for manganese EPR studies because of its well-resolved spectrum (Fig. 5).

The EPR spectra of various manganese-substrate complexes with creatine kinase can be satisfactorily fitted by the theory outlined in Sec. 2.1. By adding anions such as nitrate, formate, thiocyanate, or chloride, the EPR spectrum of the dead-end complex CK·Mn·ADP·creatine becomes much better resolved, probably because the spread in D and E values is reduced in the transition state analog complexes produced by the binding of anions. Experiments with regiospecifically ^{17}O-labeled ADP show a line broadening effect for [α-^{17}O]ADP as well as for [β-^{17}O]ADP in the CK·Mn·ADP·formate·creatine complex corresponding to a superhyperfine coupling constant of 0.3-0.4 mT [22]. A comparable influence on the spectrum can be produced by the addition of ^{17}O-labeled formate ions. Substituting normal water by ^{17}O-enriched water results in a spectral broadening. A lower limit of three water molecules bound to the metal ion in the transition state analog complex can be obtained by a simple deconvolution procedure: the probability p for finding a manganese complexed with exclusively H$_2^{16}$O decreases with the number of H$_2$O ligands; if the number of ligands assumed is too low, subtracting fraction p of the spectrum for the sample in normal water from the spectrum for the sample in ^{17}O-enriched water causes gaps in the difference spectrum [22]. The use of the diastereomers of ADP, S_p-, and R_p-[α-^{17}O]ADP leads to a line broadening only for the S_p diastereomer in the transition state analog complex with nitrate ions, meaning that the pro-S oxygen of the α-phosphate group is complexed to the metal ion in this complex [36]. Using ATP ^{17}O-labeled at the α-, β-, or γ-phosphate group together with 1-(carboxymethyl)-2-iminoimidazolin-4-one, a nonreactive analog for creatine, a broadening effect can be observed for each of the positions of ^{17}O. This leads to the conclusion that

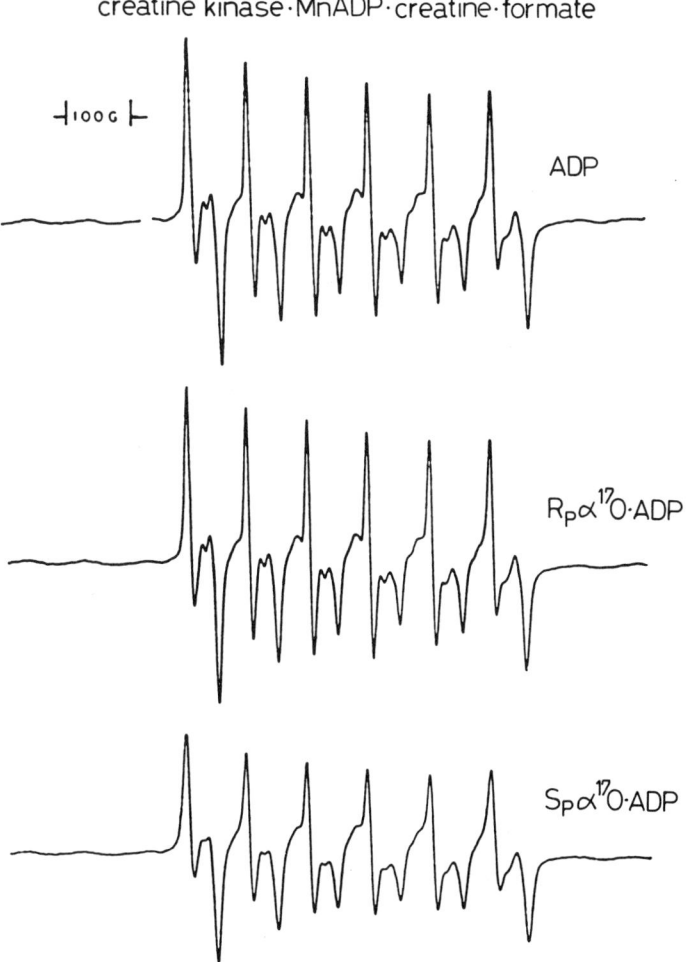

FIG. 5. Q-band EPR spectrum of the CK·Mn·ADP·creatine·formate complex with normal ADP and with the two diastereomers of $[\alpha\text{-}^{17}O]$ADP. (Reproduced by permission from Ref. 36.)

the metal ion is bound as a α,β,γ-tridentate complex of ATP to the enzyme [37]. From these results the mode of coordination can be derived for the different metal-nucleotide complexes occurring during the catalytic process (Fig. 6).

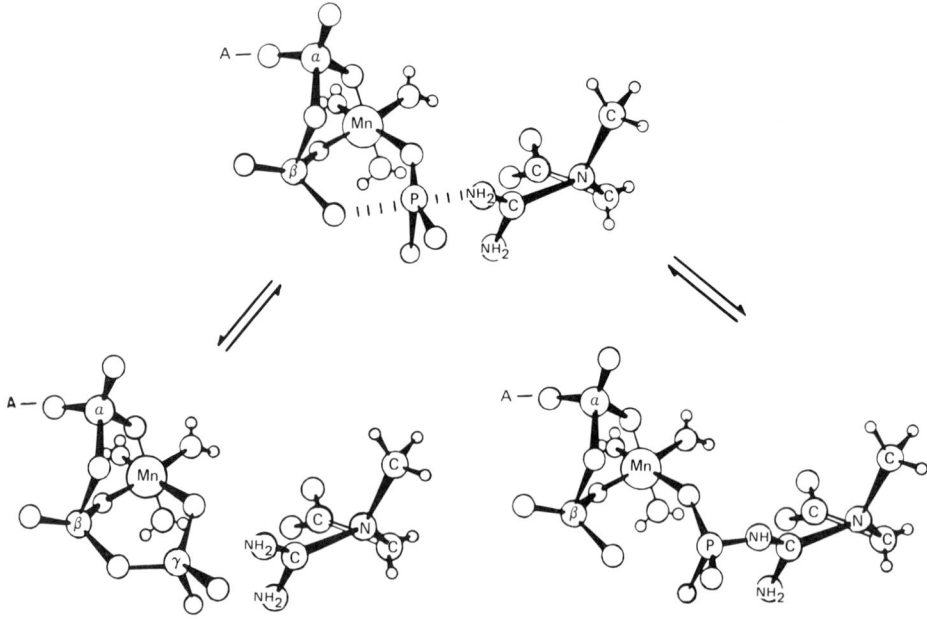

FIG. 6. Schematic representation of the metal-substrate complexes at the "active center" of creatine kinase. (Reproduced by permission from Ref. 37.)

3.3. Adenylate Kinase

Adenylate kinase (AK, EC 2.7.4.3) catalyzes the reaction:

$$\text{AMP} + \text{ATP} \xrightleftharpoons{\text{Mg}^{2+}} 2\text{ADP}$$

Mammalian muscle adenylate kinase has been studied by many different methods including ^{31}P and ^{1}H NMR (e.g., [38-41]) as well as x-ray crystallography [42]. X-ray diffraction analysis has so far failed to reveal the exact arrangement of the substrates and the metal ion in the protein structure.

EPR spectroscopy on manganese-nucleotide complexes with porcine muscle adenylate kinase shows the usually found pattern of six hyperfine lines centered around g = 2. Even at Q band the

IDENTIFICATION OF OXYGEN LIGANDS

hyperfine lines are relatively broad (approximately 2.4 mT) [23], which probably represents the upper limit for the observation of significant broadening by the interaction with a ^{17}O nuclear moment.

ATP labeled at the oxygens of the β-phosphate group leads to a line broadening in the AK·Mn·ATP spectrum which corresponds to a coupling constant of 0.22 mT under the assumption of only one Mn^{2+}-^{17}O interaction, but neither for $[\alpha$-$^{17}O_3]$ATP nor for $[\gamma$-$^{17}O_3]$ATP could a significant line broadening be observed. This means that the metal ion is probably bound only to the β-phosphate group on the enzyme. With ATP stereospecifically labeled at the β-phosphate group only the R_p isomer has an effect on the spectrum corresponding to a superfine coupling constant of 0.23 mT. From experiments with $H_2^{17}O$ it can be concluded that probably three to four water ions are coordinated to the metal ion in the enzyme-bound complex. The results from the combined ^{17}O and thio analog studies are shown in Figure 7.

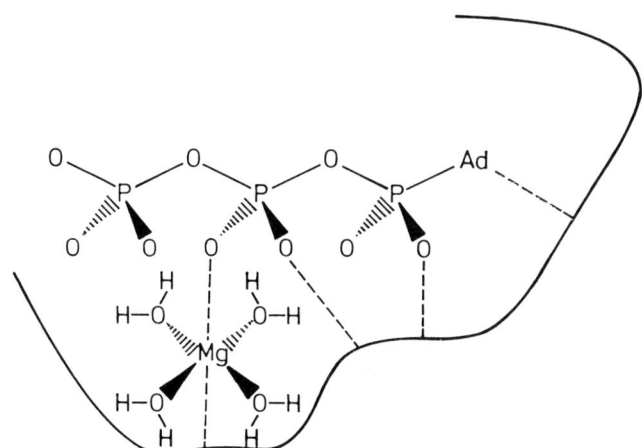

FIG. 7. Structure of the metal-nucleotide-AK complex as postulated from ^{17}O EPR and thio analog studies. (Reproduced by permission from Ref. 23.)

3.4. Myosin Subfragment 1

Myosin is a main component of the contractile filaments of muscle cells. During the contraction cycle ATP is bound to the myosin head and broken down to ADP and P_i (for a review, see [43-46]). Mg^{2+} is cofactor of the hydrolysis reaction. The ATP binding domain is located on a proteolytic fragment of myosin, myosin subfragment 1 (S1).

By using ADP labeled at the α- or β-phosphate group, Webb et al. [47] showed that manganese is coordinated to one of the oxygens of the β-phosphate group but not to the α-phosphate group. By replacing normal water with $H_2^{17}O$-enriched water they concluded that at least two water molecules are bound to the metal ion in the S1·Mn·ADP complex [47].

3.5. 3-Phosphoglycerate Kinase

3-Phosphoglycerate kinase (PGK, EC 2.7.2.3) promotes the transfer of a phosphoryl group of 1,3-bisphosphoglycerate to ADP in the presence of Mg^{2+} (for a review, see [48,49]). The crystal structure of two isoenzymes, the horse muscle [50] and the yeast [51] phosphoglycerate kinase, has been solved. In the x-ray structure of the horse enzyme manganese is bound to the α- and β-phosphate groups of ADP as well as to the carboxylate group of an aspartate residue [50].

EPR experiments with ^{17}O-labeled ADP analogs showed that the results obtained in the crystalline phase are valid also in solution [52]: both $[\alpha{-}^{17}O_3]$ADP and $[\beta{-}^{17}O_3]$ADP cause a line broadening in the PGK·Mn·ADP spectrum. Of the two diastereomers of the α-phosphate group, only the R_p isomer has an effect on the linewidth of the manganese complex, whereas the spectrum with the S_p isomer is identical to that from unlabeled ADP. The broadening by $H_2^{17}O$ could be explained by two or three water molecules in the first coordination sphere of the metal ion in the complex. The resulting structure is depicted in Figure 8.

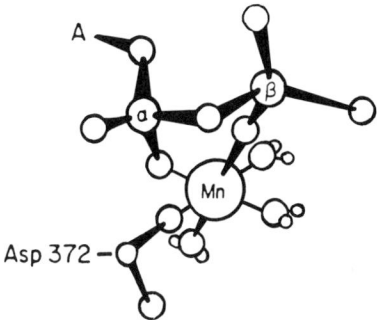

FIG. 8. Proposed structure of the metal-nucleotide-PGK complex. (Reproduced by permission from Ref. 52.)

4. CONCLUSIONS AND OUTLOOK

The observation of Mn^{2+}-^{17}O superhyperfine coupling by EPR spectroscopy of manganese protein complexes with regio- and/or stereospecifically labeled nucleotides is probably the most elegant, most reliable, and most generally applicable approach for determining the metal-nucleotide coordination at the active center of enzymes currently available. Moreover, the method can give information about the number of slowly exchanging water ligands of the complex, information which can be obtained only in special cases by other methods. Concerning the amount of biological material needed (~1 pmol) per experiment and the experimental time needed to come to a definite answer to the problem (for practical purposes around 1-2 weeks), it is comparable to methods routinely used in biochemical work.

However, every experimental method has its flaws and limits: it is not a priori justified (although probable) that in all studied complexes the coordination of Mg^{2+} can be assumed to be identical to that of its analog Mn^{2+}. In addition, as long as the experimental data basis is not large enough for estimating the possible range of magnitude of the components of the superhyperfine coupling tensor, it is not permissible to conclude definitely that a missing line-broadening effect by ^{17}O implies a lack of coordination. Even if the coupling is sufficiently large, H_2O molecules exchanging rapidly compared with the coupling constant, i.e., faster than 10^{-6} sec^{-1} may not be detected. The method is only applicable if the linewidth of the individual hyperfine components is not too large compared with the superhyperfine component, a limitation which could be overcome by measuring at higher frequencies than the commercially available Q band (see Sec. 2.1).

Nevertheless, these limitations are small relative to those of the alternative methods which all have to rely on even more stringent assumptions. Of course, the best way is always to check all results by independent experimental methods—for example, in our case by studying the enzyme kinetics with thio analog-metal complexes [1,2] which can confirm the assumed nucleotide-metal coordination and give complementary information about the interaction between the phosphate group and the enzyme itself.

ABBREVIATIONS AND DEFINITIONS

ADP	adenosine 5'-diphosphate
ATP	adenosine 5'-triphosphate
AK	adenylate kinase
CK	creatine kinase
D	axial zero field splitting parameter
E	rhombic zero field splitting parameter
EF-Tu	elongation factor Tu
EPR	electron paramagnetic resonance
GDP	guanosine 5'-diphosphate
GTP	guanosine 5'-triphosphate
H	Hamilton operator
NMR	nuclear magnetic resonance
PGK	phosphoglycerate kinase
S1	subfragment 1
τ	correlation time

REFERENCES

1. E. K. Jaffe and M. Cohn, *J. Biol. Chem.*, *254*, 10839-10845 (1979).
2. F. Eckstein, *Ann. Rev. Biochem.*, *54*, 367-402 (1985).
3. D. Dunaway-Mariano and W. W. Cleland, *Biochemistry*, *19*, 1506-1515 (1980).
4. W. W. Cleland and A. S. Mildvan, *Adv. Inorg. Biochem.*, *1*, 163-191 (1979).
5. W. W. Cleland, *Methods Enzymol.*, *87*, 159-179 (1982).
6. G. H. Reed and G. H. Markham, *Biol. Magn. Reson.*, *6*, 73-142 (1984).
7. H. Hori, M. Ikeda-Saito, G. H. Reed, and T. Yonetani, *J. Magn. Reson.*, *58*, 177-185 (1984).
8. M. Korkmaz and B. Aktas, *Phys. Stat. Sol. (b)*, *130*, 743-748 (1985).
9. A. Abragam and B. Bleaney, in *Electron Paramagnetic Resonance of Transition Ions* (W. Marshall and D. H. Wilkinson, eds.), Clarendon Press, Oxford, 1970.
10. J. J. Villafranca, D. E. Ash, and F. C. Wedler, *Biochemistry*, *15*, 544-553 (1976).
11. D. Newman and W. Urban, *Adv. Phys.*, *24*, 793-844 (1975).
12. G. D. Markham, B. D. Nageswara Rao, and G. H. Reed, *J. Magn. Reson.*, *33*, 595-602 (1979).
13. E. Meirovitch and R. Poupko, *J. Phys. Chem.*, *82*, 1920-1925 (1978).
14. A. Wittinghofer, R. S. Goody, P. Rösch, and H. R. Kalbitzer, *Eur. J. Biochem.*, *124*, 109-115 (1982).
15. H. R. Kalbitzer, R. S. Goody, and A. Wittinghofer, *Eur. J. Biochem.*, *141*, 591-597 (1984).
16. L. Burlamacchi, *J. Chem. Phys.*, *55*, 1205-1212 (1971).
17. L. Burlamacchi, G. Martini, M. F. Ottavani, and M. Romanelli, *Adv. Mol. Relax. Inter. Proc.*, *12*, 145-186 (1978).
18. R. Poupko and Z. Luz, *Mol. Phys.*, *36*, 733-752 (1978).
19. G. H. Reed, J. S. Leigh, Jr., and J. E. Pearson, *J. Chem. Phys.*, *55*, 3311-3316 (1971).
20. M. Rubinstein, A. Baram, and Z. Luz, *Mol. Phys.*, *20*, 67-80 (1971).
21. M. S. Zetter, G. Y.-S. Lo, H. W. Dodgen, and J. P. Hunt, *J. Am. Chem. Soc.*, *100*, 4430-4436 (1978).

22. G. H. Reed and T. S. Leyh, *Biochemistry*, *19*, 5472-5480 (1980).
23. H. R. Kalbitzer, R. Marquetant, B. A. Connolly, and R. S. Goody, *Eur. J. Biochem.*, *133*, 221-227 (1983).
24. D. L. Miller and H. Weissbach, in *Molecular Mechanisms of Protein Biosynthesis* (H. Weissbach and S. Pestka, eds.), Academic Press, New York, 1977, pp. 323-373.
25. Y. Kaziro, *Biochem. Biophys. Acta*, *505*, 95-127 (1978).
26. A. Parmeggiani and S. Sander, *Mol. Cell. Biochem.*, *35*, 129-158 (1981).
27. C. M. Leupold, R. S. Goody, and A. Wittinghofer, *Eur. J. Biochem.*, *135*, 237-241 (1983).
28. B. Antonsson, H. R. Kalbitzer, and A. Wittinghofer, *Hoppe-Seyler's Z. Physiol. Chem.*, *362*, 735-743 (1981).
29. J. F. Eccleston, M. R. Webb, D. E. Ash, and G. H. Reed, *J. Biol. Chem.*, *256*, 10774-10777 (1981).
30. G. E. Wilson and M. Cohn, *J. Biol. Chem.*, *252*, 2004-2009 (1977)
31. G. L. Kenyon and G. H. Reed, *Adv. Enzymol. Relat. Areas, Mol.* *54*, 367-426 (1983).
32. J. B. Walker, *Adv. Enzymol. Relat. Areas Mol. Biol.*, *50*, 177-242 (1979).
33. B. D. Nageswara Rao and M. Cohn, *J. Biol. Chem.*, *256*, 1716-1721 (1981).
34. P. R. Rosevear, P. Desmeules, G. L. Kenyon, and A. S. Mildvan, *Biochemistry*, *20*, 6155-6164 (1981).
35. A. C. McLaughlin, J. S. Leigh, Jr., and M. Cohn, *J. Biol. Chem.*, *251*, 2777-2787 (1976).
36. T. S. Leyh, R. D. Sammons, P. A. Frey, and G. H. Reed, *J. Biol. Chem.*, *257*, 15047-15053 (1982).
37. T. S. Leyh, P. J. Goodhart, A. C. Nguyen, G. L. Kenyon, and G. H. Reed, *Biochemistry*, *24*, 308-316 (1985).
38. B. D. Nageswara Rao, M. Cohn, and L. Noda, *J. Biol. Chem.*, *253*, 1149-1158 (1978).
39. K. V. Vasavada, J. I. Kaplan, and B. D. Nageswara Rao, *Biochemistry*, *23*, 961-968 (1984).
40. H. R. Kalbitzer, R. Marquetant, P. Rösch, and R. H. Schirmer, *Eur. J. Biochem.*, *126*, 531-536 (1982).
41. P. Rösch and K.-H. Groß, *J. Mol. Biol.*, *182*, 341-345 (1985).
42. G. E. Schulz, M. Elzinga, F. Marx, and R. H. Schirmer, *Nature*, *250*, 120-123 (1974).
43. H. G. Mannherz and R. S. Goody, *Ann. Rev. Biochem.*, *45*, 428-465 (1976).

44. W. F. Harrington and M. E. Rodgers, Ann. Rev. Biochem., 53, 35-73 (1984).
45. R. S. Goody and K. C. Holmes, Biochem. Biophys. Acta, 726, 13-39 (1983).
46. A. Ribeiro, J. Parello, and O. Jardetzky, Prog. Biophys. Mol. Biol., 43, 95-160 (1984).
47. M. R. Webb, D. E. Ash, T. S. Leyh, D. R. Trentham, and G. H. Reed, J. Biol. Chem., 257, 3068-3072 (1982).
48. J. L. Vandeberg, Isoenzymes Curr. Top. Biol. Med. Res., 12, 133-187 (1985).
49. C. C. Blake and D. W. Rice, Phil. Trans. R. Soc. London, 293, 94-104 (1981).
50. R. D. Banks, C. C. F. Blake, P. R. Evans, R. Haser, D. W. Rice, G. W. Hardy, M. Merett, and A. W. Philipps, Nature, 279, 773-777 (1979).
51. H. C. Watson, N. P. C. Walker, P. J. Shaw, T. N. Bryant, P. L. Wendell, L. A. Fothergill, R. E. Perkins, S. C. Conroy, M. J. Dobson, M. F. Tuite, A. J. Kingsman, and S. M. Kingsman, EMBO J., 1, 1635-1640 (1982).
52. J. M. Moore and G. H. Reed, Biochemistry, 24, 5328-5333 (1985).

3
Use of EPR Spectroscopy for Studying Solution Equilibria

Harald Gampp
Institut de Chimie
Université de Neuchâtel
Avenue de Bellevaux 51
CH-2000 Neuchâtel, Switzerland

1.	INTRODUCTION	105
2.	ANALYSIS OF EPR LINE INTENSITIES	107
	2.1. Brief History	107
	2.2. Experimental Techniques	108
	2.3. Mathematical Methods	109
3.	EXAMPLES	112
	3.1. The Cu(II)-DANA System	112
	3.2. The Cu(II)-AMIN System	120
4.	CONCLUSIONS	124
	ABBREVIATIONS AND DEFINITIONS	125
	REFERENCES	126

1. INTRODUCTION

Since electron paramagnetic resonance (EPR) spectra of paramagnetic ions are determined by the number and nature of the ligating atoms and by the geometry of the ligand field [1], EPR spectroscopy allows the assignment of the structures of paramagnetic coordination compounds. Indeed, in this regard the method has found numerous applications, e.g., in biochemical or bioinorganic studies [2,3]. One advantage of the technique is that it can be used not only for solid samples, i.e., crystals or frozen solutions, but also in the fluid phase.

Consequently, EPR should also be well suited for investigating complex equilibria in solution. Until recently, however, the method was only occasionally applied, and only to simple systems which may be described by a single equilibrium constant [4-6]. Examples are the association of iodide with radical anions [7], complexes of Mn(II) with simple anions [8,9], ion pair formation between radical anions and alkali metal ions [10,11], and the formation of a ternary complex of a Ni(III) tripeptide with 2,2',2"-terpyridine [12].

The EPR parameters that may be used for determining stability constants are the coupling constants [7], the linewidths [8,9], or the g values [11]; the latter can only be employed in those cases where the complexation kinetics are fast on the EPR time scale and where therefore only a single time-averaged signal is found. If, on the other hand, different species give rise to separate signals, only the analysis of line intensities (spin concentrations) allows one to determine stability constants [10]. In the following the focus will be exclusively on this last mentioned case, as it is usually realized in biological and bioinorganic systems.

If the species occurring in an equilibrium system have spectra consisting of sets of sharp and narrow lines, the different lines in the EPR spectrum of a mixture can be readily assigned. In this case, which is most frequently observed for organic radicals [1], the concentrations of the different species are obtained by simply integrating over the respective lines [13]. The situation is quite different for many transition metal ion complexes due to the much broader spectra [1]. For example, for the complexes of the paramagnetic Cu(II), where the dependence of the EPR parameters on the coordinating atoms, geometry, and charge is well understood [14,15], the spectra of the complexes present in an equilibrium system are strongly overlapping [16,17]. Especially in the case of minor species, their spectra are not known beforehand and the simple data treatment as mentioned above for organic radicals is no longer applicable; meaningful information about a system can only be obtained if complete spectra are evaluated, i.e., a wide range of field strengths has to be taken into account.

It is the purpose of this chapter to outline which kind of problems occur and to describe how these may be solved, in order to apply EPR spectroscopy to the investigation of complex equilibria in solution. Some of the underlying mathematical concepts will be presented. Ultimately, it will be shown that EPR is perfectly suited for studying complicated systems, including those where minor species occur and where the individual spectra are strongly overlapping. Due to recently developed algorithms, the application of the method is straightforward and even complicated biochemical equilibria may now be attacked.

2. ANALYSIS OF EPR LINE INTENSITIES

2.1. Brief History

The investigation of complicated solution equilibria by analysis of EPR line intensities in the case of overlapping spectra was first reported in 1983 by Kittl and Rode [18]. In the meanwhile, Rode and coworkers [19-23] studied in aqueous solution the formation of binary complexes between Cu(II) and dipeptides [18,22,23], tripeptides [21,23], or ATP [19], and the stability of ternary complexes between Cu(II), ATP, and dipeptides [20,22]. Their work shows that the EPR method is a useful tool for investigating complicated equilibria of biological interest, that EPR titrations are in many cases superior to pH titrations, and that the structural assignment of complexes in solution is possible. However, the calculations were done on a mainframe computer using a grid search which converges only slowly and is thus expensive in computer time [24]. The numerical procedure used [18] has the further disadvantage that for n simultaneously occurring complexes n - 1 species spectra have to be known. This limits its practical use to systems with only two species present in a certain mixture. Thus the algorithm is not easily applicable to unknown systems and considerable experience is necessary to use the program.

Recently, a more rigorous approach was successfully applied to EPR titration data [25,26]. As will be illustrated in the following sections, this algorithm is generally applicable, easy to handle, even usable on desk computers, and does not require any information about the chemical system prior to the analysis.

2.2. Experimental Techniques

In order to study a given equilibrium system a mixture of metal ion(s) and ligand(s) is titrated and after each addition of reagent the EPR spectrum is recorded. In the frequent case of pH-dependent equilibria, a potentiometric pH titration will be carried out, but any other type of titration, e.g., with a ligand at fixed pH as in [12], can be carried out in the same way.

When conventional EPR equipment is used, the recording of a spectrum may take several minutes and the complete EPR titration requires between 1 and 2 hr. If there is the danger that some of the equilibrium species may decompose during this time, after each titration step discrete samples have to be taken, sealed, and frozen immediately. After the proper titration each sample is individually defrozen and its EPR spectrum is recorded at once [18].

In cases where each species is stable, a flow system as described in Ref. 25 can be used where the titrated solution is continuously pumped through the cell of the EPR spectrometer. This method has several advantages: (1) All the spectra are recorded without changing neither the cell nor its position, i.e., the data quality is improved since no individual baseline corrections are required. (2) In the case of air-sensitive compounds there is less danger of contamination as the titration system is kept closed

USE OF EPR SPECTROSCOPY FOR STUDYING SOLUTION EQUILIBRIA 109

during the experiment. (3) With the availability of microprocessor-controlled spectrometers [27,28], the whole procedure can be made fully automatic as has been shown for spectrophotometric titrations [29]. Apart from saving considerable time, such an automatic device improves the data quality, e.g., by avoiding the manual digitizing of the spectral data which is a potential error source [30].

2.3. Mathematical Methods

The mathematical methods used for the analysis of spectroscopic titration data have been described in detail elsewhere [31-33]. This section will briefly outline how the problem of determining both stability constants and EPR spectra of the species present in solution is formulated in mathematical terms. It will be shown what kind of numerical problems occur and how they are eliminated step by step so that finally the complete data evaluation becomes straightforward and can be easily handled by the chemist himself.

The set of $\underline{S}$ EPR spectra, usually obtained as the derivative of the intensity I with respect to the magnetic field strength H (dI/dH), has to be digitized at $\underline{W}$ discrete values of H and is arranged into a matrix $\underline{Y}$. Thus, each column of $\underline{Y}$ contains one of the measured spectra, each row represents an EPR titration curve obtained at a particular value of H.

Assuming one has a pH-dependent equilibrium system with $\underline{P}$ different complexes, their concentrations depend only on the known total concentrations of metal ion and ligand, on the pH of the respective solution, and on the set of $\underline{P}$ - 1 equilibrium constants to be determined. Hence, for a given set of stability constants the concentrations of the species in the different mixtures can be calculated (e.g., by the Newton-Raphson method [34]) and arranged into a matrix $\underline{C}$. These stability constants are initially provided estimates which are then iteratively refined by the nonlinear least-squares program.

Provided that each measured spectrum is a linear combination of the spectra of the P complexes, the data matrix $\underline{\underline{Y}}$ can be represented by a calculated matrix $\underline{\underline{Y}}^c = \underline{\underline{E}} \cdot \underline{\underline{C}}$. Each column of $\underline{\underline{E}}$ contains one of the (still unknown) EPR spectra of the pure complexes. The next step is to find the set of $P \times W$ elements of $\underline{\underline{E}}$ and of $P - 1$ stability constants that leads to the minimum sum of squares of the differences between the corresponding elements of $\underline{\underline{Y}}$ and $\underline{\underline{Y}}^c$. In principle this can be achieved by applying any least-squares minimization procedure (cf. [24]), but the practical problems become obvious at once—assuming one has an equilibrium system with $P = 4$ species, where $S = 30$ spectra at $W = 50$ magnetic field strengths have been selected. Therefore, 203 parameters (200 elements of $\underline{\underline{E}}$ and 3 equilibrium constants) have to be simultaneously refined and fitted to 1,500 data points, a task which is impossible in practice.

This problem can be considerably reduced, namely, to the refinement of only $P - 1$ stability constants [31,32,35,36]. The elements of $\underline{\underline{E}}$ are linear parameters, i.e., if the stability constants were known $\underline{\underline{E}}$ could be obtained by simple linear regression [32,35]. One can benefit of this property by replacing $\underline{\underline{E}}$ in the minimization procedure by its least-squares estimate obtained for a given set of stability constants. Hence, only $P - 1$ parameters have to be refined iteratively, i.e., the elements of $\underline{\underline{E}}$ do not show up during the minimization procedure and are calculated from the final set of equilibrium constants. In other words, the calculation of stability constants does not require any knowledge of the spectra of the complexes. These spectra are obtained at the end of the calculation at little extra cost.

Due to the large size of the data matrix $\underline{\underline{Y}}$, the amount of data to be handled during the least-squares calculation may easily exceed the capacity of the computer memory. This problem can be overcome by performing an eigenvector analysis (factor analysis [37]) of $\underline{\underline{Y}}$ [31,33]: Using the method of vector iteration [38], $\underline{\underline{Y}}$ is decomposed into the product of the matrix of its eigenvectors $\underline{\underline{V}}$ and the matrix of the corresponding linear coefficients $\underline{\underline{L}}$. The latter matrix is only of the dimension $\underline{P} \times \underline{S}$ and can replace $\underline{Y}$ in the least-squares procedure without any loss of significant information [33]. For the above-mentioned example, the initial problem of fitting 203 parameters to 1,500 data points has thus finally been reduced to the refinement of 3 parameters using 120 data points.

A computer program based on the presented ideas was recently published [39]. The user has to provide the data which are first subjected to the eigenvector decomposition. The calculated number of significant eigenvectors is an unbiased estimate of the number of absorbing species and greatly helps to imagine a chemical model. Next, the user has to define the type of complexes occurring in the equilibrium system and to give rough estimates of their stability constants. The program which is based on the Newton-Gauss-Marquardt algorithm [40] automatically calculates the best set of equilibrium constants as well as the spectra of the complexes.

Very recently, a new application of factor analysis, the so-called evolving factor analysis (EFA), was developed which ultimately will lead to a completely model-free analysis of spectroscopic titration data [41,42]. EFA makes use of the spectral differences obtained between successive titration steps: for the sets of the first 2, 3, ... $\underline{S}$ spectra the corresponding eigenvalues are calculated. In this forward EFA the eigenvalues are plotted as a function of the progressing titration (e.g., of pH). If in the course of the titration a new complex is formed, this is reflected by the appearance of a new nonzero eigenvalue (= factor). The backward EFA begins at the end of the titration, starting with the last 2, 3, ... $\underline{S}$ spectra, and indicates the decay of species. A given eigenvalue, however, does not directly correspond to a certain complex or to its concentration. The actual values depend on the dissimilarity of the spectra of the species and on the differences between their concentration profiles. Nevertheless, the concentrations and the eigenvalues may be correlated as described in [41], and the combination of the results of the forward and backward EFA provides model-free species distribution curves.

3. EXAMPLES

The investigation of solution equilibria as outlined above will be illustrated with two examples, namely, the complex formation between Cu(II) and 3,7-diazanonanedioic acid diamide (DANA) or 4,7,10-triazatridecane-1,13-diamine (AMIN).

$H_2NOC-CH_2-NH-(CH_2)_3-NH-CH_2-CONH_2$ DANA

$NH_2-(CH_2)_3-NH-(CH_2)_2-NH-(CH_2)_2-NH-(CH_2)_3-NH_2$ AMIN

Each complex system is reasonably complicated, i.e., contains four species with strongly overlapping spectra, and each has been studied independently by potentiometry and by spectrophotometry as well. This allows one to judge the quality of the EPR results and to point out the advantages of the EPR method for studying equilibrium systems.

3.1. The Cu(II)-DANA System

An EPR titration was carried out in aqueous solution by adding base to an equimolar mixture of Cu(II) and DANA·2HCl. $\underline{S}$ = 44 increments of base changed the pH from 2 to 11.5. After each addition of reagent the pH was measured and the EPR spectrum was recorded between 2,750 and 3,450 G [25].

For the calculation each spectrum was resolved into $\underline{W}$ = 50 points by digitizing at equidistant intervals of 10 G (Fig. 1). Eigenvector analysis showed that the spectral data matrix $\underline{Y}$ could be represented by $\underline{P}$ = 4 eigenvectors, which is to be expected since by potentiometry [43] and by spectrophotometry [30] four different species were detected in the Cu(II)-DANA system.

The overall standard deviation between the experimental data and their eigenvector representation was 0.53 au which corresponds to 0.5% of the average measured signal and equals the error introduced by digitizing the spectra. This reflects the excellent data quality obtained by using a flow cell. With the alternative method

USE OF EPR SPECTROSCOPY FOR STUDYING SOLUTION EQUILIBRIA

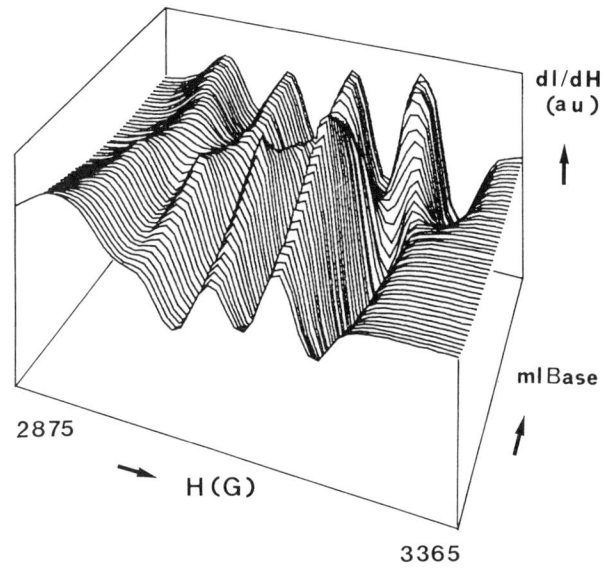

FIG. 1. Plot of the EPR data resulting from the titration of Cu(II) and DANA with NaOH (I = 0.5 M, KCl; 25°C; $[Cu^{2+}]_{tot}$ = $[DANA]_{tot}$ = 0.02 M). (Reproduced from Ref. 25 by permission of the American Chemical Society.)

of taking discrete samples (cf. Sec. 2.2), errors of ±1% are introduced only by repositioning the sample and by using different EPR tubes [18].

Due to the eigenvector representation, the calculation of stability constants and spectra could be done on a data set of 44 x 4 instead of the original 44 x 50 spectral data. This tremendous reduction of more than 90% allows one to do the complete calculation on a desk computer.

As known from the literature [30,43], the complexes formed in aqueous solution are $Cu(L)^{2+}$, $Cu(L-1H)^{+}$, and $Cu(L-2H)$. Using the potentiometrically determined ligand deprotonation constants (log $K^{H}_{H_2L}$ = 6.55, log K^{H}_{HL} = 8.40 [43]), the stability constants calculated

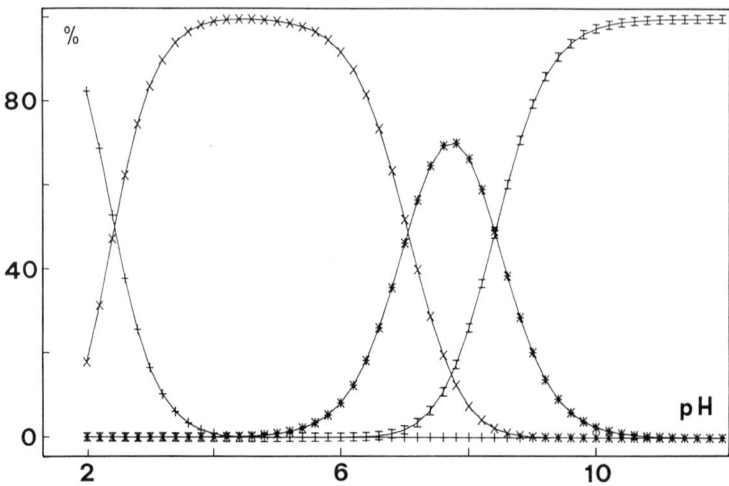

FIG. 2. Species distribution in the Cu(II)-DANA system as a function of pH. The results are given as the percentage of the total copper present: +, Cu^{2+}; x, $Cu(L)^{2+}$; *, $Cu(L-1H)^{+}$; I, $Cu(L-2H)$. (Reproduced from Ref. 25 by permission of the American Chemical Society.)

from the EPR data are log $K^{Cu}_{Cu(L)}$ = 12.06, log $K^{H}_{Cu(L)}$ = 7.05, log $K^{H}_{Cu(L-1H)}$ = 8.41 and agree well with the literature values [30,44].

The distribution curves (Fig. 2) show that $Cu(L)^{2+}$ is already formed to more than 50% at pH below 2.5 and that $Cu(L-1H)^{+}$ is never formed to more than 70%. The first observation explains why $K^{Cu}_{Cu(L)}$ cannot be determined potentiometrically [43].

The EPR titration curves (Fig. 3) demonstrate the excellent fit of the experimental data points by the calculated curves. The standard error of fit is 0.72 au, i.e., only slightly larger than the one obtained from the model-free eigenvector representation.

The calculated EPR spectra of the three complexes (Fig. 4) show four lines each, which is consistent with the interaction of the unpaired electron with a copper nucleus of nuclear spin I = 3/2 [1]. Each spectrum has an asymmetric shape with peak heights increasing with the magnetic field. Such a presence of anisotropic terms in the Hamiltonian indicates a restricted rotation of the

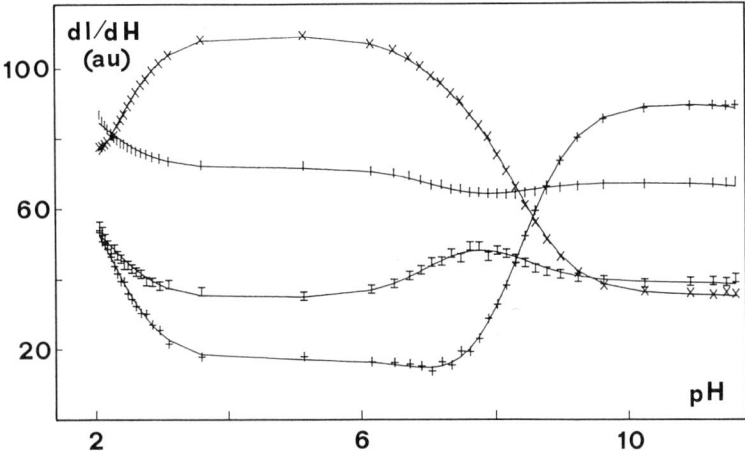

FIG. 3. EPR titration curves for the Cu(II)-DANA system. Experimental points: | (2925 G); x (3185 G); I (3205 G); + (3235 G). The solid lines represent the calculated curves. (Reproduced from Ref. 25 by permission of the American Chemical Society.)

complexes and is frequently observed for room temperature spectra of cupric complexes in solution [1,18-23].

From the calculated spectra averaged values of g and of the hyperfine splitting constant A can be obtained either graphically or by using a fitting procedure [45]. The values of g_{av} and of A_{av} allow one to assign the structures of the complexes, e.g., in the case of Cu(II) by using the empirical relations between g and A compiled for different types of ligand fields [14,15]. In our case the analysis of the EPR spectral parameters confirmed that the Cu-DANA complexes are square-planar and that the coordination sphere changes from CuN_2O_2 over CuN_3O to CuN_4 upon deprotonation [25].

As is evident from Figure 4, the spectra of the complexes overlap strongly and useful information can only be obtained if multifield strength data are evaluated. For the calculation of stability constants the data selection and the number of points per spectrum is not very critical, provided the data are taken over the whole range of H where spectral changes occur. The same stability

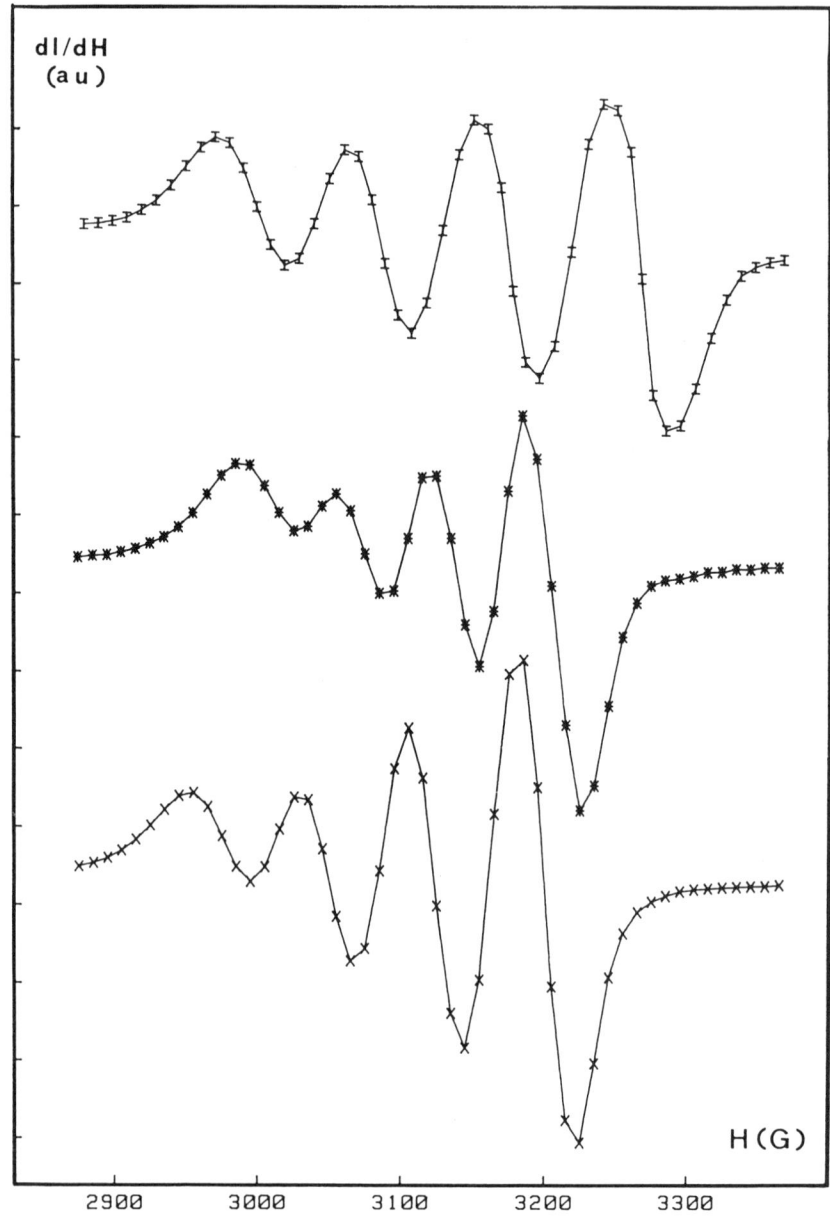

FIG. 4. Calculated EPR spectra of the Cu(II)-DANA complexes: x, Cu(L)$^{2+}$; *, Cu(L-1H)$^{+}$; I, Cu(L-2H); (g_{av} = 2.125, 2.111, 2.100; A_{av} = 0.00734, 0.00631, 0.00872 cm^{-1}). (Reproduced from Ref. 25 by permission of the American Chemical Society.)

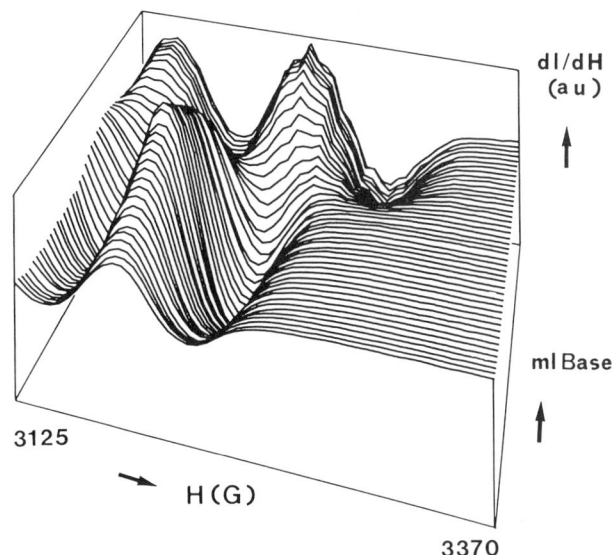

FIG. 5. Plot of the EPR data resulting from the titration of Cu(II) and DANA with NaOH; only high magnetic field strength data are included. The conditions are as given with Fig. 1. (Reproduced from Ref. 25 by permission of the American Chemical Society.)

constants were obtained (see [25]) when the equidistance interval between successive points varied between 5 and 25 G. Figure 5 shows a data set where only the upper half of the magnetic field strength region is considered, i.e., where $Cu(L)^{2+}$ and $Cu(L-1H)^{+}$ have almost identical spectra (cf. Fig. 4). Even from this deliberately poorly chosen data set, the same stability constants were calculated and only the value of log $K^{Cu}_{Cu(L)}$ was off by 0.2 log units [25].

If, on the other hand, one is interested in the spectral parameters of the complexes, each measured spectrum should be resolved into as many points as possible in the region of interest. Owing to the 10-G intervals chosen for the data set used in the calculation, no super hyperfine splitting can be seen either in the experimental data (Fig. 1) or in the calculated spectra (Fig. 4). As is evident from Figure 5, an additional structure becomes visible at high pH when digitizing at 5-G intervals is carried out. Figure 6 shows that the calculated curve satisfies the experimental points in

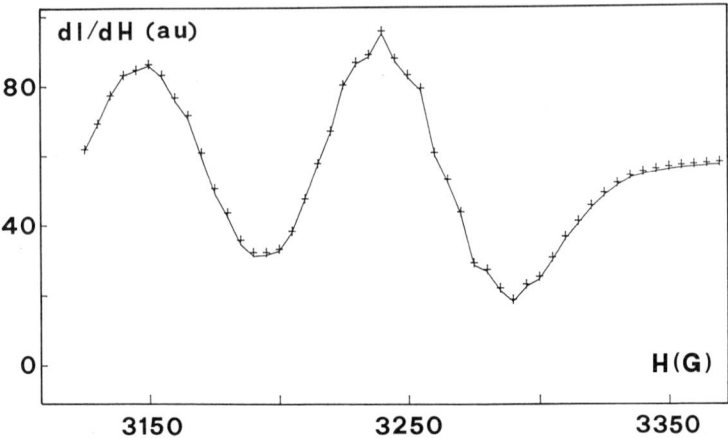

FIG. 6. Comparison of the experimental (+) and calculated (—) EPR spectrum of the Cu(II)-DANA system at pH 11.5. (Reproduced from Ref. 25 by permission of the American Chemical Society.)

any detail, and this without having made any assumption regarding the spectra of the complexes.

The results of the calculation can be tested independently by determining the spin concentration for each complex, i.e., by doubly integrating the calculated EPR spectra. As there is one unpaired electron in each complex, the double integration should give the same result in each case. The calculated values for the three Cu(II)-DANA complexes indeed differ by less than 5% [25]. As the algorithm used does not require any spectral information, this excellent agreement with the theoretical expectation confirms the results of the calculation.

Very recently, the model-free evolving factor analysis (EFA) was applied to spectroscopic titrations [41]. The results for the EPR data of the Cu(II)-DANA system are displayed in Figure 7. Forward EFA [Fig. 7(a)] shows the appearance of new factors at pH 2, 7, and 8 (factor 1 which is present from the beginning is omitted in the figure). As no further factors are found up to pH 11.5, this indicates a number of four absorbing species in the data set. In the present example factors 2, 3, and 4 clearly can be associated

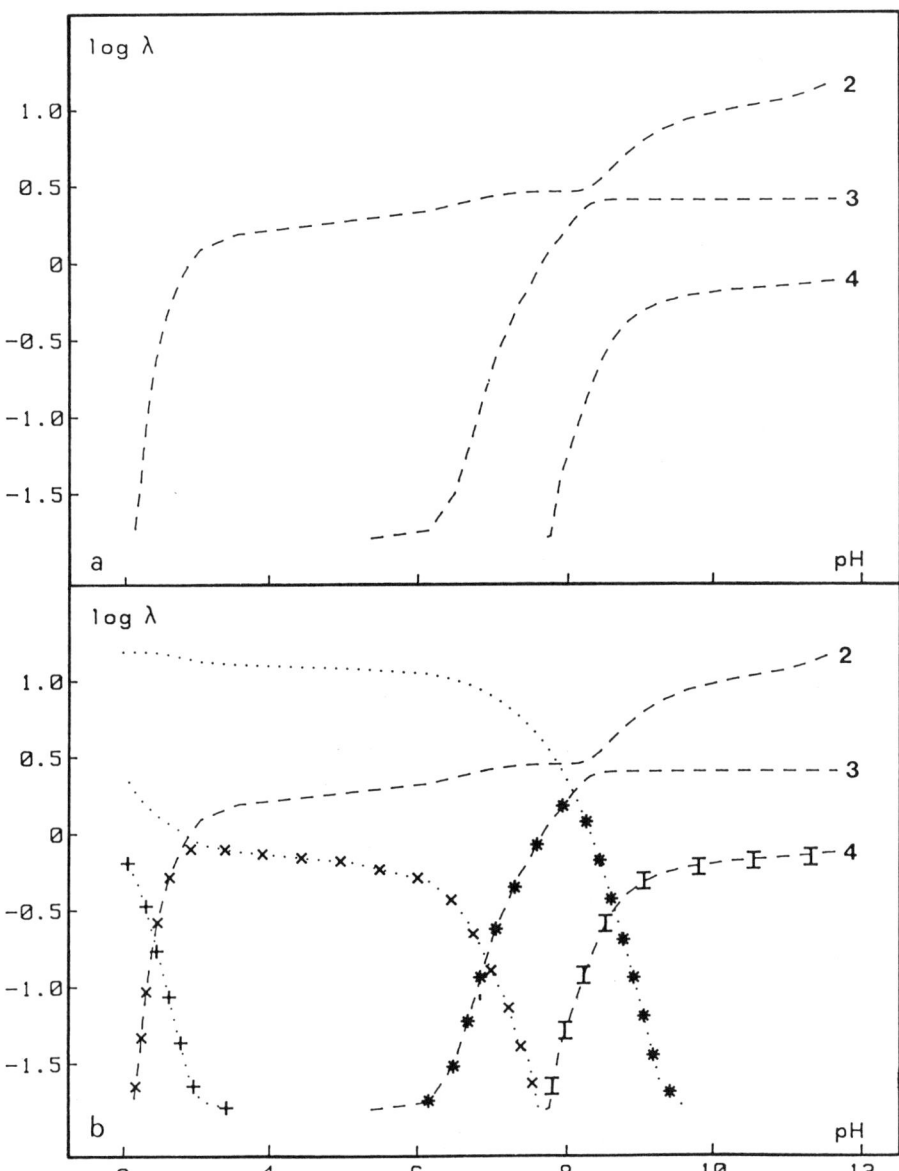

FIG. 7. EFA applied to the EPR data from Fig. 1: (a) forward, (b) forward (-----), and backward (•••••) EFA. Abstract species distribution obtained by connecting the ith forward eigenvalue with the $(P + 1 - i)$th backward eigenvalue $(i = 2 \cdots P)$. (Reproduced from Ref. 41 by permission of Pergamon Press.)

with $Cu(L)^{2+}$, $Cu(L-1H)^+$, and $Cu(L-2H)$, respectively. As has been pointed out in Sec. 2.3, the magnitude of an eigenvalue is not a direct measure of a species concentration. This can be seen from Figure 7(a) at pH 7 and 8 where the formation of $Cu(L-1H)^+$ and $Cu(L-2H)$ not only leads to the appearance of factors 3 and 4, respectively, but also influences factor 2.

The results of forward and backward EFA are displayed in Figure 7(b). Combination of the values as indicated in the figure yields a species distribution curve which has been obtained without referring to a chemical model and which closely resembles the one shown in Figure 2.

Clearly, in the case of unknown equilibrium systems EFA will be extremely helpful, both for selecting a chemical model and for obtaining estimates of the stability constants needed in the least-squares treatment.

3.2. The Cu(II)-AMIN System

An EPR titration corresponding to the one described in Sec. 3.1 yielded a set of $\underline{S}$ = 31 spectra each of which was digitized at $\underline{W}$ = 50 equidistant points between 2,830 and 3,320 G as shown in Figure 8 [26]. Eigenvector analysis indicated $\underline{P}$ = 4 absorbing species in accordance with the literature [46]. The overall standard deviation between experimental data and their eigenvector representation of 0.34 au, corresponding to 0.5% of the average measured signal, again reflects the high data quality.

Using the potentiometrically determined ligand deprotonation constants [46], the stability constants calculated from the EPR data are $\log K^{Cu}_{Cu(L)} = 21.63$, $\log K^{H}_{Cu(H_2L)} = 3.20$, and $\log K^{H}_{Cu(HL)} = 8.95$, in good agreement with potentiometric and spectrophotometric results [46].

The species distribution (Fig. 9) shows that $Cu(H_2L)^{4+}$ is present to not more than 30% over a narrow pH range only. By visual

USE OF EPR SPECTROSCOPY FOR STUDYING SOLUTION EQUILIBRIA   121

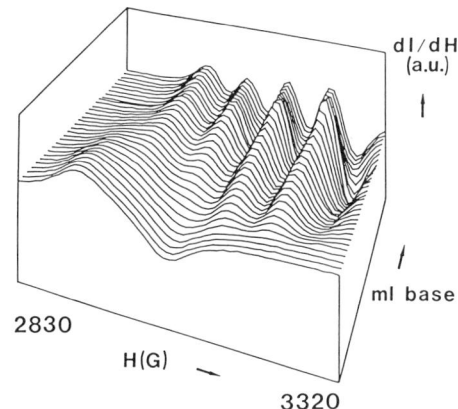

FIG. 8. Plot of the EPR data resulting from the titration of Cu(II) and AMIN with NaOH (I = 0.5 M; KCl; 25°C; $[Cu^{2+}]_{tot}$ = $[AMIN]_{tot}$ = $[HCl]_{tot}$ = 0.02 M). (Reproduced from Ref. 26 by permission of the Swiss Chemical Society [Helvetica Chimica Acta].)

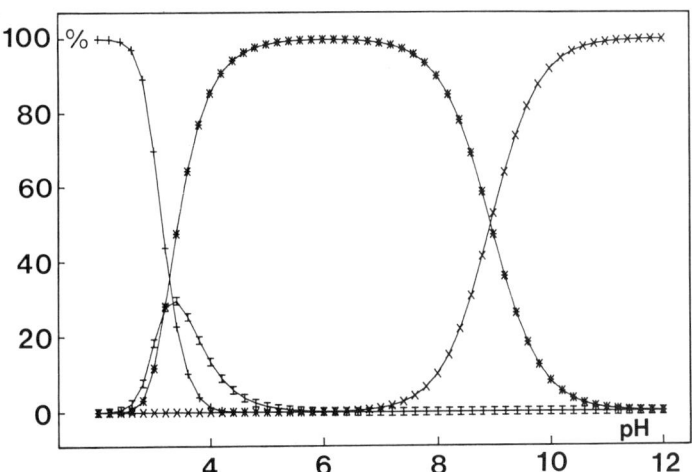

FIG. 9. Species distribution in the Cu(II)-AMIN system as a function of pH. The results are given as percentage of the total copper present: +, Cu^{2+}; I, $Cu(H_2L)^{4+}$; *, $Cu(HL)^{3+}$; x, $Cu(L)^{2+}$. (Reproduced from Ref. 26 by permission of the Swiss Chemical Society [Helvetica Chimica Acta].)

inspection of Figure 8 it would never be possible to predict the existence of this complex.

The standard error of fit is 0.37 au, i.e., only insignificantly larger than the one obtained from the model-free eigenvector representation. The titration curves shown in Figure 10 demonstrate the excellent fit of the data.

The spectra of the Cu(II)-AMIN complexes which are displayed in Figure 11 resemble those obtained in the Cu(II)-DANA system (Fig. 4) but are much stronger overlapping. In this case too, the double integrals of the three spectra agree to ±7%, which confirms the results of the calculation.

Based on the solution EPR spectral parameters and using additional information from anisotropic EPR and ligand field spectra, a detailed structural picture of the complexes was obtained (see [26]), which is summarized in Figure 12.

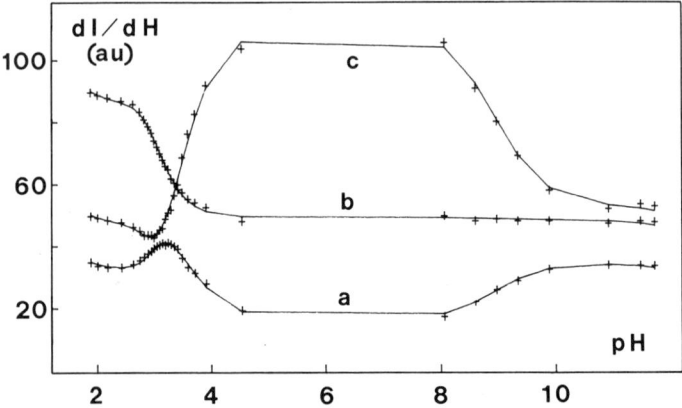

FIG. 10. EPR titration curves for the Cu(II)-AMIN system. Experimental points (+) at 3120 G (a), 2920 G (b), 3240 G (c). The solid lines represent the calculated curves. (Reproduced from Ref. 26 by permission of the Swiss Chemical Society [Helvetica Chimica Acta].)

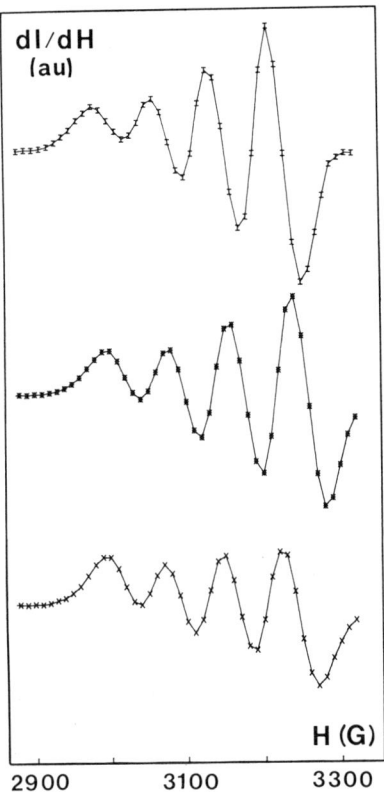

FIG. 11. Calculated EPR spectra of the Cu(II)-AMIN complexes: I, $Cu(H_2L)^{4+}$; *, $Cu(HL)^{3+}$; x, $Cu(L)^{2+}$; (g_{av} = 2.112, 2.102, 2.106; A_{av} = 0.00749, 0.00795, 0.00777 cm^{-1}). (Reproduced from Ref. 26 by permission of the Swiss Chemical Society [Helvetica Chimica Acta].)

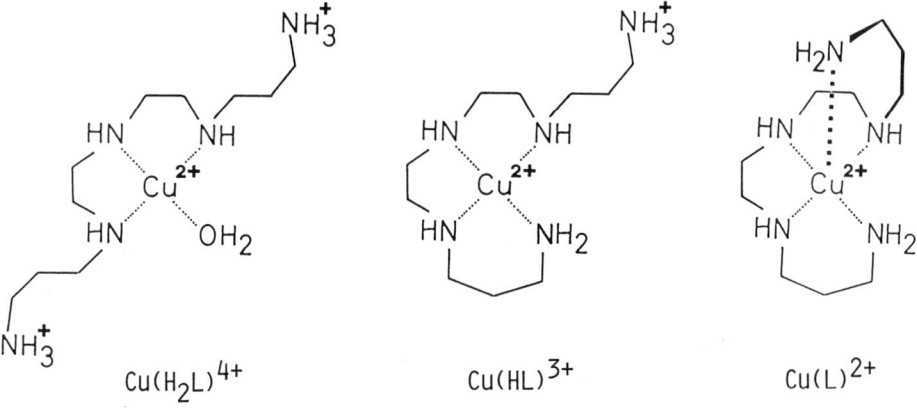

FIG. 12. Structures of the complexes occurring in the Cu(II)-AMIN system in the pH range 2-12; see Fig. 9.

4. CONCLUSIONS

As demonstrated for the complexes formed between Cu(II) and the ligands DANA [25] and AMIN [26], respectively, EPR spectroscopy is an excellent tool for investigating solution equilibria. The method can be used in complicated cases where several species are simultaneously present and where certain complexes are formed to a minor extent only.

Due to an algorithm [39] which is especially tailored for the common situation that the spectra of the complexes are unknown and strongly overlapping, the analysis of EPR titration data is straightforward and can even be done on a desk computer. For a given chemical model only initial estimates for the stability constants are necessary while no information at all with respect to the spectra of the species is required. No special care has to be taken to select the data such that only distinct complexes give rise to a signal as the calculation is done on the complete data set simultaneously.

The EPR method is superior to the commonly used potentiometric pH titration because it can also be applied in strongly acidic or basic solutions. With respect to spectrophotometry, EPR spectroscopy has the advantage that due to specific hyperfine interactions a more detailed description of the metal ion coordination sphere is possible. Moreover, the calculation of the spin concentration of each complex allows an independent test of the reliability of the results of the data analysis.

The recently developed evolving factor analysis [41] may ultimately lead to a completely model-free analysis of spectroscopic titration data. At present this technique yields abstract species distribution curves which are of great help for establishing the correct chemical model and which at the same time provide stability constants. Even more recently it was shown that EFA allows one to calculate the spectra of the species occurring in an equilibrium system without assuming a chemical model [42].

The outspreading availability of computer-controlled EPR spec-

USE OF EPR SPECTROSCOPY FOR STUDYING SOLUTION EQUILIBRIA

trometers will lead to a rapid and automatic acquisition of data of high quality.

The final conclusion to be drawn is that EPR has become a powerful method for investigating solution equilibria due to the straightforward data analysis now available. This method therefore is well suited for studying complicated biochemical equilibria involving paramagnetic metal ions like Cu(II) or Mn(II). Moreover, because the composition of the coordination sphere is reflected in a clear-cut way in the EPR spectra also detailed structural information is obtained, thus allowing studies of complicated polydentate ligands as they are common in nature.

ABBREVIATIONS AND DEFINITIONS

AMIN	4,7,10-triazatridecane-1,13-diamine
au	arbitrary units
ATP	adenosine 5'-triphosphate
$\underline{C}$	matrix of the concentrations; an element $C_{p,s}$ contains the concentration of the pth species at the sth titration point
DANA	3,7-diazanonanedioic acid diamide
$\underline{E}$	matrix of the spectra of the species; an element $E_{w,p}$ contains the value at the wth magnetic field strength of the pth species
EFA	evolving factor analysis
EPR	electron paramagnetic resonance
H	magnetic field strength in Gauss (G)
$\underline{L}$	matrix of the linear coefficients in the eigenvector representation of $\underline{Y}$; its dimension is $\underline{P} \times \underline{S}$
$\underline{P}$	number of absorbing species occurring in a given equilibrium system = number of nonzero eigenvalues = number of eigenvectors
$\underline{S}$	number of titration points = number of spectra per titration
$\underline{\underline{V}}$	matrix of the eigenvectors, its dimensions are $\underline{W} \times \underline{P}$
$\underline{W}$	number of points per spectrum
$\underline{\underline{Y}}$	matrix of the measured EPR data; an element $Y_{w,s}$ contains the data point obtained at the wth magnetic field strength of the sth spectrum
$\underline{\underline{Y}}^c$	matrix of the corresponding calculated elements

REFERENCES

1. B. A. Goodman and J. B. Raynor, *Adv. Inorg. Chem. Radiochem.*, *13*, 135 (1970).
2. H. M. Swartz, J. R. Bolton, and D. C. Borg (eds.), *Biological Applications of ESR*, Wiley-Interscience, New York, 1972.
3. "Extended Abstracts of Papers Presented at 1st International Conference on Bioinorganic Chemistry, Florence, Italy," *Inorg. Chim. Acta, 79, B7* (1983).
4. P. D. Sullivan, *Magn. Reson. Rev.*, *4*, 197 (1978).
5. R. D. Allendoerfer, *Magn. Reson. Rev.*, *5*, 175 (1980).
6. F. R. Hartley, C. Burgess, and R. M. Alcock, in *Solution Equilibria*, Ellis Horwood Ltd., Chichester, 1980.
7. A. E. Alegria, R. Conception, and G. R. Stevenson, *J. Phys. Chem.*, *79*, 361 (1975).
8. D. C. McCain and R. J. Myers, *J. Phys. Chem.*, *72*, 4115 (1968).
9. L. Burlamacchi, G. Martini, and E. Tiezzi, *J. Phys. Chem.*, *74*, 3980 (1970).
10. G. R. Stevenson and A. E. Alegria, *J. Phys. Chem.*, *77*, 3100 (1973).
11. G. R. Stevenson and A. E. Alegria, *J. Phys. Chem.*, *79*, 1042 (1975).
12. T. L. Pappenhagen, W. R. Kennedy, C. P. Bowers, and D. W. Margerum, *Inorg. Chem.*, *24*, 4356 (1985).
13. R. T. Vollmer and W. J. Caspary, *J. Magn. Reson.*, *27*, 181 (1977).
14. J. Peisach and W. E. Blumberg, *Arch. Biochem. Biophys.*, *165*, 691 (1974).
15. A. W. Addison, in *Copper Coordination Chemistry: Biochemical and Inorganic Perspectives* (K. D. Karlin and J. Zubieta, eds.), Adenine Press, New York, 1983, p. 109.
16. D. C. Gould and H. S. Mason, in *The Biochemistry of Copper* (J. Peisach, P. Aisen, and W. E. Blumberg (eds.)), Academic Press, New York, 1966, p. 35.
17. D. C. Gould and H. S. Mason, *Biochemistry*, *6*, 801 (1967).
18. W. S. Kittl and B. M. Rode, *J. Chem. Soc. Dalton Trans.*, 409 (1983).
19. E. R. Werner and B. M. Rode, *Inorg. Chim. Acta*, *80*, 39 (1983).
20. E. R. Werner and B. M. Rode, *Inorg. Chim. Acta*, *91*, 217 (1984).
21. M. J. A. Rainer and B. M. Rode, *Inorg. Chim. Acta*, *92*, 1 (1984).
22. E. R. Werner and B. M. Rode, *Inorg. Chim. Acta*, *93*, 27 (1984).

23. M. J. A. Rainer and B. M. Rode, *Inorg. Chim. Acta, 107,* 127 (1985).
24. P. Gans, *Coord. Chem. Rev., 19,* 99 (1976).
25. H. Gampp, *Inorg. Chem., 23,* 1553 (1984).
26. H. Gampp, *Helv. Chim. Acta, 67,* 2164 (1984).
27. S. E. O'Connor, T. A. Spraggins, and C. M. Grishan, *Comp. Chem., 5,* 181 (1981).
28. P. N. T. Lindberg and B. M. Peake, *J. Magn. Reson., 47,* 365 (1982).
29. G. Hänisch, T. A. Kaden, and A. D. Zuberbühler, *Talanta, 26,* 563 (1979).
30. A. D. Zuberbühler and T. A. Kaden, *Talanta, 26,* 1111 (1979).
31. M. Maeder and H. Gampp, *Anal. Chim. Acta, 122,* 303 (1980).
32. H. Gampp, M. Maeder, and A. D. Zuberbühler, *Talanta, 27,* 103 (1980).
33. H. Gampp, M. Maeder, C. J. Meyer, and A. D. Zuberbühler, *Talanta, 32,* 95 (1985).
34. H. Margenau and G. M. Murphy, in *The Mathematics of Chemistry and Physics,* Van Nostrand Reinhold, Princeton, 1968, p. 492.
35. E. A. Sylvestre, W. H. Lawton, and M. S. Maggio, *Technometrics, 16,* 353 (1974).
36. G. H. Golub and V. Pereyra, *SIAM J. Appl. Math., 10,* 413 (1973).
37. M. F. Delaney, *Anal. Chem., 56,* 261R (1984).
38. H. R. Schwarz, H. Rutishauser, and E. Stiefel, in *Numerik Symmetrischer Matrizen,* Teubner, Stuttgart, 1972.
39. H. Gampp, M. Maeder, C. J. Meyer, and A. D. Zuberbühler, *Talanta, 32,* 257 (1985).
40. D. W. Marquardt, *J. Soc. Ind. App. Math., 11,* 431 (1963).
41. H. Gampp, M. Maeder, C. J. Meyer, and A. D. Zuberbühler, *Talanta, 32,* 1133 (1985).
42. H. Gampp, M. Maeder, C. J. Meyer, and A. D. Zuberbühler, *Chimia, 39,* 315 (1985).
43. T. A. Kaden and A. D. Zuberbühler, *Helv. Chim. Acta, 57,* 286 (1974).
44. R. Blum, Ph.D. thesis, University of Basel, 1977.
45. M. Noack, G. F. Kokoszka, and G. Gordon, *J. Phys. Chem., 54,* 1342 (1971).
46. H. Gampp, D. Haspra, M. Maeder, and A. D. Zuberbühler, *Inorg. Chem., 23,* 3724 (1984).

4

Application of EPR Saturation Methods to Paramagnetic Metal Ions in Proteins

Marvin W. Makinen and Gregg B. Wells
Department of Biochemistry and Molecular Biology
The University of Chicago
Cummings Life Science Center
Chicago, Illinois 60637

1.	INTRODUCTION	130
2.	RELAXATION THEORY	131
	2.1. Spin-Lattice Relaxation	131
	2.2. Spin-Spin and Cross-Relaxation	136
	2.3. Spectral Diffusion	140
3.	METHODS FOR COLLECTING ELECTRON SPIN-LATTICE RELAXATION DATA	144
	3.1. The Continuous-Wave Saturation Technique	144
	3.2. The Pulse Saturation and Recovery Technique	154
4.	MICROWAVE POWER SATURATION STUDIES OF PARAMAGNETIC METAL ION CENTERS IN PROTEINS AND SMALL-MOLECULE COMPLEXES	159
	4.1. Differential Saturation Effects Dependent on the Mechanism of Modulation of Electron-Nuclear Dipolar Interactions	159
	4.2. Differential Saturation Effects Dependent on Spectral Diffusion	162
	4.3. Characterization of Multi-Metal Ion Clusters in Proteins	166
	4.4. Determination of the Splitting Between the Two Lowest Kramers Doublets of Paramagnetic Metal Ions with Ground State Spin Multiplets	171

4.5. Power Saturation for Estimating Distances Between Two Interacting Spins	185
4.6. Fractal Dimension and Protein Structure from Raman Spin-Lattice Relaxation Rates	190
ABBREVIATIONS	199
REFERENCES	199

1. INTRODUCTION

Saturation of a paramagnetic spin system with incident microwave power has been widely applied in electron paramagnetic resonance (EPR) studies to evaluate both electronic and structural properties of spin systems. The method essentially consists of irradiating the spin system with high levels of microwave power and observing the response of the system to reestablish thermal equilibrium of the spins with the lattice, and can be employed with both continuous wave (cw) and pulse saturation and recovery techniques. In the former, the spin system is progressively irradiated with increasingly higher levels of incident microwave power during which time the spectrum is collected to monitor the change in signal intensity; in the latter, the system is subjected to an intense pulse of microwave power of short duration after which the kinetic response of the system to reestablish thermal equilibrium is followed. Application of either method requires thorough familiarity with the physical principles and theory underlying relaxation processes as well as with the limitations of the technique with respect to instrumentation. Therefore, in this chapter we first discuss the salient features of magnetic relaxation theory that are necessary to describe and interpret results from experiments using saturation techniques in EPR studies, and we briefly outline the basic requirements for instrumentation, particularly with a view on limitations in data collection. Because this chapter is directed to structural analysis of paramagnetic metal ion sites in proteins, the discussion is restricted to relaxation theory of transition metal ions in ionic solids. Furthermore, for all of the transition metal ions discussed in this review, we need concern

APPLICATION OF EPR SATURATION METHODS   131

ourselves only with magnetic resonance theory applied to Kramers ions. Readers are directed to reviews and monographs on EPR spectroscopy and relaxation phenomena [1-6] for more detailed or specialized discussions of the concepts outlined here.

2. RELAXATION THEORY

2.1. Spin-Lattice Relaxation

The exchange of energy between a system of paramagnetic spins and thermally excited lattice vibrations is known as spin-lattice relaxation and is central to magnetic resonance. Spin-lattice relaxation can be classified either by the physical phenomenon that couples lattice phonons to the spin system or by the characteristics of the phonons that relax the spin system. The term "relaxation mechanism" describes the former concept while the term "relaxation process" applies to the latter. The importance of spin-lattice relaxation is that it provides a pathway connecting the two levels between which excitation of the spin system occurs. Without the exchange of energy with the lattice, the energy absorbed from the radiation field would heat the system to an infinite temperature whereupon no further transitions would be observed. The spin-lattice relaxation time thus becomes a characteristic property of the system, providing a measure of the rate at which the populations of the levels are restored to their Boltzmann thermal distribution.

The first detailed calculation of the spin-lattice relaxation time of paramagnetic ions of the iron group was carried out by van Vleck [7,8]. This was generalized subsequently by Mattuck and Strandberg [9] and by Orbach [10,11] to include ions with large orbital moments in their ground state. According to this formulation, the nearest ligand neighbors contribute to the crystalline potential acting on the paramagnetic electrons. The relaxation process is analyzed in terms of elementary transitions in which each spin flip is accompanied by the emission or absorption of a definite number of phonons of the host lattice, the basic interaction being phonon modulation of the ion-ligand potentials. The mechanism of spin-lattice coupling in ionic solids, therefore, may provide information both about the electronic structure of the metal ion (from spin-orbit coupling) and about the vibrational characteristics of the matrix (from orbit-lattice coupling). Other mechanisms of spin-lattice relaxation, many of which are specific for a particular spin species or arrangement of crystalline field levels, including rotational oscillations, phonon modulation of the transferred hyperfine field, and modulation of the zero field splitting tensor, are important in nonionic solids (cf. [12]) and will not be considered here.

According to the number of phonons involved in the spin flip, three types of relaxation processes can be described, known as the direct, Orbach, and Raman processes. These three processes differ in the characteristics of the phonon spectrum associated with the relaxation, resulting in three types of temperature-dependent behavior of the spin-lattice relaxation time for relaxation transitions between time-reversed states of half-integral spin. The three processes are schematically illustrated in Figure 1. At very low temperatures (T < 3 K) the magnetic field mixes excited states into the ground doublet so that relaxation occurs with absorption or emission of a single phonon, the energy of which equals that between the ground and excited spin states. Only those phonons that are of the same energy as that of the resonance frequency can take part in relaxation via the direct process, constituting an extremely small fraction of the total number of phonons, as shown by Figure 2. The spin-lattice relaxation rate is given by Eq. (1):

$$\frac{1}{T_{1d}} = A \coth\left(\frac{h\nu}{2kT}\right) \propto H^4 T \tag{1}$$

where A is a constant. Eq. (1) yields $2kT/h\nu$ as a limiting value for $h\nu/kT \ll 1$. Thus, the spin-lattice relaxation rate becomes directly proportional to the temperature under these conditions; it also exhibits a field dependence. The relaxation rate is also dependent on the occupancy of phonons of appropriate energy that may participate in the transition. Since this number is very small, it is

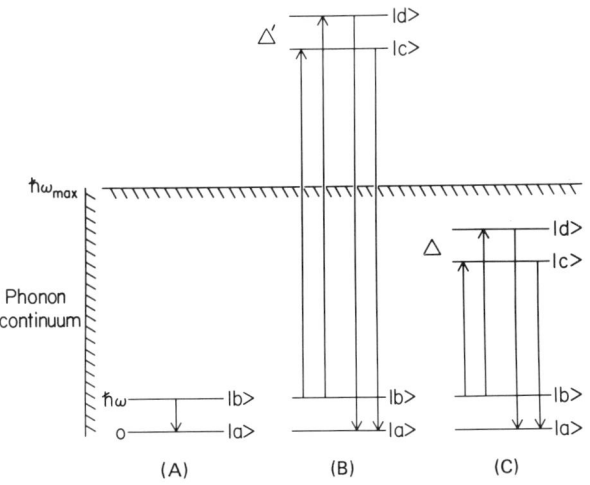

FIG. 1. Three spin-lattice relaxation processes for Kramers ions. (A) Direct process. Relaxation from $|b\rangle$ to $|a\rangle$ is accompanied by emission of a phonon of energy $h\nu$. For Kramers ions, $|a\rangle$ and $|b\rangle$ usually are time conjugate states which are split by the applied magnetic field. (B) Second-order Raman process. Relaxation from $|b\rangle$ to $|a\rangle$ proceeds through transitions to a pair of time conjugate states $|c\rangle$ and $|d\rangle$. These transitions are virtual because the energy Δ' of $|c\rangle$ and $|d\rangle$ lies outside the continuum of allowed phonon energies. (C) Orbach (resonant Raman) process. Relaxation again proceeds through the pair of states $|c\rangle$ and $|d\rangle$, except now their energy lies within the continuum of allowed phonon energies. This indirect process may be faster than the direct process because of the larger density of phonons at energy Δ compared to the density at $h\nu$.

readily seen why the spin-lattice relaxation time becomes very long near $T \sim 2$ K. Typical values of T_{1d}, for instance, of Fe^{3+} and Co^{2+} in octahedral sites are of the order of 0.05-0.01 sec at $T < 2$ K [13,14] while that for Cu^{2+} is ~ 20 sec [15]. In contrast, T_{1d} for Nd^{3+} is 0.4 sec near 2 K [16].

The two-phonon Raman process is generally operative under conditions $kT \gg h\nu$. In this relaxation process, all phonons can participate, resulting in a strongly temperature-dependent form for the spin-lattice relaxation rate as given in Eq. (2) for a Kramers ion:

$$\frac{1}{T_{1R}} = CT^9 \qquad (2)$$

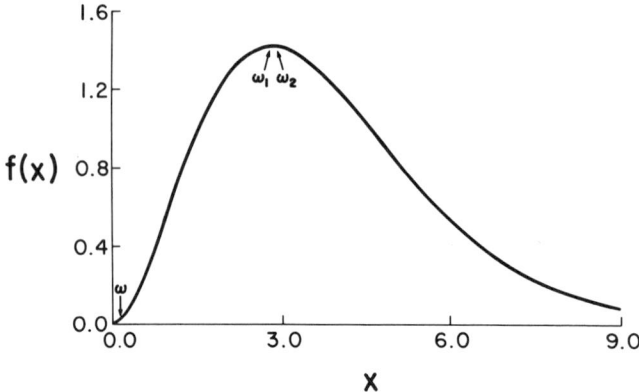

FIG. 2. Plot of $f(x) = x^3/(e^x - 1)$ showing the energy density of phonons as a function of $x = h\nu/kT$. The arrow at $x = 0.1$ indicates the energy density of phonons available for the direct process. The arrows ω_1 and ω_2 illustrate one pair of energy densities available for the Raman process. Relatively few phonons are available for the direct process; in contrast; any pair ω_1 and ω_2 which are separated by the appropriate energy ($h\nu$) can participate in the Raman process. A similar diagram calculated for the function $f(x)$ is illustrated on p. 555 of Ref. 1.

The spin flip occurs through inelastic scattering of a phonon with a change in phonon frequency and may be regarded as the virtual absorption of a phonon of energy $h\nu_1$ and the emission of a phonon of energy $h\nu_2$ so that $h\nu_1 - h\nu_2 = h\nu$ where $h\nu$ is the change in energy of the spin. Any two phonons can participate in the transition, providing their frequency difference is equal to the resonance frequency of the spin system. The relative contribution of this mechanism for spin-lattice relaxation is evident through Figure 2, which shows a plot of the energy density of phonons as a function of $h\nu/kT$. The dependence of spin-lattice relaxation on the occupancy of phonons of appropriate energy for direct and Raman processes provides the physical basis through which studies of spin-lattice relaxation can yield information about the vibrations of the entire metalloprotein. We shall discuss in Sec. 4.6 how spin-lattice relaxation studies of low-spin Fe^{3+} heme proteins have led to correlations of polypeptide structure with the density of vibrational states responsible for the Raman process.

The Orbach (resonant Raman) process consists of two kinds of spin-phonon interactions resulting in a spin flip, as illustrated in Figure 1(C). The first event is the absorption of a phonon to excite the spin system from level $|b\rangle$ to a much higher level at an energy Δ above the ground doublet; this is followed by emission of another phonon of slightly different energy to bring the system to level $|a\rangle$ whereupon the magnetic ion is transferred from one level to the other of the ground doublet [11,17]. This process is determined by the number of phonons of energy Δ available to excite the system. The temperature dependence of spin-lattice relaxation via the Orbach process for relaxation within the ground state manifold is given by Eq. (3):

$$\frac{1}{T_{10}} = \frac{B}{\exp(\Delta/kT) - 1} \simeq B\exp(-\Delta/kT) \tag{3}$$

for the condition $\Delta \gg kT \gg h\nu$. When Δ is less than the maximum phonon energy of the lattice, an exponential dependence of the spin-lattice relaxation rate on temperature is thus observed for spin-lattice relaxation via the Orbach process within the ground state manifold. On the other hand, for relaxation from the excited state to the ground state, as has been shown by Manenkov and Prokhorov [18], the spin-lattice relaxation rate via the Orbach process varies as $[1 - \exp(-\Delta/kT)]^{-1}$. Under the condition $\Delta \gg kT$, the process described by Eq. (3) exhibits a strong temperature dependence and generally masks the ordinary Raman process whenever the two are in competition while relaxation from the excited state for $\Delta \gg kT$ leads to a spin-lattice relaxation rate that is essentially temperature-independent for relaxation via the Orbach process. The importance of these different conditions in paramagnetic metal ions with ground state spin multiplets is discussed later (Sec. 4.4.2).

For a paramagnetic metal ion with an excited doublet lying below the Debye limit, the value of Δ corresponds to the splitting between the two lowest Kramers doublets. For paramagnetic ions with an orbitally nondegenerate ground state, Δ then corresponds directly to the zero field splitting. In the case of a paramagnetic metal ion with a large orbital moment associated with the ground state, Δ

corresponds to the splitting of the two lowest levels as determined by split-orbit coupling. By measurement of the temperature dependence of T_{10}, it is possible to determine the value of Δ and thus to obtain direct information about the electronic structure of the metal ion. Instances of such measurements are discussed below (Sec. 4.4), in particular how the value of the zero field splitting of high-spin Co^{2+} provides a signature of coordination number [19-21].

The complete expression for the spin-lattice relaxation rate $(1/T_1)$ becomes the sum of the three components in Eqs. (1)-(3), i.e.,

$$\frac{1}{T_1} = AT + B \exp\left(\frac{-\Delta}{kT}\right) + CT^9 \qquad (4)$$

in which the various terms A-C contribute simultaneously but to varying extents depending on the temperature (for detailed discussion of the derivation of these terms, cf. [1,3,5,6,10,11]). The temperature dependence of $1/T_1$ for (aquo)metmyoglobin exhibiting spin-lattice relaxation via the direct and Orbach processes is illustrated in Figure 3. In this case the Orbach process dominates at $T > 2$ K and masks the Raman process that is observed only at $T > 10$ K [22,23]. The results in Figure 3 from the studies of Scholes et al. [24] represent the first application of the pulse saturation and recovery method to a metalloprotein to study spin-lattice relaxation.

2.2. Spin-Spin and Cross-Relaxation

Progressive cw microwave saturation and pulse saturation and recovery methods have been applied primarily to estimate spin-lattice relaxation times. This is in principle the only relaxation process for a simple $S = 1/2$ two-level system in which the spins are completely isolated from each other. However, in general the width of an EPR absorption line is much greater than expected on the basis of T_1 due to a variety of broadening mechanisms. In this section, we shall briefly identify those processes that contribute to line broadening and may give rise to relaxation events identifiable through saturation

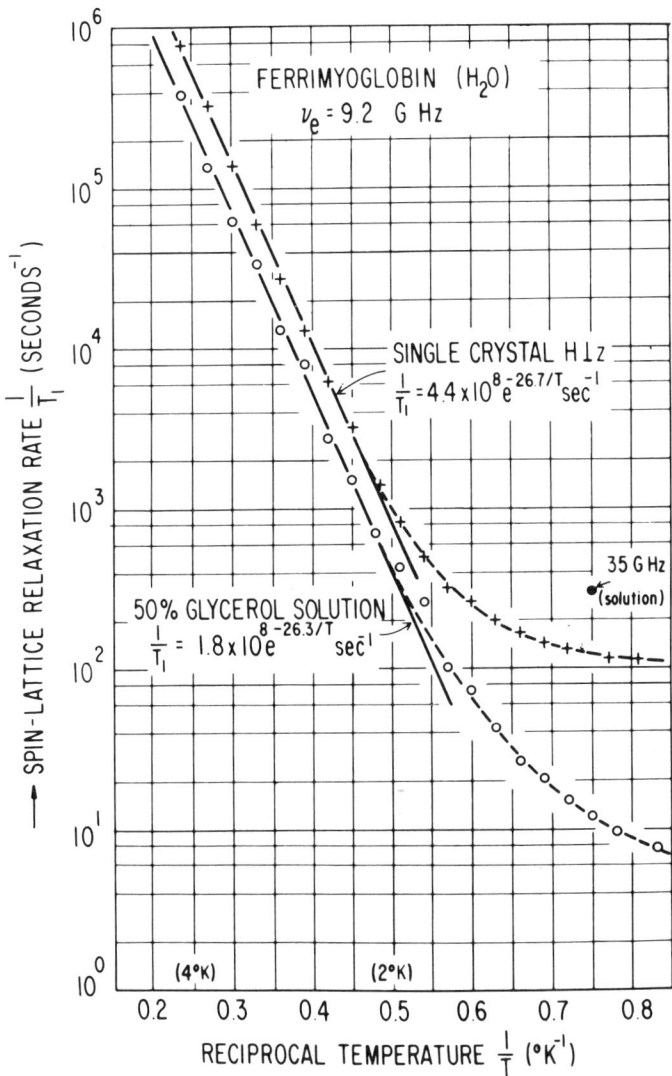

FIG. 3. Spin-lattice relaxation rate for the $g \sim 6$ transition for aquometmyoglobin vs. reciprocal temperature. These data, obtained by the pulse saturation and recovery method, show the dominance of the direct process below 2 K and of the Orbach process above that temperature. The zero field splitting (2$\underline{D}$) was found from the temperature dependence of the Orbach process to have similar values in both single crystals (18.6 cm^{-1}) and frozen glycerol-water solution (18.3 cm^{-1}). (Reproduced from Ref. 24 by permission of Elsevier Biomedical Press.)

studies. Because of the systems to be discussed in Sec. 4, we shall consider primarily inhomogeneously broadened resonance lines.

The first of these relaxation processes to be treated is that characterized by the spin-spin relaxation time, identified classically as T_2. Frequently the average distance between paramagnetic ions is such that the paramagnetic ion produces at the site of another ion a sizable magnetic field through its relaxing electron. For an ion with a moment of 1 Bohr magneton, the field is roughly proportional to β/r^3, and the net local field from a number of neighbors adds vectorially to the external field. The size of this component varies from site to site, giving a random displacement to the resonance frequency of each ion. This effect is similar to that produced by inhomogeneity in the external magnetic field and therefore the effect is known as inhomogeneous broadening. Although the resonance frequency of each ion is displaced from the average frequency of the broadened line, the lifetime of each ion in a given quantum state (T_2) is not altered. It is classically the average duration for a wave train to be emitted or absorbed by an ion in the process of paramagnetic resonance. The spin-spin relaxation time T_2 thus is the time of dephasing of the precessing spins in the magnetic field due to the dipolar site-site interactions occurring by virtue of fluctuating local magnetic fields. This type of relaxation process is, furthermore, adiabatic since energy is redistributed only within the spin system and is not transferred to the lattice. In pulse saturation studies this relaxation process gives rise to free induction decay and is discussed below (Sec. 3.2).

If we further define for each ion a resonance line the width of which is due to the processes that shorten or control the lifetime between these quantum states, we then have a series of homogeneously broadened lines, one for each type of ion displaced in the resonance frequency from the mean frequency. The inhomogeneously broadened line envelope is now considered to consist of an envelope of a continuous distribution of homogeneous resonance lines $g(\omega-\omega_0)$, each centered by a different separation $(\omega'-\omega_0)$ from the mean frequency ω_0. This essentially summarizes the spin-packet concept of an inhomogeneously broadened EPR absorption line, as originally introduced by Portis [25,26] wherein the homogeneous line with linewidth $1/T_2$ corresponds to sets of electron spins that interact with different nuclear configurations, all of which superimpose to form the overall, inhomogeneously broadened line with linewidth $1/T_2^*$.

The first discussion of cross-relaxation is found in the paper of Bloembergen, Shapiro, Pershan, and Artman [27] in which they considered the energy transfer that occurs between two spin systems of similar frequencies ν_{12} and ν_{13}. They showed that under the condition that the absorption lines overlap there is a finite probability that a transition by a spin from level 3 to level 1 may be accompanied by a simultaneous transition of the other spin system from level 1 to level 2. The transition probability for this event defines the cross-relaxation time T_{21}, the imbalance in the energy emitted and absorbed being taken up by the internal energy of the spin system. The important characteristic of this process is that the energy level separations are similar but need not be precisely identical. This process is clearly of the electron spin-electron spin type and provides a route through which the populations of the levels may be altered, in contrast to the process giving rise to T_2 described above, and the rate equations to describe the spin system must be modified accordingly. For instance, if cross-relaxation occurs in a system such that α spins of transition energy ν_{12} alters the populations of β spins of energy ν_{34} so that $\alpha\nu_{12} = \beta\nu_{34}$ the resultant rate equations become [6]:

$$n_{12} - n_{012} = R_{12} \exp\left(\frac{-t}{T_1}\right) + R'_{12} \exp\left(\frac{-t}{\tau}\right)$$

$$n_{34} - n_{034} = \left(\frac{\alpha}{\beta}\right) R_{12} \exp\left(\frac{-t}{T_1}\right) - \left(\frac{\beta}{\alpha}\right) R'_{12} \exp\left(\frac{-t}{\tau}\right)$$

(5)

where the cross-relaxation time T_{21} is defined by the relationship $1/\tau = 1/T_1 + 1/T_{21}$. Cross-relaxation events can be identified by recovery curves that are composed of two exponential terms. Examples of such cross-relaxation processes are described in the solid state literature (cf., chapter 3 in [3]). Except for our own preliminary

results of pulse saturation and recovery data described briefly
below (Sec. 3.2), no other recovery data have been reported in sufficient detail to discuss relaxation processes in metal ion systems of
biological importance other than that of spin-lattice relaxation.
It is probable that the enhanced relaxation processes observed on
the basis of both progressive cw saturation and pulse saturation and
recovery methods in multi-metal ion clusters in proteins (cf. Sec.
4.3) are ascribable in part to such cross-relaxation processes.
Unfortunately, the emphasis of analysis of saturation data has been
directed invariably to estimates of spin-lattice relaxation times
only.

2.3. Spectral Diffusion

Central to the understanding of the saturation behavior of an EPR
resonance line is the rate at which the spin system tends to an
internal thermodynamic equilibrium distribution that is describable
by a temperature. The original spin packet model of an inhomogeneously broadened line introduced by Portis [25] considered the spin
packets as groups of spins of the same frequency but with each displaced from the mean frequency of the line. These spins were, therefore, in the language of Bloembergen et al. [27], on "speaking terms"
with each other, and energy was therefore rapidly transferred within
an individual spin packet but not between spin packets. The basic
assumption, then, is that each spin packet saturated as a homogeneously broadened line according to Eq. (6):

$$\chi'' = \frac{M_0 g(\omega - \omega_0)}{1 + \gamma^2 H_1^2 T_1 g(\omega - \omega_0)} \tag{6}$$

where χ'' is the absorption component of the resonant transverse susceptibility, M_0 is the equilibrium magnetization in the absence of
H_1, γ is the magnetogyric ratio, and $g(\omega - \omega_0)$ describes the line
shape of the spin packet. This was essentially the theory developed

by Bloembergen, Purcell, and Pound [28] to describe saturation in liquids but applied to spin systems in solids. The problem with application of the model of Portis [25,26] is that the probability of a spin assigned to a given spin packet to participate in a flip-flop transition through a cross-relaxation event with a neighboring spin of only similar but not identical transition energy is much greater than that with a distant spin of identical transition energy. If we assume the theory of Portis [25,26], the spin packet is defined as an isolated spin with a Lorentzian lineshape of width $1/T_2$. The theory applies for the saturation behavior of a resonance line for only completely homogeneous and completely inhomogeneous cases. The intermediate cases are of more direct concern. While the first attempt to treat the intermediate case was that of Castner [29], who used the spin packet linewidth as an adjustable parameter, the theory of saturation and spectral diffusion in solids described by Clough and Scott [30] probably most generally applies to the saturation behavior of real materials.

This theory considers a crystal containing a low concentration of free electron spins in a strong field H_0, each spin interacting with neighboring nuclei, with resultant unresolved hyperfine structure broadening the resonance line. The field at each spin is represented as the vectorial addition of an external field and a local field whose average value is zero. A central feature of this theory is the part of the Hamiltonian H_{ss} that represents the dipole-dipole (and possibly other) interactions such that transitions of the spins are flip-flops in which an "up" spin j and a "down" spin k reverse directions. According to this model, the spin packets representing the sets of spins, each interacting with its own nuclear environment, may transfer energy instantaneously across the line in a single step [30]. The role of H_{ss} is to cause the precession frequency of an electron magnetic moment to fluctuate in a random manner so that its correlation function $G(t)$ is described by the exponential:

$$G(t) = \exp \frac{-t}{T_3} \tag{7}$$

where T_3 is the correlation time of the local field. The state toward which the spin system tends as a result of the local field fluctuations is then a steady-state balance between the effect of the correlation time T_3 tending to bring the Zeeman and local field temperatures to a common value and the spin-lattice relaxation that establishes thermal equilibrium with the lattice. The model then describes the saturation behavior of a resonance line according to Eqs. (8) and (9):

$$\gamma H_1 M_y = \frac{M_0 - M_z}{T_1} \tag{8}$$

$$M_y = \frac{M_0 g(\omega - \omega_0)}{T_3} \cdot \frac{x}{(1 + x^2)^{1/2} + ax^2} \tag{9}$$

where $x = \gamma H_1 T_3$, $a = \pi g(\omega - \omega_0)(1 + \delta^2) T_1/T_3^2$, and $\delta^2 \simeq (\omega - \omega_0)^2/\omega_L^2$ with ω_L representing the angular frequency of the local field. The theory yields saturation curves similar to those described by Castner [29], but through equations with different meaning. Also the term a acquires the same value as the ratio of the spin packet linewidth to the overall linewidth in Castner's theory. The theory of Clough and Scott [30] yields the result of Castner [29] in the limit of slow spectral diffusion, i.e., for $a \ll 1$, the saturation factor is described as $(1 + \gamma^2 H_1^2 T_1 T_3)^{-1}$. These conditions imply that $T_3 \gg T_1$. On the other hand, if $a \gg 1$, corresponding to fast spectral diffusion, $T_1 \gg T_3$ and the theory reduces to that described through Eq. (6) originally derived on the basis of the theory of Bloembergen, Purcell, and Pound [28] for completely homogeneous broadening. The theory thus reduces according to boundary conditions to the results derived earlier by Portis [25] for noninteracting spin packets, and of Bloembergen et al. [28], and also of Provotorov [31] for completely homogeneously broadened lines as special cases. More importantly, it treats the intermediate case consistently where previous theories were inadequate, the main feature of this model being that the rate of spectral diffusion depends on both the level of saturation and the intrinsic correlation time of the local field.

Except for the theoretical formalism to describe the saturation behavior of EPR resonance lines developed by Korb and Maruani for slow [32,33] and fast [34,35] spectral diffusion, the model of Clough and Scott [30] has not been explicitly considered in progressive cw or pulse saturation and recovery studies. For paramagnetically dilute systems of metal ions with separations greater than approximately 20 Å, the local field contributed by nuclear moments may compete favorably in magnitude with the fluctuating field provided by the relaxing unpaired spins of neighbor sites. Moreover, there is reasonable experimental justification for the appropriateness of the model. The early studies of Feher and Gere [36] and of Bekauri et al. [37] indicate that multispin cross-relaxation processes between unresolved hyperfine structure components are an important source of spectral diffusion and require spins with only similar difference energy levels between the quantum states of the transitions, and the studies of Boscaino and coworkers [38] have provided important evidence that the spin packet linewidth is determined by the correlation time of the local field. Also, prior to the theory of Clough and Scott, Kiel [39] derived expressions for the spectral diffusion time T_D within the formalism for cross-relaxation in an inhomogeneously broadened line in aggreement with the earlier theory of Bloembergen et al. [27] showing that $T_D \simeq (T_2)^4/(T_2^*)^3$. Both the spin packet linewidth $(1/T_2)$ and the overall linewidth $(1/T_2^*)$ are influenced by the extent of unresolved hyperfine broadening. The process described by T_D in Kiel's theory is contained within T_3 according to the model of saturation derived by Clough and Scott [30].

The theory of Clough and Scott [30] shows that the spin-lattice relaxation is modulated by the correlation time of the local field since the energy imbalance from the cross-relaxations between spin packets is absorbed by the spin system. On this basis, it is probably better to describe the spin-lattice relaxation by an apparent spin-lattice relaxation time T_1' such that $1/T_1' = 1/T_1 + 1/T_D$ where T_D represents the spectral diffusion time between spin packets within an inhomogeneously broadened line that contributes to the correlation time of the local field. Some experimental justification for this relationship has been provided on the basis of electron-electron double-resonance experiments [40,41]. These considerations imply that if the environment of the spins defined by the distribution of magnetic nuclei contributing to the unresolved hyperfine structure is altered through isotopic enrichment without a change in their spatial relationships to each other, the saturation behavior of the EPR resonance line may be altered. We shall present (cf. Sec. 4.2) some experimental verification of this expectation on the basis of progressive cw saturation and pulse saturation and recovery studies [42-44].

3. METHODS FOR COLLECTING ELECTRON SPIN-LATTICE RELAXATION DATA

3.1. The Continuous-Wave Saturation Technique

3.1.1. Collection and Analysis of Saturation Data

The most direct means of obtaining information about spin-lattice relaxation processes is by application of the resonance saturation technique. The first application of a cw microwave saturation method was by Bloembergen et al. [28] to determine nuclear relaxation times, and in 1951 Schneider and England [45] applied the cw resonance saturation method to study Mn^{2+} in zinc sulfide phosphors. In this study they estimated the spin-lattice relaxation time of Mn^{2+} ions at liquid air temperatures. Subsequent early applications of cw saturation effects in EPR studies to estimate spin-lattice relaxation times were made by Eschenfelder and Weidner [46] and by Feher and Scovil [47]. The attraction of the cw resonance saturation method in EPR is that conventional spectrometers can be readily applied for collection of data.

Let us consider first a simple two-level system characterized by a homogeneous lineshape $g(\omega - \omega_0)$ and an inhomogeneous distribution function $h(\omega - \omega_0)$. We define the saturation function $[1 + \gamma^2 H_1^2 T_1 g(\omega - \omega_0)]^{-1}$ such that $H_{1/2}$ is the microwave magnetic field H_1 when $\gamma^2 H_1^2 T_1 g(\omega - \omega_0) = 1$. Correspondingly, $P_{1/2}$ and P are

APPLICATION OF EPR SATURATION METHODS 145

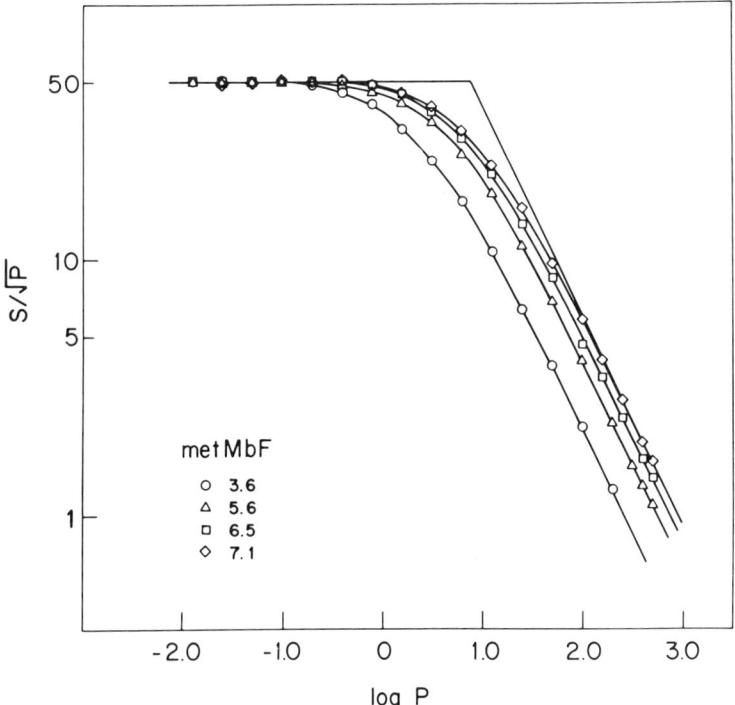

FIG. 4. The temperature dependence of the cw saturation of the $g \sim 2$ transition of metmyoglobin fluoride. The plotting technique shown here is used to characterize cw saturation behavior. The intersection of the asymptotic slopes at low power and high power defines $P_{1/2}$. In this figure, these slopes are drawn only for the data at 7.1 K for purposes of clarity. The temperature independence of the slope at high power indicates that $P_{1/2}$ depends only on T_1 in this temperature range. (Reproduced from Ref. 22 by permission of Academic Press.)

the microwave powers. For cw saturation measurements the sample is placed in the cavity in a region of maximum H_1 and minimum E fields, and the signal amplitude is monitored (at constant modulation amplitude and frequency) as a function of increasing microwave power. As shown in Figure 4, a plot of the data in the form of $S/P^{1/2} \propto (1 + P/P_{1/2})^{-b/2}$ yields the saturation curve that is then analyzed. In practice, because of the sample size, the microwave field H_1 experienced by different regions of the sample will vary from its maximum value at the cavity center to zero at each end of the cavity, and parts of the sample will saturate at different power levels. This circumstance has the effect of distorting the saturation curve, and the actual signal will be the integrated average over the sample [48,49]. At low microwave power levels, the shape of the saturation curve will not be greatly affected, although the absolute magnitude of the low-power signal is diminished by a factor of 2 in comparison with conditions where H_1 is uniform over the sample. At high power, i.e., $P/P_{1/2} > 1$, the distortion is important and Blum and Ohnishi [49] have calculated the effect on the EPR absorption spectrum of a paramagnetic site with anisotropic g values. This is illustrated in Figure 5. These considerations show that small sample volumes should be employed to minimize distortion of the saturation curves. In addition, it is obvious that the sample volumes and geometries must be maintained constant if saturation effects of a related series of samples are to be compared.

To analyze the saturation curve, we first recall that the detected signal is given by $S \propto \chi''H_1$ for the direct absorption mode, where χ'' is the imaginary part of the absorption susceptibility under conditions of crystal diode detection. Alternatively, for the first harmonic (phase-sensitive detection) mode the detected signal $S' = [(\partial\chi''/\partial H)H_{mod}]H_1$. The maximum value of the absorption or derivative functions must be then found. These can be directly solved for the limiting conditions of a completely homogeneously or a completely inhomogeneously broadened line, leading to the expressions given in Table 1. Correspondingly, theoretical saturation curves for the limiting and intermediate conditions are given in Figure 6. This type of plot of progressive saturation data is useful for analysis because $S/P^{1/2}$ is constant at low levels of incident microwave power. The plot provides a convenient graphical estimate of $P_{1/2}$ as the intersection of the two asymptotes of the saturation curve while the slope provides an estimate of b. It is seen from the plot that $S/P^{1/2}$ falls off more rapidly at $P/P_{1/2} > 1$ for a homogeneously broadened line so that at $P = P_{1/2}$, the value of $S/P^{1/2}$ is 0.5. Correspondingly, for an inhomogeneously broadened line $S/P^{1/2} = 0.707$ for $P = P_{1/2}$.

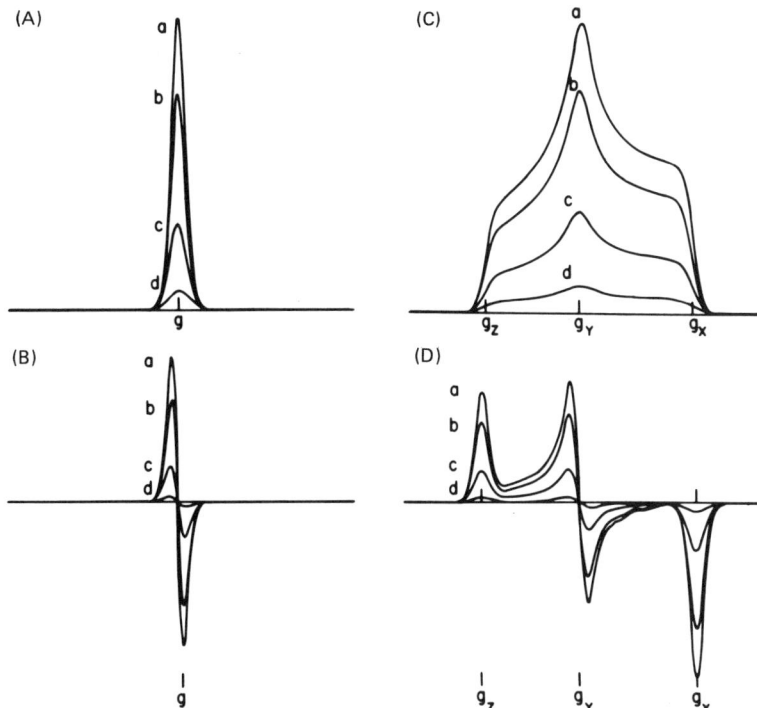

FIG. 5. Simulated EPR absorption spectra as a function of increasingly higher levels of microwave power to illustrate effects of correction for nonuniform H_1 along the sample length. (A) and (B) illustrate spectra for isotopic absorption at $g = 2$. (C) and (D) illustrate spectra for a rhombic site with $g_x = 1.9$, $g_y = 2.1$, and $g_z = 2.3$. The upper traces represent spectra in the direct absorption mode; the lower figures represent the corresponding first-derivative traces. The vertical axis shows the signal amplitude as a function of $S/P^{1/2}$, and spectral simulations are calculated for a Gaussian resonance line width of 20 Gauss with $T_2 = 10^{-8}$ sec. Values of $P/P_{1/2}$ are for each trace: (a) 0, (b) 1, (c) 10, and (d) 100. (Reproduced from Ref. 49 by permission of Elsevier Biomedical Press.)

Comparison of the plots in Figure 6 also shows that the distortion at high-power levels due to a nonuniform H_1 field along the sample tends to displace the progressive saturation curve toward the inhomogeneous limit for a given value of b.

The first analysis of progressive saturation data was carried out by Portis [25] to study the relaxation of F centers in solid

TABLE 1

Comparison of the Function $S_{max}/P^{1/2}$ for Continuous-Wave Progressive Microwave Power Saturation of the Resonance Absorption of a Paramagnetic System According to Mode of Signal Detection and Type of Line Broadening[a,b]

	Signal detection mode[c]	
Type of broadening	Zeroeth harmonic	First harmonic
Homogeneous	$C_1/(1 + P/P_{1/2})^{2/2}$	$C_3/(1 + P/P_{1/2})^{3/2}$
Inhomogeneous	$C_2/(1 + P/P_{1/2})^{1/2}$	$C_4/(1 + P/P_{1/2})^{1/2}$

[a]S_{max} respresents the maximum signal intensity corrected for spectrometer gain. See text for definitions of P and $P_{1/2}$.
[b]$P_{1/2}$ is here defined as the intersection of the asymptotic limits of the saturation curve as illustrated in Fig. 4.
[c]The value of b is directly indicated for $S/P^{1/2} \propto (1 + P/P_{1/2})^{-b/2}$. The terms $C_1, \ldots, C_4$ represent proportionality constants.

potassium chloride. This model assumes a completely inhomogeneously broadened line with no interaction of the spin packets. It is rigorously valid only in the limit of the ideal inhomogeneous case for zeroeth harmonic detection. Castner [29] extended this model to include interaction of spin packets, i.e., spectral diffusion, but this method is also valid only for zeroeth harmonic detection. Furthermore, a critical assumption of these earlier models is that the lineshape is Gaussian. This condition is quite restrictive for first harmonic detection since there is generally significant overlap between different EPR lines at the points of maximum slope, and saturation tends to occur more rapidly for a Lorentzian distribution of spin packet linewidths than for a Gaussian [50]. In this respect, second harmonic detection may be more useful for analysis of relaxation times since there is little overlap at the EPR line centers [51]. Furthermore, all of the models assume absence of spectral diffusion or cross-relaxation, and the classical study of Lloyd and Pake [52] pointed out that in saturation studies a single relaxation time cannot

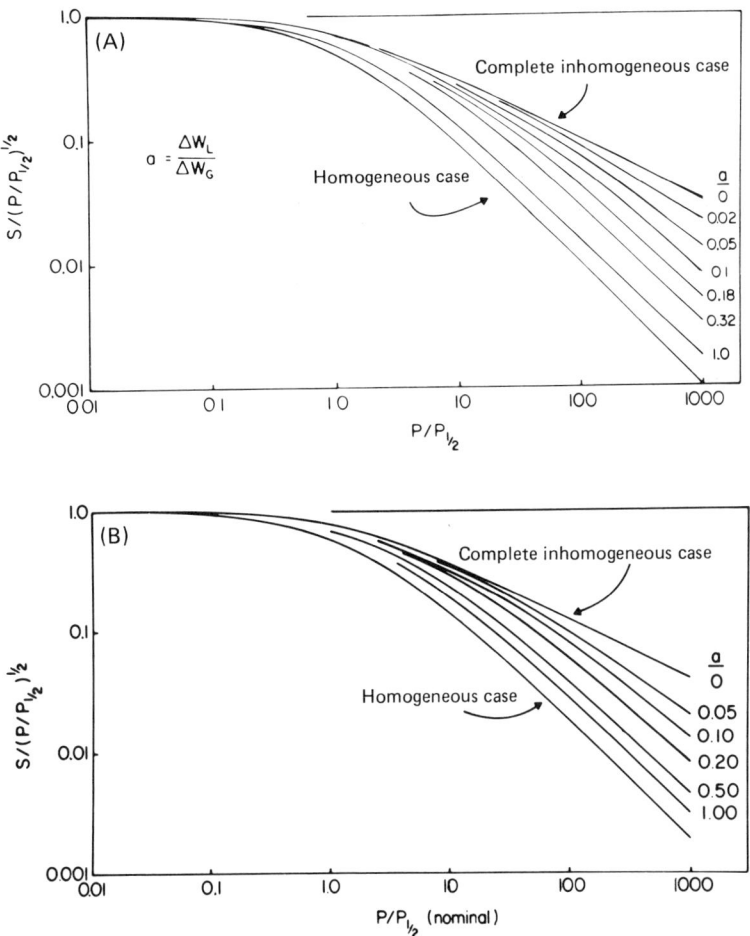

FIG. 6. Comparison of power saturation curves in (A) the idealized case of constant magnetic field over the entire sample and in (B) the case of a geometry where the sample extends completely through the microwave cavity. In (B) the calculated curves have been corrected for geometry. This correction is most important for $P/P_{1/2} > 1$. The parameter $a = \Delta\omega_L/\Delta\omega_G$ measures the degree of inhomogeneous broadening. (Reproduced from Ref. 49 by permission of Elsevier Biomedical Press.)

be extracted for more complicated relaxation schemes. Although progressive saturation curves of multilevel systems can be meaningfully interpreted [53], the analysis is difficult and only a relaxation probability corresponding to a sum of exponential terms is obtained. Nonetheless, progressive microwave power saturation curves have been generally employed to estimate relaxation times. For accurate estimates of T_1 there are considerable problems, and the monograph by Standley and Vaughan [3] should be consulted for a discussion of the various assumptions inherent in the different methods of estimating H_1 at the sample. These assumptions together with the results of Blum and Ohnishi [49] to evaluate distortion at high power due to a nonuniform H_1 field along the sample indicate that only order-of-magnitude estimates of T_1 can be obtained at best.

Relative values of T_1 are accurately determined when measured under identical conditions with samples of nearly identical geometry, volume, and dielectric properties. For instance, from the definition of the cavity quality factor Q = 2 x (energy stored/average energy lost/sec), it follows that $Q = H_1^2 V/2P$ where V is the cavity volume and P is the power dissipated in the cavity. Then under the condition that $P = P_{1/2}$, i.e., $\gamma^2 H_1^2 T_1 T_2 = 1$,

$$P_{1/2} = \frac{1}{2} \cdot \frac{V}{Q} \cdot \frac{1}{\gamma^2 T_1 T_2} \tag{10}$$

There are two conditions under which it is possible to demonstrate that the susceptibility is a function of only T_1. The first applies to the condition that T_2 is constant. For all the models formulated to analyze progressive cw saturation curves [25,29,50-53], the slope at $P/P_{1/2} > 1$ is proportional to the ratio of the spin packet linewidth ($1/T_2$) to the overall (unsaturated) linewidth at half maximum intensity ($1/T_2^*$). For the data illustrated in Figure 4, for instance, it was observed [22] that the overall linewidth near the $g \sim 2$ absorption of the metmyoglobin-fluoride complex did not change with temperature. Since the slope at $P/P_{1/2} > 1$ does not change with increasing temperature, it follows that T_2 also remains constant, and $P_{1/2}$ is then a function of only T_1 for a series of related measurements. The second condition obtains under high microwave power levels where $T_1 > T_2 \gg 1/(\gamma H_1)$. Hyde [54] demonstrated that for inhomogeneously broadened lines under these conditions the susceptibility is governed directly only by T_1 because of a breakdown in the Bloch equations at high microwave powers. We have employed these criteria in cw saturation studies of a variety of high- and low-spin metalloprotein complexes to demonstrate that the parameter $P_{1/2}$ is an accurate monitor of relative changes in spin-lattice relaxation time [19-22,42-44,55,56]. More detailed discussion of the salient aspects of these results will be presented below. Furthermore, in a number of studies of Fe^{3+} heme proteins, it has been shown that the temperature dependence of $P_{1/2}$ for relaxation by the Raman process is identical to that of $1/T_1$ determined by the pulse saturation and recovery method [57-63]. These results will also be discussed later (Secs. 4.4 and 4.6).

3.1.2. Temperature Measurement and Calibration

In general, cw saturation studies of transition metal ions must be carried out at cryogenic temperatures because of their associated short relaxation times, and cryogenic temperatures are a necessary condition when investigations are directed to relaxation by the Orbach and Raman processes. For cryogenic temperatures conventional liquid helium dewars are readily employed for cw saturation studies. In our laboratory, we have made extensive use of the Oxford Instruments ESR10 cryostat [64] in which sample temperatures ≥3.5 K can be routinely obtained. In this cryostat the temperature of the liquid helium at the outlet needle positioned just beneath the sample is monitored directly with a gold-chromel thermocouple probe. It is therefore necessary to determine the temperature gradient between the sample placed in the cavity and the helium gas at the position of the outlet in the cryostat.

For calibration of the sample temperature, we have employed the following procedure: The gold-chromel thermocouple probe located at the level of the helium outlet inside the cryostat was first calibrated at the temperature of boiling nitrogen and helium and at the λ point of liquid helium by direct observation through the quartz dewar of the cryostat. The sample temperature was determined with a carbon resistor thermometer [64] or an Air Products (Allentown, Pennsylvania) gold-chromel thermocouple digital temperature indicator placed in the sample, both probes having been previously calibrated at the temperatures of boiling nitrogen, hydrogen, and helium. By control of the flow of helium and heat applied to the liquid helium outlet in the ESR10 cryostat, the temperature gradient between the helium at the outlet and the sample was then determined over the 3 to 20 K range. We have found that the temperature gradient between the helium outlet and the sample could be reproducibly set to within ±0.3 K over the entire range in this manner simply by thermostating and controlling the flow of helium gas [21,22].

Another method of determining sample temperature is to employ an internal standard in the EPR sample tube. Normally, a thermocouple probe cannot be left in the sample during measurements because of the inherent decrease in the cavity quality factor. Slappendel et al. [65] for this reason employed lyophilized horse heart ferricytochrome c as an internal standard to determine the zero field splitting of high-spin Fe^{3+} in soybean lipoxygenase. The EPR absorption of the low-spin Fe^{3+} cation of the ferricytochrome c acts as a temperature indicator because excited states can be neglected in the 3 to 25 K temperature range. The signal exhibits Curie law behavior, i.e., intensity (I) x temperature = constant [66]. Since the linewidth of the $g \sim 3$ signal of ferricytochrome c is constant below 25 K, the peak-to-peak amplitude can be employed as a measure of the signal intensity. For accurate determination of the signal amplitude of the ferricytochrome c, the EPR absorption spectrum must be recorded with the incident microwave power well below saturation. Representative results from this study are provided in Figure 7. It is clear that this method of sample temperature determination requires that the added solid substance employed for temperature determination does not interact chemically with the test system and that the EPR

APPLICATION OF EPR SATURATION METHODS 153

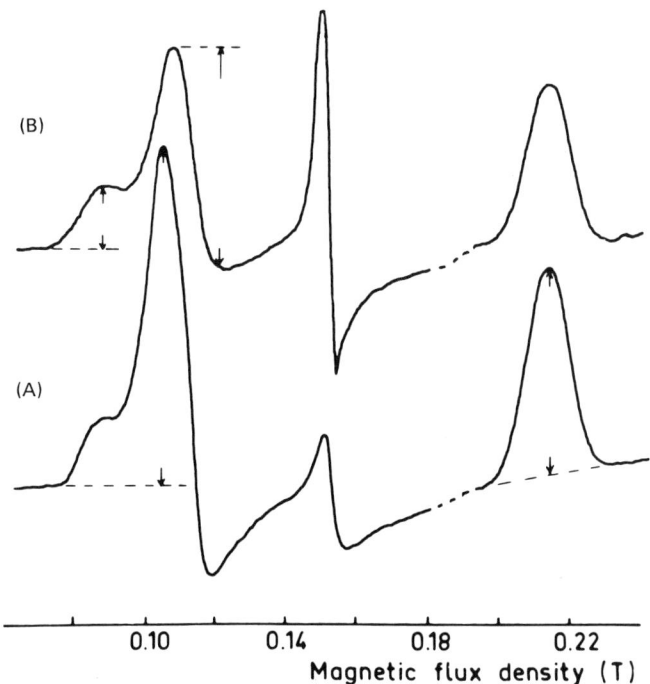

FIG. 7. Use of ferricytochrome c as a monitor of sample temperature. The sample temperatures of two lipoxygenases (A) and (B) below 0.18 T were determined from the amplitude of the g = 3 transition of ferricytochrome c, shown at 0.21 T. The arrows indicate the way of measuring the amplitudes of the signals. This method is useful when a significant thermal gradient exists between the sample and the thermocouple of a gas flow system. (Reproduced from Ref. 65 by permission of Elsevier Biomedical Press.)

absorption lines of the two systems do not overlap. Although in the studies of Slappendel et al. [65] solid ferricytochrome c was employed, in principle any paramagnetic system can be used that obeys the Curie law. It may also be possible to keep a small sample of the reference material in a sealed quartz capsule suspended in the EPR sample tube to ensure noninteraction of the two chemical systems. This, however, will absorb helium gas since quartz becomes permeable to gaseous helium just below 77 K.

The dewars employed for these cryogenic studies are designed for convenient access to the sample and for efficiency in carrying out EPR studies of a number of samples over a relatively short period of time. This construction requires calibration of the sample temperature since the sample is not enclosed by surrounding helium as in dewars employed for single-crystal EPR and pulse saturation and recovery studies [58-63]. In the latter types of dewars the sample temperature is defined directly by the temperature of the helium in contact with the microwave cavity and can be controlled often to within a few milli-Kelvin.

3.2. The Pulse Saturation and Recovery Technique

The pulse saturation and recovery technique may be described as a step technique wherein the sample is subjected first to a pumping pulse of intense microwave irradiation for a sufficiently long period for the system to come to saturation equilibrium, followed by turning off the pumping power to a low level to monitor the dynamic processes by which the spin system returns to thermal equilibrium with the lattice. The origin of the technique derives from the nuclear magnetic resonance studies of Bloembergen et al. [28], in which they suggested that the steady-state population difference of a pair of levels may be perturbed by a pumping radio frequency field followed by monitoring the time-dependent return to equilibrium. The salient features of a pulse saturation and recovery experiment for a simple two-level spin system, then, are described by the rate equation:

$$\frac{dn}{dt} = \frac{n - n_0}{T_1} \tag{11}$$

where n is the instantaneous difference in populations of the two levels and n_0 is the Boltzmann-determined equilibrium difference of the populations. Since the observed EPR transition is first saturated and the return to equilibrium is monitored with low-power microwave irradiation, the signal recovery may be correspondingly described by Eq. (12):

$$S(t) = S_{LP} - \Delta S \exp\left(\frac{-t}{T_1}\right) \tag{12}$$

where S_{LP} is the signal at $t = \infty$ and ΔS is the difference in signal amplitudes at $t = 0$ and $t = \infty$. With reference to Figure 8, the signal grows at an exponential rate from $(c/d)L_P^{1/2}$ to $(a/b)L_P^{1/2}$. Because of limitations in instrumentation, specifically delivering a pulse of a sufficiently intense microwave field in times less than 10^{-7} sec, the technique is best applied to systems with spin-lattice relaxation times longer than 10^{-7} sec. Furthermore, if the magnetization after a pulse is not negligible, i.e., M_x and M_y are finite, a free induction decay signal may compete with the saturation recovery signal. This circumstance is of particular concern when T_2 and T_1 are nearly comparable. Hyde and coworkers [67-69] have discussed this aspect in great detail and the reader should consult these papers for further details of instrumentation and theoretical analysis of data. In the spin systems to be discussed here, as in paramagnetic metal ions in solids at cryogenic temperatures, this circumstance does not generally obtain and need not concern us further.

Instrumentation for pulse saturation and recovery is not commercially available, and the experimentalist therefore must construct his own. There are numerous publications describing suitable instrumentation design [62,63,67-71]. The major components for this type of time domain EPR spectroscopy include:

1. A source of microwave pulses of high power (rated 0.5-1 W continuous output) and of brief duration (10^{-6}-10^{-1} sec) that saturate the energy levels of the transition.
2. A low-power microwave source (10^{-6} W) for monitoring the return to equilibrium (this may be the same as in item 1 above).
3. Magnet, microwave cavity, and waveguide system for delivering the microwave field to the sample.
4. Detector with transient recorder, e.g., signal averager with 10-μsec dwell time and 1,024 channels.
5. Clocked controlling unit for coordination of the operation of the microwave source and signal-recording system.

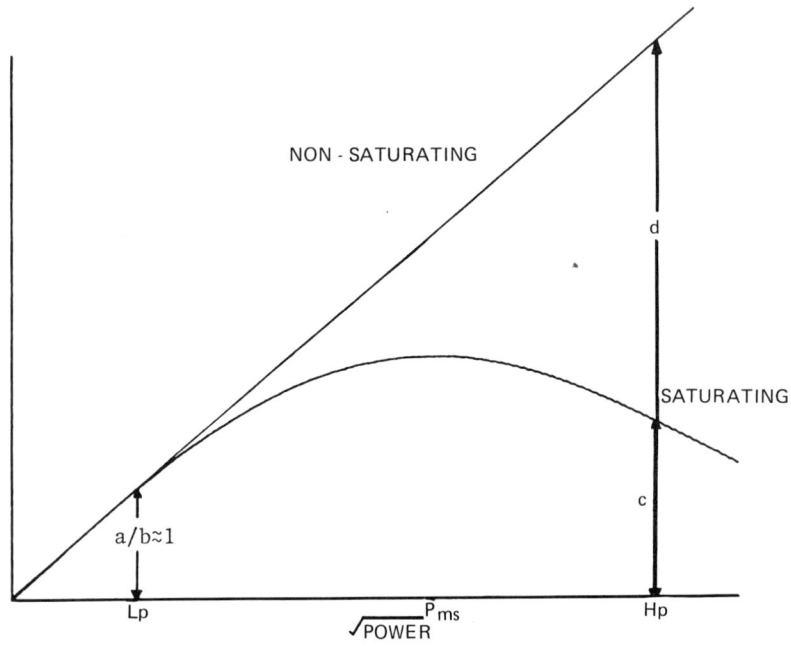

FIG. 8. Saturating and nonsaturating EPR signals as a function of microwave power. Amplitudes of the signals are plotted as a function of the square root of power (proportional to microwave magnetic field). The pulse saturation and recovery technique for measuring spin-lattice relaxation times depends on the behavior of the saturable signal relative to the nonsaturable one. Hp = high-power setting at which the spin system is saturated. Lp = low-power setting at which the return of the spin system to thermal equilibrium is monitored. (From Ref. 63 with permission of the author.)

Time domain EPR spectroscopy requires components (4) and (5). Items under number (3) are comparable with those of a cw EPR spectrometer. The operating characteristics described for components (1) and (2) are required by the special characteristics of relaxing paramagnetic systems. In addition, a crucial feature, first introduced by Isaacson [70], is the use of a boxcar integrator that repetitively samples an input signal during a variable-width gate time. The integrator output approaches the average value of the input signal over the gate time, whereby the signal-to-noise ratio is enhanced.

APPLICATION OF EPR SATURATION METHODS 157

The pulse saturation technique is best suited for determining values of T_1. For this objective it is necessary to ensure that the recovery signal that monitors spin-lattice relaxation is isolated from the effect of other time-dependent processes. Equation (11) is applicable to a simple two-level system of an EPR transition and does not account for the presence of other EPR transitions or the presence of magnetic nuclei. Both give rise to time-dependent effects through cross-relaxation and spectral diffusion, respectively. Hyde [67] has outlined general guidelines to ensure isolation of the spin-lattice relaxation event from the effects of other, rapid, time-dependent processes, and we summarize them below:

1. Monitored signals should show single exponential behavior, otherwise some complication is occurring.
2. Measurements of spin-lattice relaxation constants are more reliable when based on the late part of recovery signals rather than on the early part.
3. Relaxation constants should be independent of the duration of pumping.
4. Relaxation constants should be independent of the energy of the pumping.
5. Relaxation constants should be independent of observing power.

Effects due to free induction decay are sensitive to the phase of the pumping microwave signal relative to a reference level. Since the saturation recovery signal due to spin-lattice relaxation is independent of the phase, this criterion can be employed to ensure isolation of and to identify accurately time-dependent signals characteristic of spin-lattice relaxation.

The extraction of a spin-lattice relaxation rate constant from the time-dependent monitoring signal is straightforward if adherence to the above rules has produced a single exponential decay, as the rate equation model through Eqs. (11) and (12) predicts. On the other hand, more complicated patterns often occur. Such patterns can be interpreted by fitting an exponential function to the late part of the data, thereby ignoring the early part of the trace since cross-relaxation and spectral diffusion generally influence only this region. This has been the procedure in most investigations of metalloproteins thus far. One may also use a multiterm function over the entire time range the parameters of which are determined by a computer program based on a nonlinear, least-squares algorithm, such as the Marquardt algorithm [72]. The use of a more complicated model than that of a single exponential is illustrated in Figure 9. In this

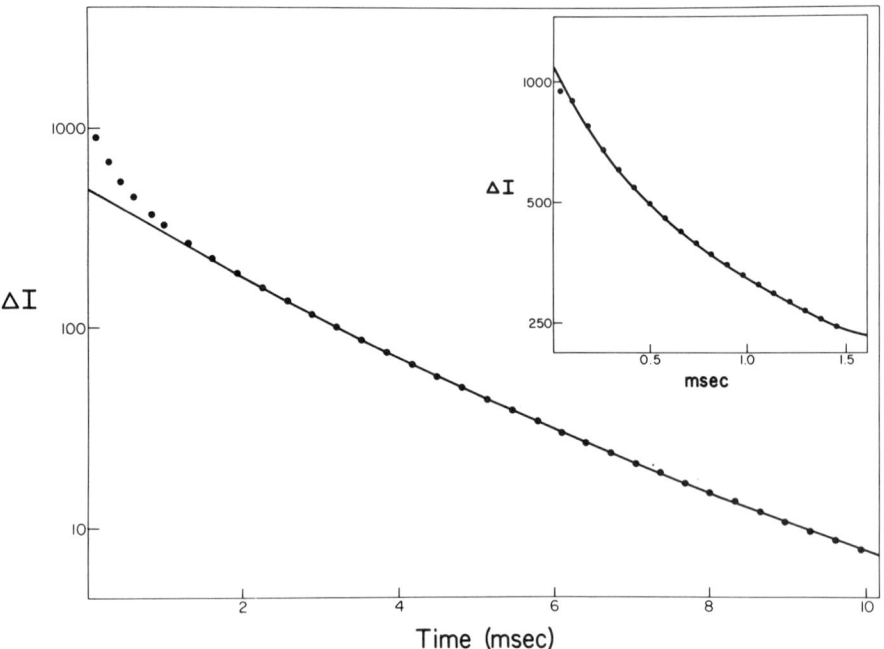

FIG. 9. Pulse saturation and recovery for the $g \sim 2$ signal of aquomethemoglobin at 1.90 K. The multiple exponential recovery behavior is demonstrated by two different models fitted to the observed data, with use of a nonlinear least squares algorithm [72] to find analytical expressions to fit the data (●). In the main panel, a two-exponential term function ($y = 383 \exp(-608t) + 115 \exp(-283t)$, t = time) fits the long times but cannot fit data from time < 1.5 msec. In the inset, a three-exponential term function [$y = 541 \exp(-3805t) + 359 \exp(-772t) + 203 \exp(-330t)$] satisfactorily fits the data at t < 1.5 msec, and the fit for long times is comparable to that for the two-exponential function. Such multiexponential expressions permit quantitative analysis of time domains where other processes giving rise to relaxation in addition to spin-lattice relaxation are significant.

figure we have illustrated the recovery signal of the $g \sim 2$ EPR absorption of aquomethemoglobin under conditions in which two fast relaxation events are observed, in addition to the spin-lattice relaxation that is predominantly responsible for the tail of the recovery data. To model these data required a function with a minimum of three exponential components. In this manner, the temperature dependence of all three relaxation processes could be determined from data collected over the 1.5 to 2.9 K range. Use of the late portion of the recovery data to estimate the spin-lattice relaxation rate only yielded values essentially equivalent to those extracted with the three-component model. This type of modeling of data collected according to the guidelines outlined above to ensure isolation of relaxation processes from instrumentation artefacts has allowed us to identify interactions of magnetic nuclei with paramagnetic ions underlying spectral diffusion [44]. Other examples of multiterm modeling of pulse saturation and recovery data have also been reported. For example, Standley and Vaughan [73] fitted recovery traces of Cr^{3+} in ruby with a sum of two exponentials based on a model of cross-relaxation superimposed on spin-lattice relaxation [cf., Eq. (5)] and Kyhl and Nageswara-Rao [74] attributed $t^{-1/2}$ behavior to coupling between the electron dipole-dipole reservoir and the aluminum nuclear Zeeman reservoir in Cr^{3+} doped ruby. For biological systems, however, the pulse saturation and recovery technique has made its major contribution to studies of spin-lattice relaxation with analysis based on the expected single exponential behavior of the long-time portion of the recovery data.

4. MICROWAVE POWER SATURATION STUDIES OF PARAMAGNETIC METAL ION CENTERS IN PROTEINS AND SMALL-MOLECULE COMPLEXES

4.1. Differential Saturation Effects Dependent on the Mechanism of Modulation of Electron-Nuclear Dipolar Interactions

Application of the cw microwave power saturation technique to paramagnetic systems in ionic solids was first carried out to estimate spin-lattice relaxation times. These studies and the attendant problems in data collection and analysis have been discussed above, and the subject has been well reviewed by Standley and Vaughan [3]. Since estimation of spin-lattice relaxation times is more accurately made by application of pulse saturation and recovery methods, we shall not review further other cw power saturation studies of paramagnetic metal ion salts. In this section we shall review studies in which application of the progressive cw power saturation method yields structural and electronic information that could not be as directly gained through the pulse saturation method.

The coupling of an unpaired electron spin with magnetic nuclei in its environment is known as hyperfine interaction. This leads to the appearance of (2I + 1) satellite lines in the EPR absorption spectrum spaced at constant intervals from the main transition, and the splitting is indicative of the strength of the hyperfine interaction. In nuclear magnetic resonance spectroscopy, Bloch [75] originally suggested that the saturation factors of multiplet lines of a transition should be different, and Shimizu [76] showed that the differential saturation behavior of the main and satellite lines of an EPR transition due to hyperfine interaction may be used to assign the mechanism of the modulation of the electron spin-nuclear spin dipolar interaction. By taking into account differences in the microwave-induced and in the lattice-induced transition probability for the main and satellite lines, he showed that their differential saturation would depend on the mechanism of spin-lattice relaxation. The two mechanisms considered are (1) radial modulation of the dipolar coupling between electron and nuclear spins by lattice modes, and (2) angular modulation of the dipolar coupling by intramolecular motions. For the first case, the intensity ratio ($I_{satellite}/I_{main}$) at high values of H_1 is twice the value at low values of H_1 while for the second case this ratio may change by a factor of 50-100.

The first application of this theory in the case of a paramagnetic metal ion complex was carried out by Schlick and Kevan [77]. They studied calcium cadmium acetate hexahydrate doped with ^{63}Cu. Although the exact concentration of Cu^{2+} in the crystallizing medium or in the crystals was not stated, it appears from the spectra [77] that the Cu^{2+} sites were sufficiently dilute so that site-site

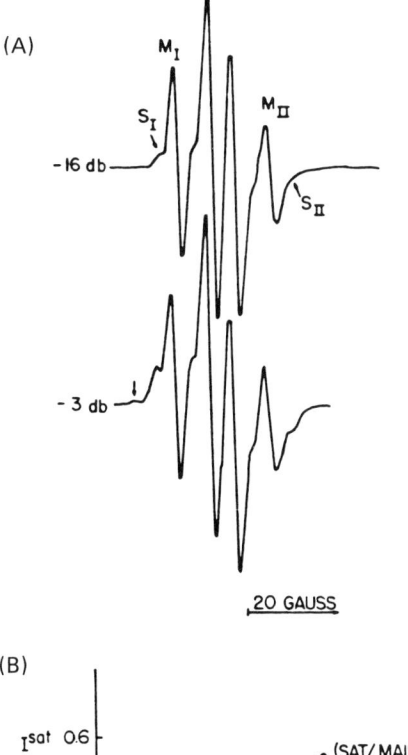

FIG. 10. Differential saturation of allowed and forbidden transitions in ^{63}Cu-doped calcium cadmium tetraacetate hexahydrate. (A) A quartet signal showing the main lines M_I and M_{II} accompanied by the satellite lines S_I and S_{II}. (B) Variation with microwave power of the intensity ratio I_{sat}/I_{main} for the pairs I and II shown in A. The ratio changes only by about a factor of 2, suggesting that radial modulation of the crystal field is the dominant mechanism for modulation of the electron spin-lattice relaxation. (Reproduced from Ref. 77 by permission of Academic Press.)

interactions were negligible. The results are illustrated in Figure 10. It is seen that the ratio ($I_{satellite}/I_{main}$) changes only by a factor of approximately 2 over the entire microwave power range used.

Provided the lineshape for the main and satellite transitions does not change, there would be only a gradual decrease in intensity of both lines at higher power with no essential change in relative intensity. In view of the highly anisotropic principal components of the g tensor for Cu^{2+} in this system, Schlick and Kevan [77] point out that the orbital moment will predominately "sense" the radial modulation of the crystalline field giving rise to spin-lattice relaxation in agreement with Shimizu's theory.

Schlick and Kevan [77] also studied the progressive power saturation of single crystals of $Na_2HPO_3 \cdot 5H_2O$ irradiated with a source of ^{60}Co. The differential saturation behavior of main and satellite lines was distinctly different from that of the Cu^{2+} system, giving evidence for angular modulation of the electron-nuclear dipolar coupling. Similarly, the saturation properties of trapped H atoms in γ-irradiated frozen glasses of H_2SO_4 indicates that the spin-lattice relaxation is dominated by angular modulation [77,78]. In these systems the g value is nearly isotropic and very close to that of the free electron with consequently little orbital contribution. For such systems relaxation via spin orbit and the orbit-lattice interaction will be slow. Torsional motions within the free radical may then be expected to affect angular modulation of the crystalline field more than radial modulation. This behavior may be characteristic of organic free radical systems in general. For instance, the description of the differential saturation behavior of γ-irradiated barbituric acid dihydrate [79] indicates that angular modulation of the relaxation is predominant, and the progressive power saturation curves of glassy 6 M H_2SO_4 γ-irradiated with ^{60}Co similarly show that the ratio ($I_{satellite}/I_{main}$) changes by a very large factor between low and high values of H_1 [80].

4.2. Differential Saturation Effects Dependent on Spectral Diffusion

In a series of spectroscopic studies to assign metal ion coordination structure in metalloenzymes, Makinen and coworkers [42-44] observed that the progressive cw power saturation of the EPR absorption of paramagnetic metal ions is sensitive to the presence of oxygen-17 enriched into inner-sphere coordinated oxygen donor ligands. The effect is illustrated in Figure 11 showing the influence of $H_2^{17}O$ on the cw microwave power saturation of the prominent low-field resonance absorption of active site specific [81] Co^{2+}-reconstituted liver alcohol dehydrogenase. It is seen that increasing the concentration of $H_2^{17}O$ results in a decrease in the value of $P_{1/2}$ with no change in slope at $P/P_{1/2} > 1$. Also, the presence of $H_2^{18}O$ has no influence on the value of $P_{1/2}$, demonstrating that the shift in $P_{1/2}$ is magnetic in origin. Similar results have been observed for Co^{2+}-substituted carboxypeptidase A [42] and a variety of structurally defined complexes of Co^{2+}, Cr^{3+}, as well as of high- and low-spin Fe^{3+} [43,44]. Moreover, the effect has also been observed in progressive cw saturation studies of metmyoglobin ^{13}C cyanide [44]. The salient features of the effect are that (1) the slope of the asymptote of the saturation curve at $P/P_{1/2} > 1$ is not altered upon introduction of the isotope; (2) the magnitude of the shift in $P_{1/2}$ from its value observed for the complex prepared with natural abundance materials decreases with increasing temperature; (3) the shift in $P_{1/2}$ may occur toward either higher or lower value; and (4) the shift in $P_{1/2}$ is observed only for complexes in which the isotopically enriched donor atom of the ligand is bound within the inner coordination sphere of the paramagnetic metal ion.

The influence of the magnetic nucleus, for instance, as illustrated by the presence of $H_2^{17}O$ in Figure 11, can be interpreted on the basis of the spin packet model of an inhomogeneously broadened EPR absorption line with spectral diffusion of the type described by Clough and Scott [30]. This theory of saturation and spectral diffusion has been discussed above (Sec. 2.3). As the EPR transition is saturated with increasing levels of microwave power, the spin packets interact through spectral diffusion to transfer saturation, providing an additional pathway to achieve internal thermal equilibration. Under conditions of high microwave power, as shown by Hyde [54], the susceptibility, i.e., $P_{1/2}$, is governed by $1/T_1$. According

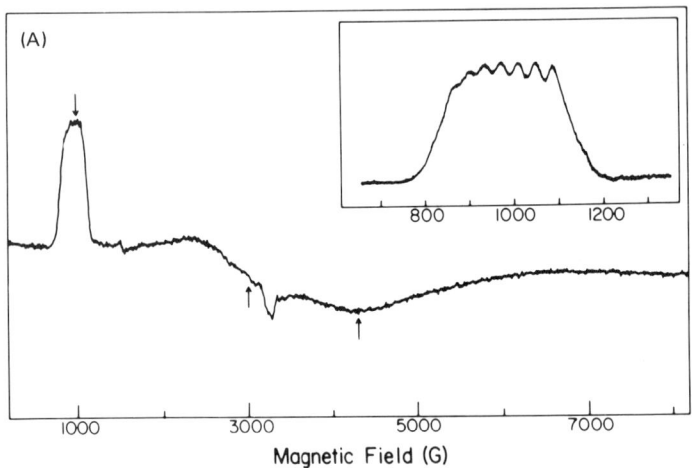
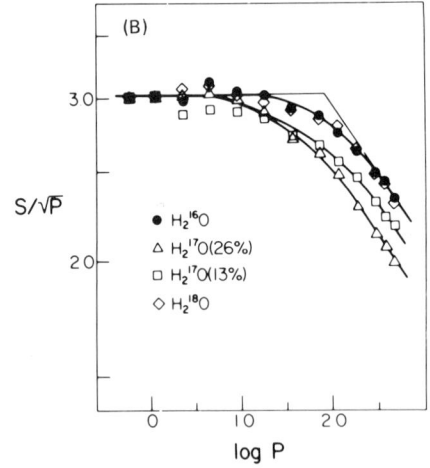

FIG. 11. First-derivative EPR absorption spectrum of active site-specific CoLADH. The inset in panel (A) shows the expanded spectrum of the low-field resonance absorption with hyperfine structure due to the (I = 7/2) ^{59}Co nucleus. In panel (B) are compared the cw saturation curves of the low-field absorption for CoLADH in media of different isotopic oxygen content. The equivalence of the saturation curves of natural abundance and oxygen-18-enriched solvent shows that the shift in $P_{1/2}$ with introduction of oxygen-17 is magnetic in origin. The log $P_{1/2}$ values for these saturation curves are (●,◊), 2.00, natural abundance or >45% $H_2^{18}O$; (■) 1.75, 13 g atoms% $H_2^{17}O$; (△) 1.53, 26 g atoms% $H_2^{17}O$. Only the slope for the enzyme in natural abundance solvent is drawn for purposes of clarity in data presentation. The sample temperature was 8 K. Typical conditions of spectral data collection have been published [19,55]. (Reproduced from Ref. 43 by permission of the VCH Verlagsgesellschaft.)

to our discussion above (Sec. 2.3), we must consider T_1 as an apparent spin-lattice relaxation time (T_1') because of the steady-state balance between spin-lattice relaxation (T_1) and the influence of the local field (T_3) through spectral diffusion. To evaluate the influence of T_1' and therefore the corresponding change in $P_{1/2}$, it is necessary to consider only the limiting conditions for the relationship $T_1' = T_1 T_D/(T_1 + T_D)$, where T_D is the characteristic spectral diffusion time and T_1 is the true spin-lattice relaxation time in the absence of spectral diffusion. These are defined by the relationships $T_1 \gg T_D$, under which condition there is strong interaction of the spin packets, and $T_D \gg T_1$, under which condition the spin packets may be characterized as independent or noninteracting. As discussed in Sec. 2.3, the introduction of an isotopically enriched ligand with a magnetic nuclear moment will modulate the apparent spin-lattice relaxation time through contact hyperfine interactions. Accordingly, the value of $P_{1/2}$ is shifted to higher or lower values, compared with that for the natural abundance system, depending on whether $T_1 > T_D$ or $T_D > T_1$. We have presented preliminary observations and discussion of this phenomenon earlier [42,43] and treat these results in more detail elsewhere [44] to discuss how the shift in $P_{1/2}$ may be dependent on the extent of covalency in the metal-ligand bond. Qualitatively this can be understood as related to the basis of the modulation pattern observed in spin-echo spectroscopy of metal ion complexes wherein the modulation pattern due to nearby ligand nuclei is observed only for weak covalent interactions, suggesting relatively effective spectral diffusion, while the modulation pattern is suppressed in the presence of strong covalent interactions, indicating noninteraction of the spin packets [82,83].

Aasa and coworkers [84] compared the progressive cw saturation properties of the dioxygen-bound intermediate of a copper-containing laccase obtained by reacting the reduced enzyme with dioxygen. The paramagnetic spin is assigned to an oxygen radical, possibly O^- [85], and in the presence of oxygen-17-enriched dioxygen they observe no change in the value of $P_{1/2}$. Since the oxygen radical appears to interact with another paramagnetic site in the protein, presumably

Cu^{2+}, this dipolar coupling must mask the influence of the oxygen-17 nucleus on the relaxation properties of the oxygen radical.

4.3. Characterization of Multi-Metal Ion Clusters in Proteins

The first application of a progressive cw power saturation technique to the study of paramagnetic metalloproteins was carried out by Beinert and Orme-Johnson [86]. In this study they compared the saturation behavior of the flavin semiquinone radical in metal ion containing flavoproteins to that in metal-free flavoproteins. They introduced the saturation curve plotted as in Figures 4-6 and noted that the slope was a function of the degree of inhomogeneous broadening. They correctly evaluated with T. Vänngård (cf. [87]) that the value of the parameter b for a completely homogeneously broadened line was 3 instead of 4 for first harmonic detection (cf. Table 1), and focused attention on employing spin-lattice relaxation properties to characterize metal ion centers in proteins. The saturation studies were carried out with samples in the 78 to 112 K range and consequently probed only Raman spin-lattice relaxation. Furthermore, Beinert and Orme-Johnson [86] noted that the relaxation properties of flavin radicals in proteins could be altered by addition of metal ions such as Ni^{2+} and that the relaxation rate of the Mo(V) center in aldehyde oxidase was maximally influenced in the course of an anaerobic reductive titration when the maximum amount of the semiquinone form of the protein-bound flavin group was formed. This observation indicated that the two spin systems were near each other, but no theory was applied to extract the distance between the two interacting spins from the cw saturation data.

Hall and coworkers subsequently applied the cw technique pioneered by Beinert and Orme-Johnson [86] to investigate the relaxation properties of iron-sulfur proteins [88] and of the electron acceptor site in photosystem I reaction centers isolated from spinach [89]. They noted that the relaxation rate as measured by $P_{1/2}$ in the oxidized form of high-potential iron-sulfur protein of *Chromatium D*

with a [4Fe-4S] center was significantly faster in the 10 to 20 K
range than in proteins with a [2Fe-2S] center and that the relaxation
rate in general was decreased upon addition of denaturing agents.
They also noted that the value of $P_{1/2}$ of the Mo(V) center of xanthine
oxidase was dependent on the extent of reduction of Fe-S center I of
the protein. While these studies, as those of Beinert and Orme-Johnson
[86], demonstrated that nearby paramagnetic centers influenced the
relaxation properties of the radical under investigation, no theory
was applied to allow further conclusions about the distance over which
this interaction occurred.

The electron spin-lattice relaxation of the [4Fe-4S] center in
the ferredoxin of *B. stearothermophilus* has been subsequently studied
by pulse saturation and recovery and the temperature dependence of
$1/T_1$ has been determined over the 1.2 to 12 K range [90]. The results
are illustrated in Figure 12 and compared with those of the [2Fe-2S]
ferredoxin of *S. maxima*. The results show that spin-lattice relaxation
time is shorter for the [4Fe-4S] protein than for the [2Fe-2S] protein,
essentially confirming the earlier conclusions of Hall, Rupp, and coworkers [88,89,91]. Bertrand et al. [90] also compared the interpretation of their data according to a so-called Tarassov model [92]
which assumes a dimensionality of 3 for relaxation via the Raman process in contrast to that of Stapleton and coworkers [58-60] involving
a dimensionality less than 3. The interpretations of the mechanism
of spin-lattice relaxation provided by Gayda and coworkers [89,90,93]
are discussed in Sec. 4.6 with reference to the model of Stapleton
and coworkers [58-60] based on fractal dimensionality.

More recent examples of the method pioneered by Beinert and
Orme-Johnson [86] are found in the investigations of Ohnishi and
coworkers [94,95]. Here the cw saturation technique has been applied
to assign the location of iron-sulfur clusters with respect to the
polypeptide subunits of *B. subtilis* succinate dehydrogenase [94] and
to estimate the distance of closest approach of added Dy^{3+} to iron-
sulfur and heme iron centers in a variety of iron-containing proteins
[95]. The theoretical basis for estimating distances by power saturation is discussed below (Sec. 4.5).

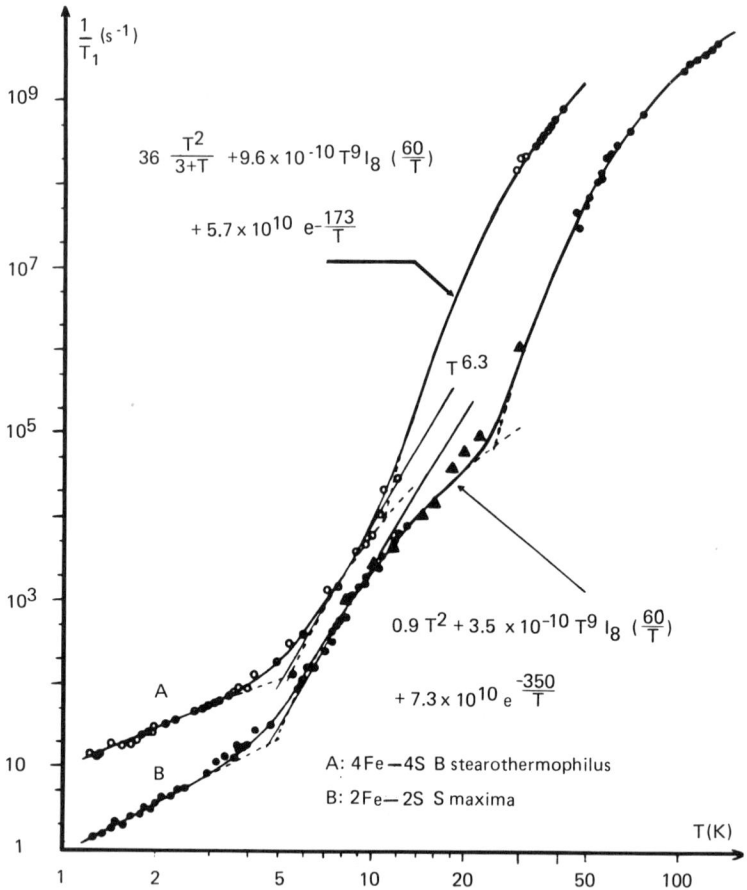

FIG. 12. Temperature dependence of electron spin-lattice relaxation time T_1 of the ferredoxins isolated from *B. stearothermophilus* (○) and from *S. maxima* (● and ▲). Experimental techniques for the *B. stearothermophilus* enzyme: below 12 K, pulse saturation-recovery; above 30 K, relaxation-broadening measurements. For the *S. maxima* enzyme: below 30 K, pulse saturation-recovery (●) and cw power saturation (▲); above 50 K, relaxation-broadening measurements. The analytical expressions fitted to the data incorporate direct (with phonon bottleneck), Raman, and Orbach processes for both enzymes. (Reproduced from Ref. 90 by permission of the American Institute of Physics.)

Mammalian cytochrome oxidase is the terminal electron transfer catalyst of the mitochondrial respiratory chain and functions by accepting electrons from ferrocytochrome c to reduce molecular dioxygen to water. It contains four inequivalent metal ion centers consisting of a heme a and heme a_3 group and two copper sites. One of these copper sites is associated with the heme a_3 group constituting the O_2-binding site. The other Cu^{2+} site (the Cu_A site) and the iron cation of the heme a group are shielded from interaction with added ligands and mediate the transfer of electrons to the site of reduction of dioxygen. In efforts to identify the ligands to the Cu_A site, Chan and coworkers [96,97] compared the progressive cw saturation properties of the Cu_A site to those of several heme complexes, copper proteins, and the cysteine radical. The latter system was included on the basis that ENDOR studies suggest that there is substantial spin density associated with the sulfhydryl group of a coordinating cysteine side chain of the Cu_A site [98]. By progressive cw power saturation, Chan and coworkers observed that the temperature dependence of $P_{1/2}$ of the Cu_A site was unlike that of any of the model systems in the 8 to 30 K range but was parallel to that of the Fe^{3+} cation of heme a in the enzyme complex. This observation suggested that the heme a and Cu_A sites interact through dipolar coupling and are separated by <10 Å. This distance is larger than that between the heme a_3 group and its associated Cu^{2+} ion. By analyzing the nitrogen superhyperfine splitting of complexes of cytochrome oxidase prepared with NO, which competes with O_2 in binding to the enzyme, these investigators have shown that the heme a_3-Cu_{a_3} sites are within 5 Å of each other [99-101].

Scholes and coworkers [102] investigated the temperature dependence of spin-lattice relaxation of the Cu_A and heme a sites separately in mammalian cytochrome oxidase in the 1.5 to 20 K range by pulse saturation and recovery and have confirmed the observations of Chan and coworkers [96,97]. Their results are summarized in Figure 13. The results show that the temperature dependence of $1/T_1$ of heme a and of Cu_A are remarkably similar in the 7 to 20 K range and

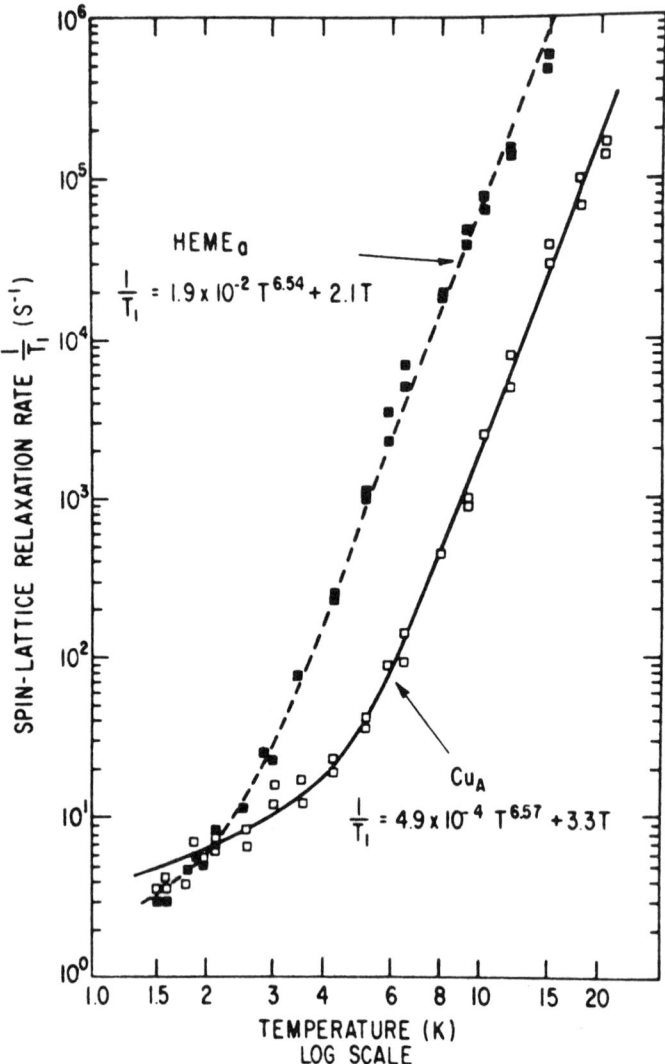

FIG. 13. Electron spin-lattice relaxation vs. temperature for heme a and Cu_A in resting cytochrome oxidase. The data were collected by pulse saturation and recovery at $g \sim 2.30$ for heme a and at $g \sim 2.02$ for Cu_A. The analytical expressions fitted to the data incorporate direct processes and Raman processes with noninteger exponents, in accordance with theories of the fractal dimensionality of proteins. (Reproduced from Ref. 102 by permission from the Biophysical Society.)

APPLICATION OF EPR SATURATION METHODS

that the spin-lattice relaxation rate is faster for heme a than for the Cu_A site. These results indicate that the heme a relaxation dominates that of the Cu_A site through dipolar interactions.

4.4. Determination of the Splitting Between the Two Lowest Kramers Doublets of Paramagnetic Metal Ions with Ground State Spin Multiplets

4.4.1. Heme Proteins and Heme Enzymes

The first application of a pulse saturation method to determine the zero field splitting of a paramagnetic metal ion in a metalloprotein was carried out by Scholes, Isaacson, and Feher [24] (cf. Fig. 3). In this study they compared the temperature dependence of $1/T_1$ of aquometmyoglobin in frozen aqueous solution, frozen 50:50 (v/v) glycerol-water, and in crystals of the protein obtained from solutions of concentrated ammonium sulfate. They found to within 1-2% accuracy that the value of D was 9.1 ± 0.2 cm^{-1} for all these systems in excellent agreement with results of magnetic susceptibility studies [103]. They also found that the value of D in metmyoglobin fluoride was 6.1 ± 0.2 cm^{-1}, similarly in excellent agreement with magnetic susceptibility results [103]. These results have been subsequently confirmed through studies by Brill and coworkers [61-63]. Also, as discussed above, essentially identical results have been obtained by progressive cw saturation studies [22].

Herrick and Stapleton [104] similarly determined the zero field splitting of the Fe^{3+} cation in cytochrome P-450$_{cam}$ by pulse saturation and recovery methods. This hemeprotein catalyzes the hydroxylation of organic aromatic and alkyl residues as a mixed function oxidase. The species of enzyme employed in this case from P. putida acts specifically on $D(+)$-camphor, the binding of which converts the Fe^{3+} cation from a low-spin to a high-spin species. Through the temperature dependence of $1/T_1$ for the associated Orbach relaxation rates, the energies of the two excited crystal field doublets of the $S = \pm 5/2$ manifold were found to lie 6.59 ± 1.94 and 20.6 ± 1.32 cm^{-1}

above the ground doublet, yielding a value for D of 3.29 cm^{-1}. This is approximately threefold smaller than that found in aquometmyoglobin [22,24,103]. The two splittings were obtained by fitting the late recovery data to the expression:

$$\frac{1}{T_1} = 5.0 \times 10^4 \left[\exp\left(\frac{\Delta_1}{kT}\right) - 1\right]^{-1} + 3.2 \times 10^9 \left[\exp\left(\frac{\Delta_2}{kT}\right) - 1\right]^{-1} \quad (13)$$

By observing that the relaxation rates at $T > 1.6$ K were identical for detection and saturation at X-band and K-band frequencies, indicating no magnetic field dependence of the relaxation rate [cf. Eq. (1)], Herrick and Stapleton [104] concluded that the data were not influenced by direct process contributions. For the $S = \pm 5/2$ manifold of a high-spin Fe^{3+} ion, theory requires that the $S = \pm 3/2$ doublet lies above the ground level by $2D$ while the $S = \pm 5/2$ doublet is separated from the ground doublet by $4D$. That the ratio of Δ_2/Δ_1 is greater than 2 suggests that there is considerable error in determining the value of D. This circumstance may be due to incomplete formation of the protein-$D(+)$-camphor complex. Also, since the analyzed data were collected only over the 1.67- to 2.46-K range, this temperature range may not suffice to adequately determine the value of D by fitting relaxation data to an expression of the form of Eq. (13).

In the reaction cycle of horseradish peroxidase, the Fe^{3+} heme group is cyclically oxidized by substrates resulting in two reaction intermediates of the enzyme that differ from the native Fe^{3+} enzyme by 2 and 1 in oxidation state. These intermediates are known as compounds I and II, respectively. The electronic structure of the heme iron in both intermediates has been conjectural, in particular whether the heme iron cation in compound I has an oxidation state of 5+ or whether this oxidation state is the result of an Fe(IV) ferryl cation and a porphyrin π-cation radical. With superheterodyne detection Aasa and coworkers [105] found evidence for an organic free radical coupled to an $(S = 1)$ Fe^{4+} moiety in compound I of horseradish peroxidase. Together with the known d^4 electronic structure of the heme iron determined through Mössbauer [106] and magnetic susceptibility [107] studies of the compound II intermediate, it is therefore probable that the heme group in compound I of peroxidase consists of an organic π-cation radical localized on the porphyrin ring with an Fe^{4+} ferryl group at the heme center. In compound II, on the other hand, the porphyrin is in a neutral state, and the heme group contains a Fe^{4+} ferryl group with d^4 electronic structure [107].

Through analysis of the temperature dependence of the Orbach spin-lattice relaxation rate in compounds I and II of horseradish peroxidase, Stapleton and coworkers [108] were able to determine that the first excited electronic state of the Fe^{3+} in native horseradish peroxidase was 28.4 ± 0.8 cm^{-1} above the ground state while the first excited electronic state of the Fe^{4+} moiety in compound I was 25.5 ± 1.7 cm^{-1} above the ground doublet. Furthermore, by comparing the strength of the Orbach process in compound I, as denoted by the value of the preexponential factors in Eq. (13) to that in the native enzyme, Stapleton and coworkers showed that the weaker Orbach process in compound I is consistent with a model of an $S = 1/2$ free radical coupled to an $S = 1$ Fe^{4+} ion at the heme center. They also showed that the zero field splitting of the Fe^{4+} ion in compound I determined earlier by a progressive cw saturation technique [109] was 20% underestimated because of underestimation of the sample temperature.

4.4.2. High-Spin Co^{2+} in Enzymes and Small Molecules

In a variety of Zn^{2+} metalloenzymes, the catalytically required Zn^{2+} can be removed and the metal-free enzyme substitutionally reconstituted with Co^{2+}. Generally such chemically modified enzyme derivatives exhibit high catalytic efficiency, often comparable to that of the native enzyme. Because of the spectroscopic complexities of the Co^{2+} ion dependent on the crystal field and spin-orbit coupling with nearby excited states, however, the use of high-spin Co^{2+} in enzymes has been largely directed only toward monitoring kinetic and equilibrium properties of ligand binding. The splitting of spectroscopic states of high-spin Co^{2+} in coordination complexes results in two general patterns. They correspond to a high-spin d^7 configuration either in an orbitally nondegenerate ground state ($^4A_2$) or in an orbitally degenerate ground state ($^4T_1$) in which the levels are split by spin-orbit coupling. The former is generally observed in tetra- and pentacoordinate complexes, while the latter is observed in hexacoordinate complexes. To develop a method of assigning the coordination geometry and structural environment of high-spin Co^{2+} substitutionally incorporated into metalloproteins, Makinen and coworkers [19-21] carried out a detailed analysis of the EPR absorption and progressive cw saturation properties of Co^{2+} in structurally defined complexes. They find that the zero field splitting (Δ) follows the order:

$$\Delta_4 < \Delta_5 \leq \Delta_6 \tag{14}$$

where the subscript designates the coordination geometry characteristic of classical tetra-, penta-, and hexacoordinate sites with monoatom ligands. Here we discuss the results of studies to characterize the coordination environment of the active site metal ion in carboxypeptidase A and liver alcohol dehydrogenase.

The splitting between the two lowest Kramers doublets of high-spin Co^{2+} is dependent on the strength of the crystal field, the relative magnitude of the distortion away from cubic symmetry, and the strength of the spin-orbit coupling. This is schematically illustrated in Figure 14, wherein the spectroscopic terms of the ground state of high-spin Co^{2+} that exhibit multiplet separation through spin-orbit coupling have been assigned by group theory for sites of (double-point group) T_d', D_{2h}', and D_3' symmetry. These correspond to crystal fields of exact tetrahedral, distorted tetrahedral, and trigonal bipyramidal geometry, respectively. The first two are found in tetracoordinate complexes while the latter corresponds to a pentacoordinate site. For high-spin Co^{2+} in pentacoordinate sites of C_4' or C_{3v}' symmetry, comparable arguments apply to those developed for D_3' sites, and the term splitting for high-spin Co^{2+} in a hexacoordinate site is the inverse of that for the T_d' site. The upper diagram in Figure 14, furthermore, shows according to group theory that the

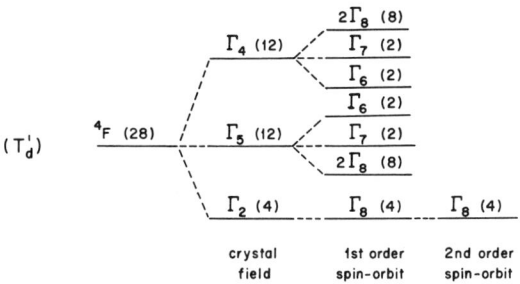
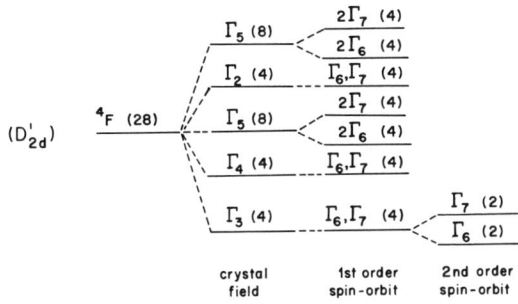
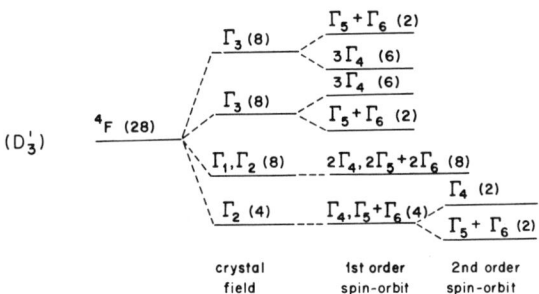

FIG. 14. Term-splitting diagram of high-spin Co^{2+} in a crystal field of T'_d, D'_{2d}, and D'_3 symmetry (the prime indicates double-point group notation). The multiplet splitting resulting from spin-orbit coupling is shown only for the ground state. The numbers in parentheses indicate the total spin and orbital degeneracies of each term. The upper part illustrates the splitting in T'_d symmetry; the central diagram corresponds to D'_{2d} symmetry, and the lower diagram corresponds to a site of D'_3 symmetry. (Reproduced from Ref. 21 by permission of the American Chemical Society.)

splitting must be equal to zero for a site of *exact* T_d' symmetry, and, as shown by Makinen and coworkers [21], the zero field splitting constant D can have negative values only in distorted tetrahedral sites.

To determine the splitting between the two lowest Kramers doublets of high-spin Co^{2+} in coordination complexes, a progressive cw microwave power saturation technique [22] was employed. Typical results are illustrated in Figure 15 for bis(triphenylphosphineoxide)-dichlorocobaltate(II). The temperature dependence of $P_{1/2}$ monitors that of $1/T_1$, and in Figure 15(C) there are three distinct regions in the plot of $P_{1/2}$ vs. reciprocal temperature. The two steepest slopes correspond to spin-lattice relaxation via the Raman and Orbach processes; near 5 K there is a gradual decrease in slope that is dominated by cross-relaxation and spectral diffusion under the modulation conditions employed [21]. The slope of the plot for Orbach relaxation yields a value of 6.6 cm^{-1} for $2D$. In Table 2 is a summary of representative values of the zero field splitting of high-spin Co^{2+} in a variety of structurally defined complexes. The results show that there are three ranges of Δ. These three ranges are approximately ≤ 13 cm^{-1} for tetracoordinate complexes, 20-50 cm^{-1} for pentacoordinate complexes, and ≥ 50 cm^{-1} for hexacoordinate complexes.

On this basis, the magnitude of the zero field splitting of high-spin Co^{2+} substitutionally incorporated into the active site of carboxypeptidase A and liver alcohol dehydrogenase has been employed as a spectroscopic signature to assign the coordination structure of the active site metal ion in catalytically competent reaction intermediates. In both metalloenzymes the Co^{2+} ion can be employed to replace the active site Zn^{2+}, with equivalent catalytic activity compared with that of the native enzyme [55,120]. Furthermore, x-ray crystallographic studies have shown that the respective native and Co^{2+}-reconstituted enzymes have essentially identical active site coordination structure [113,121]. In Figures 16 and 17 are illustrated the spectroscopic data that define the value of Δ of the active site metal ion in Co^{2+}-reconstituted carboxypeptidase A for the free enzyme and for the mixed-anhydride reaction intermediate formed with a specific ester substrate and stabilized at

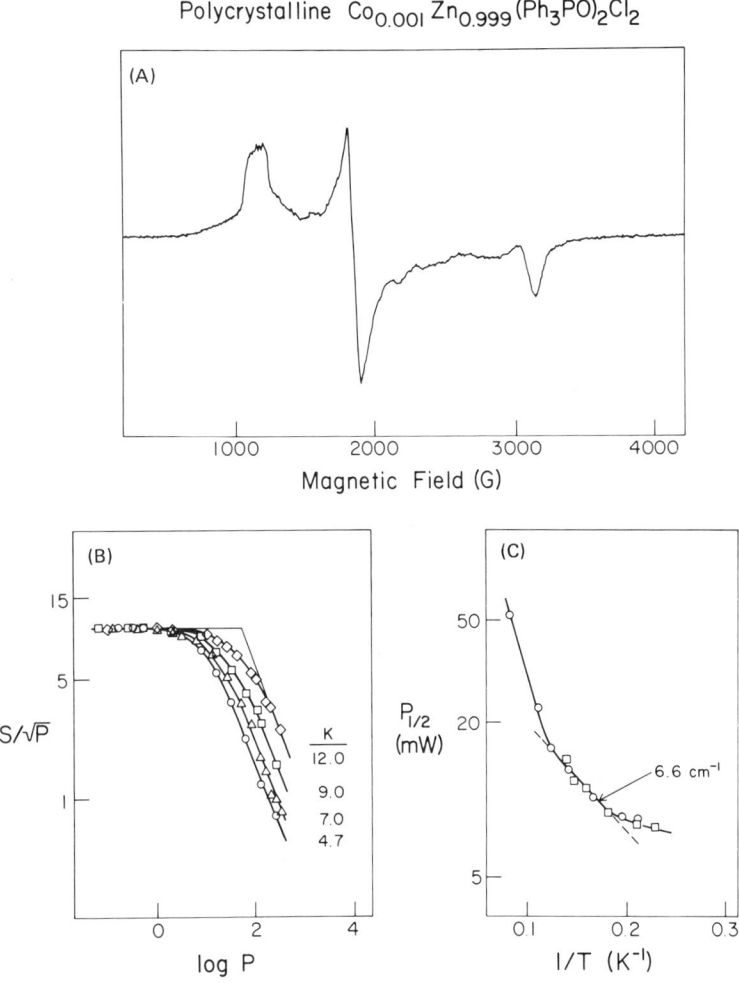

FIG. 15. First-derivative EPR absorption spectrum and cw microwave power saturation properties of bis(triphenylphosphineoxide)cobalt(II) dichloride incorporated to ≤0.1 mol% into the polycrystalline matrix of the isomorphous Zn^{2+} complex. (A) illustrates the first-derivative EPR absorption spectrum. (B) illustrates the saturation behavior of the $g \sim 3.64$ component at various temperatures. (C) shows the temperature dependence of $P_{1/2}$ for two different polycrystalline samples (the $P_{1/2}$ data of one set are frame-shifted to compensate for different volumes of the two samples). The linear region in the 4.7 to 8.0 K temperature range plot yields an estimate of 6.6 cm^{-1} for the value of the zero field splitting. (Reproduced from Ref. 21 by permission of the American Chemical Society.)

TABLE 2

Comparison of the Splitting Between the Two Lowest Kramers Doublets of High-Spin Co^{2+} in Structurally Defined Small-Molecule Complexes and Enzymes Determined by EPR, Magnetic Susceptibility, or Spin-Lattice Relaxation Methods

Structurally defined environments	Δ (cm^{-1})	Comments
A. Tetracoordinate		
Co^{2+} in CdS	1.34	a
Co^{2+} in ZnO	5.50	b
Cs_2CoCl_4	11.6	c
$Co(imidazole)_2(acetate)_2$	4.8	d
$Co(Ph_3PO)Cl_2$	6.6	d
$[(C_2H_5)_4N]_2CoCl_2$	6.52	c
$(CoBr_4)^{2-}$	11.0	d
$Co(Ph_3P)_2Cl_2$	12.9	d
Co^{2+}-carboxypeptidase A	8.3	d
Co^{2+}-LADH	9.3	e,f
Co^{2+}-LADH (+NADH + DACA)	13.0	g,h
B. Pentacoordinate		
$Co(MePh_2AsO)_4(ClO_4)_2$	19	d
$Co(2\text{-picoline-N-oxide})_5(ClO_4)_2$	25	d
acetazolamide complex of Co^{2+}-substituted carbonic anhydrase	33	i,j
CoLADH (+NAD$^+$ + CF_3CH_2OH)	26	k

C. Hexacoordinate

Co(pyridine-N-oxide)$_6$(ClO$_4$)$_2$	55	d
Co^{2+} in ZnSO$_4 \cdot$ 7H$_2$O	90	d
Co^{2+} in Al$_2$O$_3$ (site 1)	110	l
Co^{2+} in Al$_2$O$_3$ (site 2)	180	l
Co^{2+} in MgF$_2$	152	m
Co^{2+} in MgO	305	n

[a] Ref. 110.
[b] Ref. 111.
[c] Ref. 112.
[d] Ref. 21.
[e] Ref. 19.
[f] The tetracoordinate environment of the Co^{2+} ion in the active site-specific Co^{2+}-reconstituted form of liver alcohol dehydrogenase (LADH) has been defined by x-ray crystallographic studies [113].
[g] Ref. 55.
[h] This ternary complex, stabilized at pH 9.1 under nonturnover conditions, is identical to the complex of LADH formed with the inhibitor 1,4,5,6-tetrahydro-NADH and with the aldehyde substrate p-N,N-dimethylaminocinnamaldehyde (DACA) as defined in the x-ray studies of Cedergren-Zeppezauer et al. [114].
[i] Ref. 115.
[j] The pentacoordinate environment of the metal ion in this inhibitor complex is defined by the x-ray studies of Kannan et al. [116].
[k] Unpublished observations of M. B. Yim and M. W. Makinen quoted in Refs. 20 and 43; the coordination of a water molecule as the fifth ligand in this ternary inhibitor complex of liver alcohol dehydrogenase is confirmed [43] by detection of contact ligand hyperfine interactions in the presence of ^{17}O-enriched H$_2$O.
[l] Ref. 117.
[m] Ref. 118.
[n] Ref. 119.

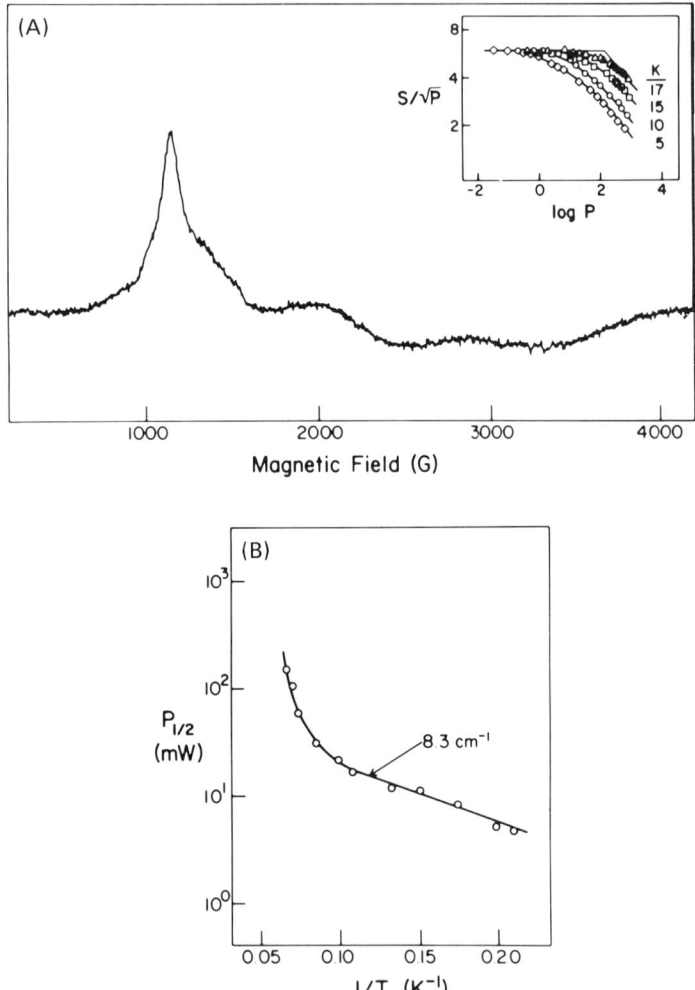

FIG. 16. First-derivative EPR absorption and cw microwave power saturation behavior of Co^{2+}-reconstituted carboxypeptidase A. The spectrum in (A) was recorded with the sample at 8 K with 2 mW incident microwave power irradiation and 25 G_{pp} modulation amplitude. The inset to (A) illustrates the general cw microwave power saturation behavior of the peak-to-peak resonance absorption centered at 1,530 G. The data illustrate that the shape at $P/P_{1/2} > 1$ remains independent of temperature over the 5 to 17 K range. (B) illustrates the temperature dependence of $P_{1/2}$. The linear portion in the 5 to 10 K region yields an estimate of 8.3 cm^{-1} for the value of Δ for the high-spin Co^{2+} ion in the metal-substituted enzyme. (Reproduced from Ref. 56 by permission of the American Chemical Society.)

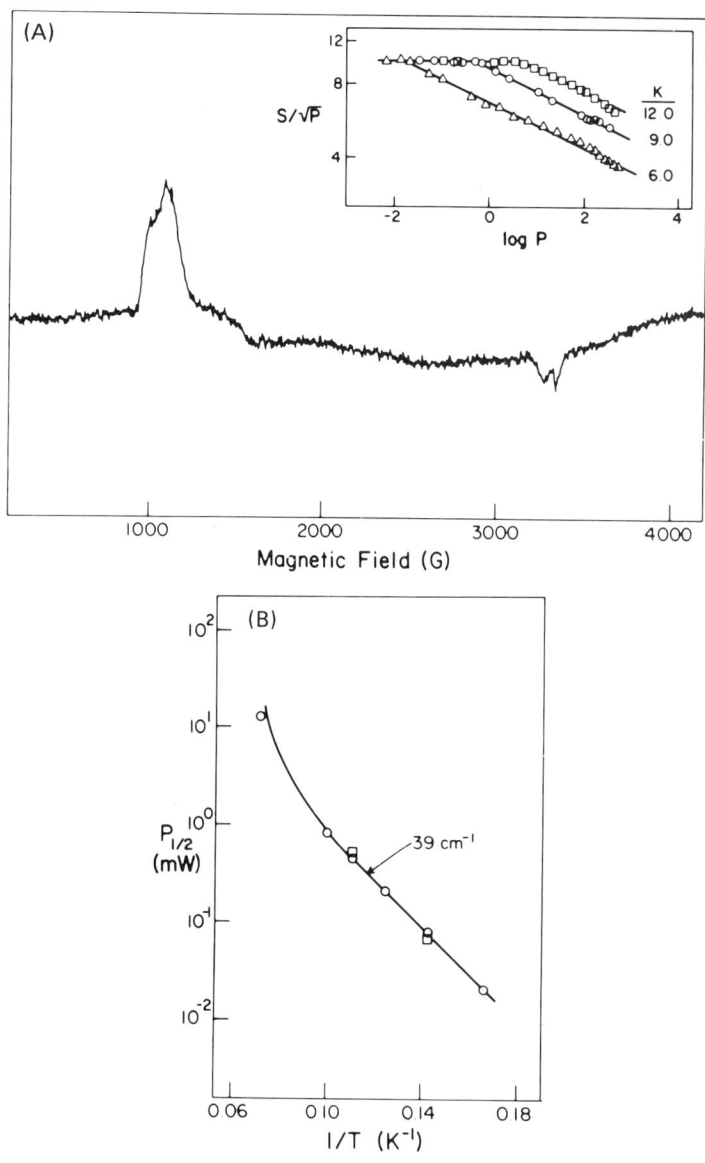

FIG. 17. First-derivative EPR absorption spectrum and cw microwave saturation behavior of the acyl enzyme intermediate of CoCPA formed with O-(p-chlorocinnamoyl)-L-β-phenyllactate. The spectrum in (A) was recorded at 7 K at 2 mW incident microwave power and 25 G_{pp} modulation amplitude. The inset illustrates the saturation behavior of the strong resonance feature centered at 1,100 G. In (B) the temperature dependence of $P_{1/2}$ of the acylenzyme intermediate in the 4 to 12 K range yields a value of 39 cm^{-1} for the splitting between the two lowest Kramers doublets of the Co^{2+} ion. The open squares indicate the $P_{1/2}$ values for the intermediate prepared in the presence of ^{17}O-enriched substrate from a separate study arate study [42]. These two values are frame-shifted to demonstrate that their relative change with temperature adheres to that defined by the open circles. (Reproduced from Ref. 56 by permission of the American Chemical Society.)

−70°C [42,120]. The results show that the splittings are ~8.3 and 39 cm^{-1} for the free enzyme derivative and for the mixed anhydride species, respectively. As seen in Table 2, these values are correspondingly compatible only with tetra-liganded and penta-liganded species of the active site metal ion. While the assignment of a tetra-liganded metal ion by spectroscopic methods is in direct agreement with the high-resolution, refined x-ray structures of both the native [122] and the Co^{2+}-reconstituted [121] enzyme, the results demonstrate that the coordination number of the metal ion is increased in the mixed anhydride intermediate to accommodate both a water molecule and the carbonyl oxygen of the scissile bond of the substrate, in addition to the three ligands from the protein. Furthermore, as discussed previously (Sec. 4.2), these interpretations are confirmed on the basis of progressive cw saturation studies in which oxygen-17 enriched materials were employed [42]. These results interpreted in light of the pH dependence of deacylation of the acyl enzyme [120] specify that breakdown of the mixed anhydride intermediate is catalyzed by metal-hydroxide nucleophilic attack. A schematic illustration of the coordination environment of the active site metal ion in this reaction intermediate of carboxypeptidase A is shown in Figure 18.

Liver alcohol dehydrogenase catalyzes the oxidation of alcohol at the level of a ternary complex formed by the enzyme, the coenzyme NAD$^+$, and the substrate. While the active site metal ion is tetracoordinate in the free enzyme [113,123], the coordination environment

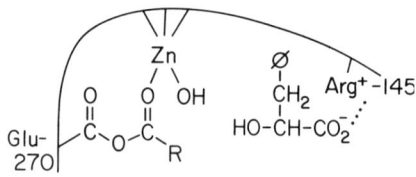

FIG. 18. Schematic drawing of the structure of the metal ion environment in the acyl enzyme (mixed anhydride) reaction intermediate of carboxypeptidase A formed with substrates.

TABLE 3

Comparison of Zero Field Splittings of High-Spin Co^{2+} in Active Site-Specific Co^{2+}-Reconstituted Liver Alcohol Dehydrogenase

Enzyme derivative	Δ (cm^{-1})	Coordination number[a]
CoLADH	9.3	4
CoLADH-CF$_3$CH$_2$OH	3.1	4
CoLADH-NADH-DACA[b]	13	4
CoLADH-NAD$^+$-CF$_3$CH$_2$OH	26	5
CoLADH-NADH-CF$_3$CH$_2$OH	22	5

[a]Coordination number is assigned on basis of ranges of Δ established in Table 2.
[b]See footnotes g and h in Table 2.
Source: From Refs. 19, 43, 44, and 55.

of the active site metal ion has not been assigned in catalytically competent reaction intermediates by direct experiment. Interpretations of the pH dependence of kinetic parameters have invoked either a tetracoordinate or pentacoordinate species as the catalytically active species [124-126]. In Table 3 are summarized the values of the zero field splitting of the high-spin Co^{2+} ion in the active site-specific metal-reconstituted enzyme for a variety of substrate and inhibitor complexes determined by progressive cw saturation methods [19,43,44]. The values of the zero field splitting of the free enzyme, the enzyme-CF$_3$CH$_2$OH complex, and the ternary enzyme-NADH-(p-N,N-dimethylaminocinnamaldehyde) complex stabilized at high pH are in direct agreement with x-ray studies (cf. [55]). However, a critical complex that has not been characterized by x-ray diffraction is the ternary enzyme-NAD$^+$-CF$_3$CH$_2$OH complex. Since the enzyme complex is crystallized from aqueous mixtures of 2,4-dimethylpentanediol, this alcohol reduces the oxidized coenzyme to NADH during crystal formation [127]. The results in Table 3 show that the zero field splitting is compatible only with a pentacoordinate metal ion in both the ternary enzyme-NAD$^+$-CF$_3$CH$_2$OH and the ternary enzyme-NADH-CF$_3$CH$_2$OH complex. These assignments of active site metal ion coordination structure in liver alcohol dehydrogenase made on the basis of zero field splitting estimates have been confirmed through use of oxygen-17 enriched water [43,128]. Thus, these results suggest that the active site metal ion coordinates both the alcoholic substrate and a water molecule in catalytically competent reaction intermediates. On this basis, Makinen and coworkers [43,55] pointed out that the catalytic function of the metal-bound water molecule may be to serve as a conduit for abstraction of the alcoholic proton prior to hydride transfer, as illustrated in Figure 19.

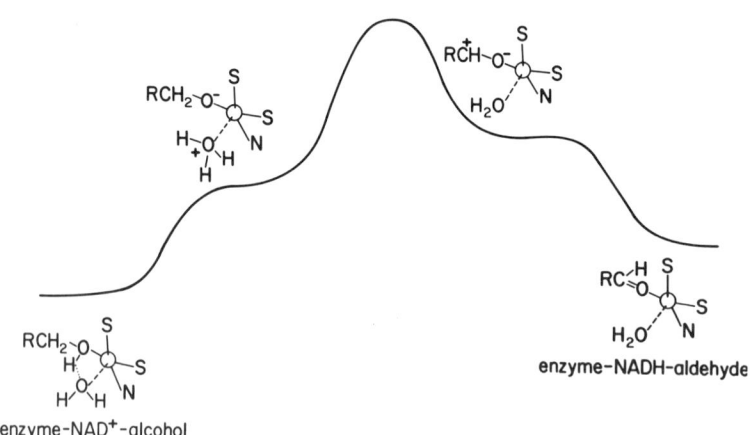

FIG. 19. Catalytic role of the active site metal-bound water molecule of horse liver alcohol dehydrogenase during the conversion of the ternary enzyme-NAD$^+$-alcohol complex to the ternary enzyme-NADH-aldehyde complex. The active site metal ion is represented by a circle, and the donor ligand atoms are those of cysteine-46, cysteine-174, and histidine-67. The pathway begins with the metal-bound alcohol forming a hydrogen bridge to the metal-bound water molecule and proceeding to transfer of the proton to form a hydronium ion that is fleetingly coordinated to the metal ion. With release of the proton to bulk solvent, abstraction of the hydride ion by the nearby pyridinium group of NAD$^+$ can occur. The process of hydride ion abstraction at this point presumably constitutes the primary origin of the activation barrier. It is not known whether the species drawn, beginning with proton abstraction, correspond to stable intermediates, i.e., troughs in the reaction coordinate scheme, or to intrinsically unstable species. Reduction of the aldehyde to alcohol can proceed directly through the same species by the reverse reaction. (Reproduced from Ref. 55.)

4.5. Power Saturation for Estimating Distances Between Two Interacting Spins

Measurement of the spin-lattice relaxation rate of a spin species influenced through the dipolar interaction of a different nearby relaxing spin can provide an estimate of the distance of closest approach between the two interacting spins. The basis of estimating this interaction distance is seen in the dipolar Hamiltonian originally derived to calculate the influence of paramagnetic metals in solids on relaxation of nearby nuclei [129,130]. For weak interactions the equation describing the spin-lattice relaxation rate for dipolar induced relaxation is:

$$\frac{1}{T_{1s}} = \frac{1}{T_{1s}^0} + J(J+1) \left[\frac{b^2 T_{2f}}{1 + (\omega_f - \omega_s)^2 T_{2f}^2} + \frac{c^2 T_{1f}}{1 + \omega_s^2 T_{1f}^2} \right.$$
$$\left. + \frac{e^2 T_{2f}}{1 + (\omega_f + \omega_s)^2 T_{2f}^2} \right] \quad (15)$$

where

$$b^2 = \frac{1}{6}\gamma_f^2 \gamma_s^2 \hbar^2 r^{-6} (1 - 3\cos^2\theta)^2$$

$$c^2 = \frac{3}{4}\gamma_f^2 \gamma_s^2 \hbar^2 r^{-6} \sin^2 2\theta$$

$$e^2 = \frac{3}{2}\gamma_f^2 \gamma_s^2 \hbar^2 r^{-6} \sin^4\theta \quad (16)$$

In these equations $\hbar$ is Planck's constant divided by 2π, s and f refer to the slow-relaxing and fast-relaxing spins, r is the distance of closest approach [131], θ is the angle between the intersite radial vector and H_0, ω is the precession frequency in the applied field, and T_{1s}^0 is the spin-lattice relaxation time in the absence of the fast-relaxing paramagnetic perturbant. A corresponding expression can be similarly derived for $1/T_{2s}$.

For inhomogeneously broadened lines under conditions of high microwave fields, Hyde [54] showed that the susceptibility is governed only by T_1 whereby

$$\chi''(\omega)H_1 = \frac{\pi}{2} \frac{\chi_0 \omega h(\omega - \omega_0)H_1}{(1 + \gamma^2 H_1^2 T_1^2)^{1/2}} \qquad (17)$$

because of a breakdown in the Bloch equations at high microwave powers. A similar conclusion was reached earlier by Redfield [132] for resonance saturation of nuclei in solids. On this basis, Hyde and Rao [133] derived essentially the same relationships as in Eqs. (15) and (16) but with T_{2f} replaced by T_{1f}. This treatment by Hyde and coworkers [54,133-136] is critical to estimating the distance of closest approach of dipolar interacting spins in pulse saturation and recovery and cw saturation experiments.

In application of these relationships, Hyde and coworkers showed [133-136] that the Dy^{3+} ion is particularly effective in inducing spin-lattice relaxation of organic free radicals because its spin-lattice relaxation time is very nearly equal to ω^{-1} where ω is the x-band microwave frequency. The resultant effect is that the c component of the dipolar Hamiltonian in Eq. (15) dominates and Eq. (17) then is given by

$$\chi''(\omega)H_1 = \frac{\pi}{2}\chi_0 \omega \gamma_s^{-1} \hbar (\omega - \omega_0) \left[\frac{4T_{1f}}{1 + \omega^2 T_{1f}^2} \right]^2 <\Delta\omega>_c^2 \qquad (18)$$

where

$$<\Delta\omega>_c^2 = \frac{3}{4}\gamma_s^2 \gamma_f^2 \hbar^2 J(J+1)N(13.07)a^{-3} \qquad (19)$$

In Eq. (19) a is the distance of closest approach and N is the concentration of the metal ions where $f = Na^3$, f being the fractional population of lattice sites occupied by the metal ions with (cubic) lattice parameter a. Hyde and coworkers [134-137] employed these relationships with Dy^{3+} as the paramagnetic perturbant to investigate cellular compartmentalization of intrinsic free radicals in biological tissue. Ohnishi and coworkers [138] applied these methods with Dy^{3+} to estimate $<a^6>^{1/6}$ for a Dy^{3+}-EDTA complex at the surface of the protein relative to the intrinsic paramagnetic iron center of a number of iron-sulfur and heme proteins.

Similarly, Kulikov and Likhtenstein [139] showed that the cw power saturation properties of nitroxide spin labels interacting with a rapidly relaxing paramagnetic ion in frozen solutions near 77 K may be employed to estimate the distance between the spin centers. In this study they employed Ni^{2+}, Mn^{2+}, and $[Fe(CN)_6]^{3-}$ as the paramagnetic metal ion perturbants. Although they did not extract directly the distance of closest possible approach of the two-spin systems for a series of spin-labeled proteins with added paramagnetic metal ions, they showed that the parameter $P_{1/2}$ was sensitive to the concentration of the metal ion and that it was possible to discern whether the spin label was buried within a protein or was attached to a surface residue. A more recent review of their studies [140] applies Eq. (15) with its corresponding counterpart for $1/T_{2s}$ for nitroxide spin-labeled proteins in frozen solution. This treatment ignores the discussion by Hyde [54] as summarized above and evaluates distances on the basis of relaxation times extracted from cw saturation curves. As discussed earlier (Sec. 3.1.1), such estimates of absolute relaxation times are of dubious accuracy.

An interesting application of the relationships described through Eqs. (15)-(17) is the estimate of the distance between the heme a and heme a_3 groups of mammalian cytochrome oxidase [141,142]. Under the conditions of the experiment, the cytochrome oxidase enzyme was poised by redox titration to 0.357 and 0.100 V with $K_3Fe(CN)_6$, so that the iron cation of the heme a group was in the Fe^{3+} state while that in the heme a_3 group was in the Fe^{2+} state and liganded to an NO molecule. In this state, the reduced NO-liganded heme complex is in an S = 1/2 state. The spin-lattice relaxation rate of the NO-liganded (Fe^{2+}) heme a_3 complex in the enzyme was found to be enhanced threefold by the ferric heme a complex. On the basis of limits placed on the ratios of the spin-lattice relaxation times for the fast- (heme a) and slow- (heme a_3) relaxing components and the principal magnetic directions of the heme a and heme a_3 groups established in oriented membrane preparations

[143,144], it was concluded that the b term of Eq. (15) is much larger than the other two, leading to the relationship:

$$r^6 = \frac{\gamma_f^2 \gamma_s^2 \hbar^2}{8\Delta\omega^2} \cdot \frac{T_{1s}}{T_{1f}} \cdot (1 - 3\cos^2\theta)^2 \qquad (20)$$

Since the value of $(1 - 3\cos^2\theta)^2$ ranges from 0 to 4, an upper limit of 16 Å could be calculated for measured ratios of T_{1s}/T_{1f} based on $P_{1/2}$ values. Since resonance x-ray diffraction studies of oriented membrane preparations of cytochrome oxidase show that the z coordinates of the iron cations (normal to the membrane surface) in the two types of heme groups differ by 8 or 38 Å [145], the limiting distance between the two heme iron cations could be further restricted to 12-16 Å. Although polarized absorption and EPR studies of oriented multilayer preparations of cytochrome oxidase have shown that both hemes are normal to the membrane surface [143,144], the relative orientation of both heme groups with respect to the a-a_3 vector is not established. Nonetheless, the combined results of x-ray, oriented membrane EPR, and cw power saturation studies have placed constraints on the geometric relationships between the two heme groups and have led to the resultant picture illustrated in Figure 20.

Chan and coworkers [146] similarly estimated distances of closest approach of the metal ion centers in mammalian cytochrome oxidase (cf. also Sec. 4.3) by cw saturation methods but have applied Eq. (15) and its counterpart for $1/T_{2s}$ assuming that the susceptibility is governed by both T_1 and T_2, as originally derived in the model by Castner [29]. Although they conclude that the a-a_3 distance is ~20 Å, their analysis ignores the treatment by Hyde and coworkers [54,133] for derivation of the dipolar Hamiltonian under conditions $T_1 > T_2 \gg 1/(\gamma H_1)$ that apply to inhomogeneously broadened lines in high microwave fields, showing that the susceptibility is governed only by T_1. Similarly, a more recent analysis of cw progressive saturation data of cytochrome oxidase by Goodman and Leigh [147] to estimate the heme a-Cu_A distance ignores the consequences of Hyde's theory.

APPLICATION OF EPR SATURATION METHODS

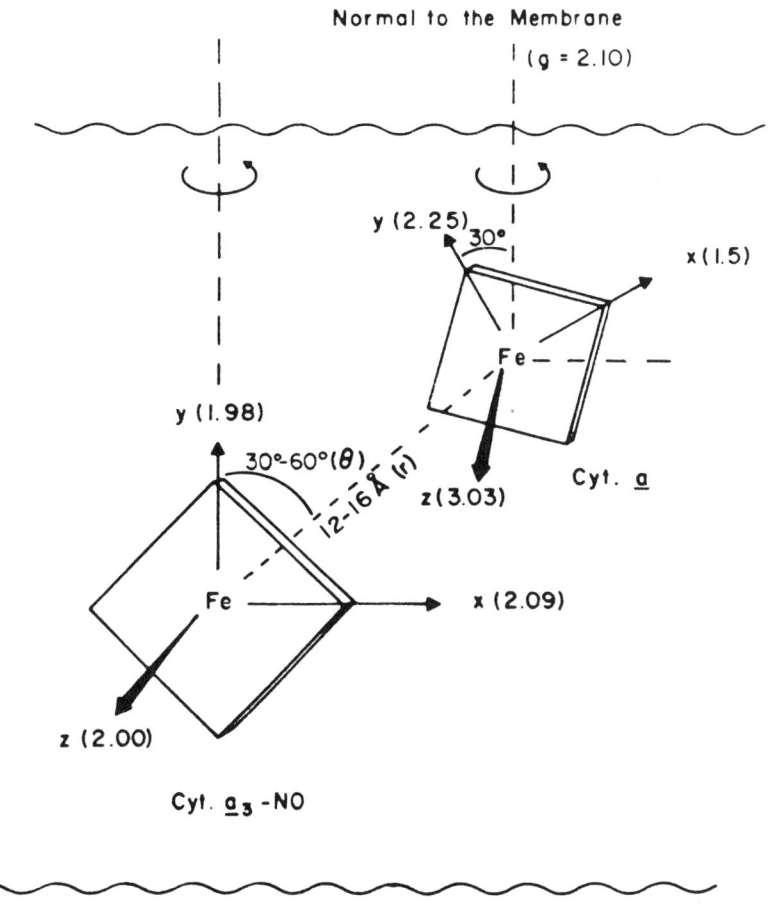

FIG. 20. Schematic illustration of the spatial relationships between the heme a and a_3 groups in mammalian cytochrome oxidase. The numbers following the labels on the axes of the two heme groups refer to the g values of the EPR spectrum. (Reproduced from Ref. 141 by permission from the authors.)

4.6. Fractal Dimension and Protein Structure from Raman Spin-Lattice Relaxation Rates

4.6.1. Overview

As shown through Eq. (2), spin-lattice relaxation by the Raman process is expected to show a T^9 dependence for Kramers ions. This circumstance occurs because of the manner in which the lattice phonons contribute to the transition probability for relaxation. However, measurements of spin-lattice relaxation rates of iron proteins for Raman relaxation have not clearly shown a T^9 process. Early studies of Mailer and Taylor [148] as reinterpreted by Stapleton and coworkers [58] yielded an apparent T^6 dependence of the Raman spin-lattice relaxation rate in single crystals of ferricytochrome c. Subsequently, Herrick and Stapleton [104] reported a temperature dependence of Raman relaxation much slower than T^9 for the low-spin form of Fe^{3+} cytochrome P-450 from *P. putida*. They fitted their data to a T^7 function, suggesting that the seventh-power dependence might be caused by the planar structure of the heme group. Stapleton and coworkers [149] later reported that the temperature dependence of the Raman process for the low-spin Fe^{3+} cation in metmyoglobin azide, ferricytochrome c, as well as in ferricytochrome P-450 from *P. putida* could be fitted by a function of the type CT^n where n equaled 6.3, quite different from the T^9 law predicted by theory applied to rare earth and transition metal salts. These results are illustrated in Figure 21.

Two interpretations have been provided to account for this unusual temperature dependence of the Raman process. One applied by Gayda and coworkers for two ferredoxins [90,93] maintains that a regular T^9 dependence is indeed present but is also accompanied by direct and Orbach processes. According to this model, the direct process is subject to an efficient phonon bottleneck for T < 4 K (cf. [1,3,5,150] for discussion of the phonon bottleneck), a relatively low Debye temperature is invoked for the Raman process, and the Orbach process involves excited energy levels 120 cm^{-1} in one ferredoxin and 243 cm^{-1} in the other ferredoxin above the ground

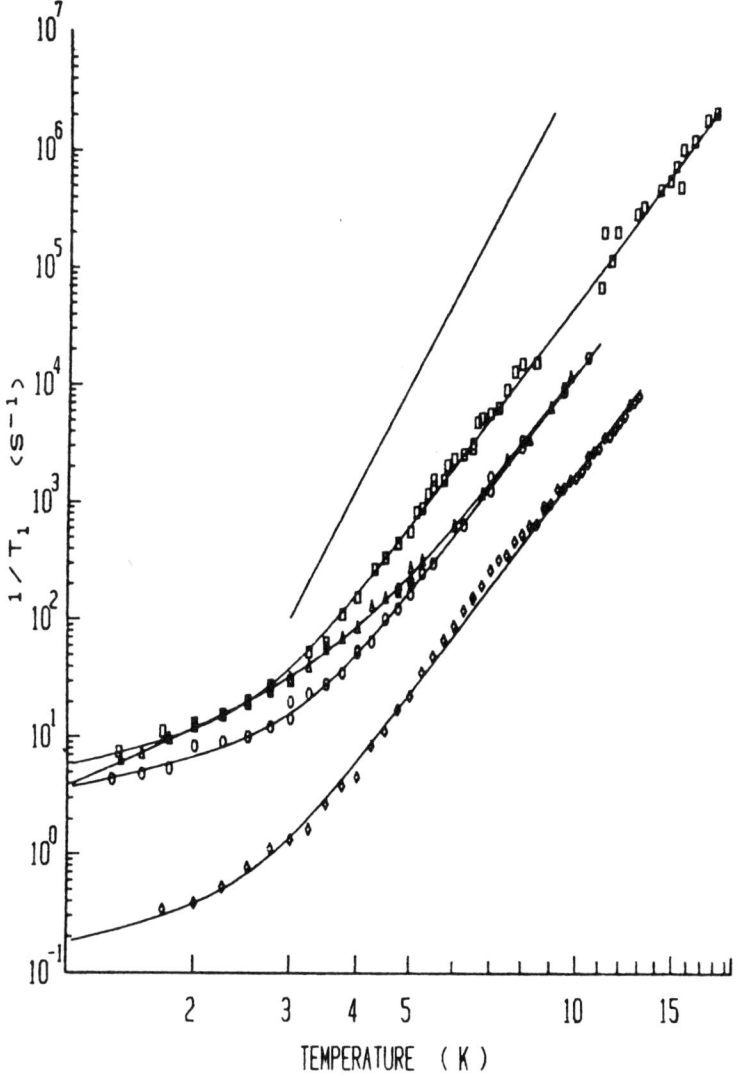

FIG. 21. Spin-lattice relaxation rates of four low-spin Fe^{3+} heme proteins. The data from ferricytochrome c ($\square$, $1/T_1 = 4.85T + 0.0221T^{6.34}$), metmyoglobin hydroxide ($\triangle$, $1/T_1 = 2.79T^2 + 0.007267T^{6.22}$), metmyoglobin azide ($\circ$, $1/T_1 = 3.11T + 0.00617T^{6.29}$), and ferricytochrome P-450 ($\diamond$, $1/T_1 = 0.152T + 0.000910T^{6.27}$) are fitted to the sum of a direct process and a T^n power law. For comparison, the straight line shows a T^9 power law as classically expected for a Kramers ion. The noninteger power terms describe Raman relaxation of objects with spectral dimension d_s, where $n = 2d_s + 3$. (Reproduced from Ref. 58 by permission from the Biophysical Society.)

doublet. As pointed out by Stapleton [150], it is improbable that such low Debye temperatures and large excitation levels obtain to apply this model to all of the observed instances of CT^n processes in the various metalloproteins studied hitherto. The other explanation, developed by Stapleton and coworkers [58,60,150], utilizes noninteger exponent values for the temperature power law of Raman relaxation by postulating the property of fractal dimensionality for proteins. This model has led to novel applications of paramagnetic relaxation data for examining protein structure and dynamics. The rest of this section shows how fractal dimension enters the theory of the Raman process and outlines the implications of fractal dimensionality for the paramagnetic relaxation behavior of iron proteins in particular and for the three-dimensional structure of proteins in general.

4.6.2. Fractal Dimension and the Temperature Dependence of Raman Spin-Lattice Relaxation

The second-order Raman process has a transition probability $\underline{w}$ given by:

$$\underline{w} = \frac{2\pi}{\hbar^2} \left| \frac{\varepsilon_1 V_1^{(1)} \varepsilon_2 V_2^{(1)}}{\Delta'} \right|^2 f(\omega) \tag{21}$$

where $V_1^{(1)}$ and $V_2^{(1)}$ are the matrix elements of $V^{(1)}$ between states $|\underline{a}\rangle$ and $|\underline{b}\rangle$ and the excited states $|\underline{c}\rangle$ and $|\underline{d}\rangle$ (cf. Fig. 1 and Sec. 2.1). $V^{(1)}$ is the coefficient of the linear term of the power series expansion of the orbit-lattice interaction in terms of the lattice strain ε, with ε_1 associated with the excitation from $|\underline{a}\rangle$ to $|\underline{c}\rangle$ and ε_2 associated with the relaxation from $|\underline{c}\rangle$ to $|\underline{b}\rangle$. The factor $f(\omega)$ is a shape function, and Δ' is the energy of $|\underline{c}\rangle$ above the ground state. For a Kramers ion, the excited state $|\underline{c}\rangle$ is accompanied by its time conjugate state $|\underline{d}\rangle$. The matrix elements $(\varepsilon_1 V_1^{(1)} \varepsilon_2 V_2^{(1)}/\Delta')$ in Eq. (21) must be summed over states $|\underline{c}\rangle$ and $|\underline{d}\rangle$, and the net result is a near-cancellation of terms resulting in an expression for the transition probability by integrating over all phonon frequencies ω_ℓ to give:

APPLICATION OF EPR SATURATION METHODS 193

$$\frac{1}{T_{1R}} = 2w \propto \iint f(\omega)\varepsilon_1^2\varepsilon_2^2\omega_\ell^2 \exp\left(\frac{\hbar\omega_2}{kT}\right) d\omega_1 d\omega_2 \qquad (22)$$

where ε_1^2 and ε_2^2 are functions of the density of phonon states ρ_{ph1} and ρ_{ph2}. In the limit Eq. (22) can be rewritten as:

$$\frac{1}{T_{1R}} \propto \int_0^{\omega_{max}} \frac{\rho_{ph}^2 \omega_\ell^4 \exp(\hbar\omega_\ell/kT)}{[\exp(\hbar\omega_\ell/kT) - 1]^2} d\omega_\ell \int_0^\infty f(\omega) d\omega \qquad (23)$$

wherein the second integral equals one and is consequently of no further interest. For one, two, and three dimensions in Euclidean space (cf., [151]), ρ_{ph} is proportional to $\omega^{d_s - 1}$ where d_s is the spectral dimension. With the substitutions $x = (\hbar\omega_\ell/kT)$ and $\theta = (\hbar\omega_{max})/k$, the integral in Eq. (23) can be rewritten for three dimensions as:

$$\frac{1}{T_{1R}} \propto \left(\frac{kT}{\hbar}\right)^9 \int_0^{\theta/T} \frac{x^8 \exp(x) dx}{[\exp(x) - 1]^2} = \left(\frac{kT}{\hbar}\right)^9 J_8\left(\frac{\theta}{T}\right) \qquad (24)$$

Here θ represents the Debye temperature, $J_8(\theta/T)$ is a transport integral (cf. [58]) and is approximately constant for $T \ll \theta$. For $T \ll \theta$, the resulting relationship in Eq. (25) obtains:

$$\frac{1}{T_{1R}} \propto T^{2d_s+3} \qquad (25)$$

The key to the introduction of fractal dimensionality into Raman relaxation is found in the dependence of the phonon density of states on a power d_s of the phonon frequency. In Euclidean space, d_s can take the integer values 1, 2, or 3. With the possibility of noninteger values, d_s more broadly represents the spectral dimension, the value of which is determined by fitting spin-lattice relaxation data to a function containing terms which represent direct, Orbach (if present), and Raman processes. The relationship between n of the T^n Raman process and d_s then becomes $n = 2d_s + 3$. An example of this fitting procedure for low-spin ferric heme protein complexes is provided in Figure 21. To relate spectral dimensions to structural properties of proteins, however, requires a discussion of the concept of fractal dimensionality and how it applies to proteins.

4.6.3. Fractal Dimensions of Proteins

The concept of noninteger dimensionality introduced by Mandelbrot [152,153] may be approached through the idea of a similarity ratio. Consider a standard object, a line segment of unit length. Suppose a segment is divided into b equal parts. Each small segment can be deduced from the whole by a similarity of ratio r such that $r = 1/b = 1/N$. Now consider a square with sides of unit length. If each side is divided into b equal parts, then the original square will contain $b^2 = N$ small squares. Now each small square is deduced from the original square by a similarity of ratio $r = 1/b = 1/N^{1/2}$. This description generalized for all d-dimensional parallelepipeds leads to the relationship $Nr^d = 1$ where N is the number of parts obtained from applying the similarity ratio r to each component of the original structure. In other words, each component is subjected to a scaling operation so that the N parts which are produced for the whole also are obtainable from the whole with a similarity of ratio r. At the center of fractal dimension is the concept of scale invariance: division into smaller and smaller parts does not change the appearance of the parts. Regardless of the resolution with which an object is examined, its appearance remains similar.

This type of geometric self-similarity is not found in nature, so a variation of the concept must be used. The theory of fractal dimensionality applies to structurally irregular objects like the backbones of proteins if they demonstrate a statistical property that exhibits self-similarity. The self-avoiding property of proteins can be used as such a statistical property. Colvin and Stapleton [59] describe two methods for calculating the fractal dimensionality of a protein from its crystallographic coordinates. The first method determines d_c, the "contour length" dimension, by comparing the length along the backbone with the end-to-end length. The second method

APPLICATION OF EPR SATURATION METHODS   195

ignores the path of the polypeptide backbone and scales the total
mass of the protein with respect to the distance in the embedding
space, thus calculating d_r, the "reentrant" dimension. The contour
length method assumes that the protein is a chain of topological
dimension one. The reentrant method views the protein as a collection of points of topological dimension zero, with the scaling
exponent related to the mass within a sphere with respect to the
radius of that sphere.

In principle, a fractal object of fractal dimension d_f possesses
a well-defined density of phonon states [154-156]. The spectral dimension d_s is a readily defined number, being a parameter relatively
easily obtained from fitting the temperature-dependent behavior of
spin-lattice relaxation to a power law function of the temperature.
However, the relationship between d_f and d_s for proteins is not always
clear. First, d_s depends on the range of temperature over which the
dimensionality is calculated and the coupling of the protein with its
solvent environment; second, d_f does not possess a fixed value for
globular proteins but rather depends on the method of calculation
[59]. Indeed, attempts to reconcile values of d_f and d_s may contribute substantially to an understanding of intramolecular phonons in
proteins and the coupling of protein vibrations to the surrounding
medium. These aspects of d_f and d_s are discussed below with respect
to the results of spin-lattice relaxation studies of iron proteins.

4.6.4. Fractal and Spectral Dimensions and Structures of Iron Proteins

Low-spin Fe^{3+} hemeproteins and iron-sulfur proteins possess paramagnetic characteristics that make them ideally suited for investigating
relationships between fractal dimensionality and protein structure.
First, Raman spin-lattice relaxation times of the Fe^{3+} ion can be
measured accurately by the pulse saturation method at temperatures
between 6 and 20 K. Second, wavelengths of acoustic phonons involved
in the Raman process are long enough to involve large sections of the
protein (localized vibrations away from the iron do not affect its
Raman relaxation) and short enough to exclude vibrations of the

solvent. Third, rate contributions from an Orbach process are absent because of the isolated S = 1/2 ground state. Analysis of the Raman process, therefore, is straightforward since it must be accompanied only by analysis of the direct process. The Orbach process itself has been used to examine the phonon density in high-spin aquometmyoglobin, but the Orbach process interferes with the measurements of the Raman process [61].

In Figure 22 are illustrated the fit to the data for spin-lattice relaxation of the reduced [2Fe-2S] ferredoxin of S. maxima by modeling a fractal Raman process. This is compared to the fit reported by Bertrand et al. [90] for a conventional Raman mechanism and an Orbach process with $\Delta' \sim 243$ cm^{-1}. It is seen that the fractal Raman model finds better agreement with the experimental observations. Further examples of interpretations based on fractal Raman processes are found in Figure 21 and in the studies of Stapleton and coworkers [58-60,104,108,149,150]. It is of interest in these studies that the spin-lattice relaxation of the paramagnetic Fe^{3+} ion has provided a probe of vibrations of the protein structure. The spin-lattice relaxation of the (S = 1/2) NO-bound Fe^{2+} heme group in nitrosylmyoglobin, in which the unpaired electron is localized primarily to the antibonding π^* orbital of the NO group with the symmetry of the $Fe(d_z2)$ orbital [157], has not proven to be a probe of spectral dimension [158]. It is not clear whether systems in which two paramagnetic centers interact through dipolar coupling, as, for instance, the Cu_A and heme a sites of cytochrome oxidase (cf. Sec. 4.3), also can serve as probes of fractal dimensionality.

The investigations of the spectral dimension of metalloproteins on the basis of the temperature dependence of the Raman spin-lattice relaxation have provided some new approaches to study vibrational modes in the protein. Measurements of Raman relaxation processes to study the effect of solvent [59] and ionic strength [60] have shown that the coupling between solvent and protein may be probed by Raman spin-lattice relaxation studies. In a study of low-spin metmyoglobin azide in 11 solvent conditions, a model based on solvent phonons with

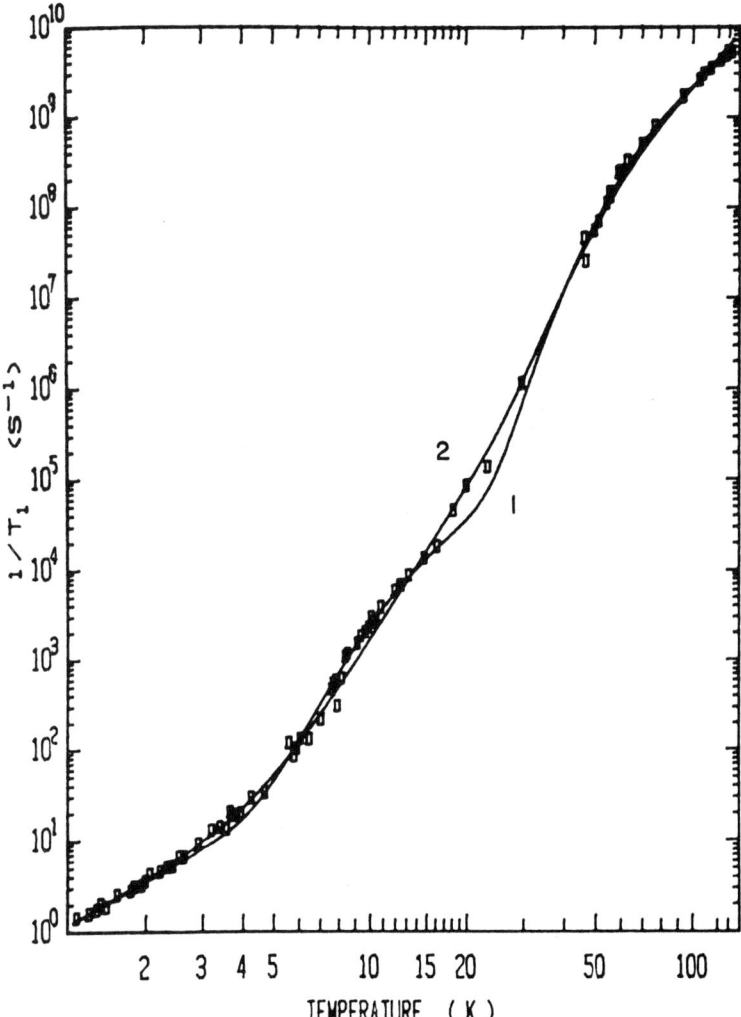

FIG. 22. Spin-lattice relaxation data from the reduced [2Fe-2S] ferredoxin of *S. maxima*. Data were obtained from pulse saturation-recovery, continuous-wave, and relaxation broadening methods (see legend of Fig. 13). Curve 1 is the original fit of the investigators, which included bottlenecked direct, Orbach, and conventional Raman processes [90]. Curve 2 is from the equation $1/T_1 = 0.874\ T^2 + 4.40 \times 10^{10} \exp(-348/T) + 0.00341\ T^{5.67}$, which substitutes a fractal Raman process (noninteger power dependence on temperature) for the conventional Raman process and does not invoke a low Debye temperature. (Reproduced from Ref. 160 by permission of the author.)

a spectral dimension of 3 and protein phonons of spectral dimension 1.61 was derived to explain the sharp rise in the effective temperature exponent that fits the data below 6 K [59]. Notably, at 6 K acoustic phonons of solids have a wavelength that is approximately the dimension of the myoglobin molecule, and strong solvent-protein coupling was cited as an explanation of small spectral dimensions. Also, certain aspects of protein secondary structure have been evaluated. Wagner et al. [60] demonstrated a correlation between fractal dimension and the fraction of tetrapeptide sequences of proteins in α-helical or β-sheet conformations. They conclude that the fractal character of proteins is determined primarily by relative contributions of compact and extended secondary structures. In addition, Colvin and Stapleton [59] showed that the calculated fractal dimension of bovine pancreatic trypsin inhibitor predicts a density of low-frequency states that is qualitatively confirmed by a normal mode calculation [159]. Thus, it may prove feasible to compare fractal dimensions with results from protein molecular dynamics calculations. Thus far investigations of Raman relaxation processes in heme and iron-sulfur proteins have shown that the fractal dimensionality of proteins is a rich source of information about structure and dynamics. Since relationships between fractal and spectral dimensions of proteins are still incompletely defined, this area of research will continue to bring new understanding to intramolecular interactions involving regions of different secondary structure and of structural domains.

ACKNOWLEDGMENTS

G. B. Wells is a predoctoral student supported by an MSTP grant of the National Institutes of Health (GM 07281). This work was supported by grants of the National Institutes of Health (GM 21900 and AA 06374).

ABBREVIATIONS

cw   continuous wave
EPR   electron paramagnetic resonance
NAD^+ oxidized nicotinamide adenine dinucleotide
NADH   reduced nicotinamide adenine dinucleotide

REFERENCES

1. A. Abragam and B. Bleaney, *Electron Paramagnetic Resonance of Transition Ions,* Clarendon Press, Oxford, 1970.
2. J. W. Orton, *Electron Paramagnetic Resonance,* Gordon and Breach, New York, 1969.
3. J. Standley and R. A. Vaughan, *Electron Spin Relaxation Phenomena in Solids,* Adam Hilger, Ltd., London, 1969.
4. C. P. Poole and H. A. Farach, *Relaxation in Magnetic Resonance,* Academic Press, New York, 1971.
5. R. Orbach and H. J. Stapleton, in *Electron Paramagnetic Resonance* (S. Geschwind, ed.), Plenum Press, New York, 1972, pp. 121-126.
6. K. W. H. Stevens, *Rep. Prog. Phys., 30,* 189 (1975).
7. J. H. van Vleck, *J. Chem. Phys., 7,* 72 (1939).
8. J. H. van Vleck, *Phys. Rev., 57,* 426 (1940).
9. R. D. Mattuck and M. W. Strandberg, *Phys. Rev., 199,* 1204 (1961).
10. R. Orbach, *Proc. Roy. Soc. (London) Ser. A, 264,* 458 (1961).
11. R. Orbach, *Proc. Phys. Soc. (London), 77,* 821 (1961).
12. M. K. Bowman and L. Kevan, in *Time Domain Electron Spin Resonance* (L. Kevan and R. N. Schwartz, eds.), Wiley, New York, 1979, pp. 67-105.
13. C. F. Weissfloch, *Can. J. Phys., 44,* 3185 (1966).
14. W. P. Unruh and J. W. Culvahouse, *Phys. Rev., 129,* 2441 (1963).
15. J. C. Gill, *Proc. Phys. Soc. (London), 85,* 119 (1965).
16. G. H. Larson and C. D. Jeffries, *Phys. Rev., 141,* 461 (1966).
17. C. B. P. Finn, R. Orbach, and W. P. Wolf, *Proc. Phys. Soc. (London), 77,* 261 (1961).

18. A. A. Manenkov and A. M. Prokhorov, Soviet Physics-JETP (Engl. Trans.), 15, 951 (1962).
19. M. W. Makinen and M. B. Yim, Proc. Natl. Acad. Sci. USA, 78, 6621 (1981).
20. M. W. Makinen, L. C. Kuo, M. B. Yim, W. Maret, and G. B. Wells, J. Mol. Catalysis, 23, 179 (1984).
21. M. W. Makinen, L. C. Kuo, M. B. Yim, G. B. Wells, J. M. Fukuyama, and J. E. Kim, J. Am. Chem. Soc., 107, 5245 (1985).
22. M. B. Yim, L. C. Kuo, and M. W. Makinen, J. Magn. Reson., 46, 247 (1982).
23. E. Wajnberg, H. Kalinowski, G. Bemski, and J. S. Helman, CBPF Notas de Fisica (Rio de Janeiro), No. 043 (1984).
24. C. P. Scholes, R. Isaacson, and G. Feher, Biochim. Biophys. Acta, 244, 206 (1971).
25. A. M. Portis, Phys. Rev., 91, 1071 (1953).
26. A. M. Portis, Phys. Rev., 104, 584 (1956).
27. N. Bloembergen, S. Shapiro, P. S. Pershan, and J. O. Artman, Phys. Rev., 114, 445 (1959).
28. N. Bloembergen, E. M. Purcell, and R. V. Pound, Phys. Rev., 73, 679 (1948).
29. T. G. Castner, Phys. Rev., 115, 1506 (1959).
30. S. Clough and C. A. Scott, J. Phys., C1, 919 (1968).
31. B. N. Provotorov, Sov. Phys.-Solid State (Engl. Trans.), 5, 411 (1963).
32. J. P. Korb and J. Maruani, J. Magn. Reson., 41, 247 (1980).
33. J. P. Korb and J. Maruani, Phys. Rev., B23, 971 (1981).
34. J. P. Korb and J. Maruani, J. Magn. Reson., 37, 331 (1980).
35. J. P. Korb and J. Maruani, Phys. Rev., B23, 5700 (1981).
36. G. Feher and E. A. Gere, Phys. Rev., 114, 1245 (1959).
37. P. I. Bekauri, B. G. Berulava, T. I. Sanadze, and O. G. Khakhanashvili, Sov. Phys.-JETP (Engl. Trans.), 25, 292 (1967).
38. R. Boscaino, M. Brai, and I. Ciccarello, Phys. Rev., B13, 2798 (1976).
39. A. Kiel, Phys. Rev., 125, 1451 (1962).
40. H. Yoshida, D. F. Feng, and L. Kevan, J. Chem. Phys., 58, 3411 (1973).
41. L. Kevan and P. A. Narayana, in Multiple Electron Resonance Spectroscopy (M. M. Dario and J. H. Freed, eds.), Plenum, New York, 1979, pp. 229-259.

42. L. C. Kuo and M. W. Makinen, *J. Biol. Chem.*, *257*, 24 (1982).
43. M. B. Yim, G. B. Wells, L. C. Kuo, and M. W. Makinen, in *Bioinorganic Chemistry* (A. V. Xavier, ed.), VCH Verlagsgesellschaft, Weinheim, FRG, 1985, pp. 562-570.
44. G. B. Wells, M. B. Yim, and M. W. Makinen, manuscript in preparation.
45. E. E. Schneider and T. S. England, *Physica*, *17*, 221 (1951).
46. A. H. Eschenfelder and R. T. Weidner, *Phys. Rev.*, *92*, 869 (1953).
47. G. Feher and H. E. D. Scovil, *Phys. Rev.*, *105*, 760 (1957).
48. C. Mailer, T. Sarna, H. M. Swartz, and J. S. Hyde, *J. Magn. Reson.*, *25*, 205 (1977).
49. H. Blum and T. Ohnishi, *Biochim. Biophys. Acta*, *621*, 9 (1980).
50. O. P. Zhidkov, V. I. Muromtsev, I. G. Akhvldiani, S. N. Safronov, and V. V. Kopylov, *Soviet Physics-Solid State (Engl. Trans.)*, *9*, 1095 (1967).
51. M. K. Bowman, H. Hase, and L. Kevan, *J. Magn. Reson.*, *22*, 23 (1976).
52. J. P. Lloyd and G. E. Pake, *Phys. Rev.*, *94*, 579 (1954).
53. J. Maruani, *J. Magn. Reson.*, *7*, 207 (1972).
54. J. S. Hyde, *Phys. Rev.*, *119*, 1492 (1960).
55. M. W. Makinen, W. Maret, and M. B. Yim, *Proc. Natl. Acad. Sci. USA*, *80*, 2584 (1983).
56. L. C. Kuo and M. W. Makinen, *J. Am. Chem. Soc.*, *107*, 5255 (1985).
57. R. Calvo and G. Bemski, *J. Chem. Phys.*, *64*, 2264 (1976).
58. J. P. Allen, J. T. Colvin, D. G. Stinson, C. P. Flynn, and H. J. Stapleton, *Biophys. J.*, *38*, 299 (1982).
59. J. T. Colvin and H. J. Stapleton, *J. Chem. Phys.*, *82*, 4699 (1985).
60. G. C. Wagner, J. T. Colvin, J. P. Allen, and H. J. Stapleton, *J. Am. Chem. Soc.*, *107*, 5589 (1985).
61. A. S. Brill, F. G. Fiamingo, D. A. Hampton, P. D. Levin, and R. Thorkildsen, *Phys. Rev. Lett.*, *54*, 1864 (1985).
62. F. G. Fiamingo, Ph.D. thesis, University of Virginia, Charlottesville, 1981.
63. R. Thorkildsen, Ph.D. thesis, University of Virginia, Charlottesville, 1980.
64. S. J. Campbell, I. R. Herbert, C. B. Warwick, and J. H. Woodgate, *Rev. Sci. Instrum.*, *47*, 1172 (1976).

65. S. Slappendel, G. A. Veldink, J. F. G. Vliengenthart, R. Aasa, and B. G. Malmström, *Biochim. Biophys. Acta, 642,* 30 (1980).

66. A. Tasaki, J. Otsuka, and M. Kotani, *Biochim. Biophys. Acta, 140,* 284 (1967).

67. J. S. Hyde, in *Time Domain Electron Spin Resonance* (L. Kevan and R. N. Schwartz, eds.), Wiley, New York, 1979, pp. 1-30.

68. M. Huisjen and J. S. Hyde, *Rev. Sci. Instrum., 45,* 669 (1974).

69. P. W. Percival and J. S. Hyde, *Rev. Sci. Instrum., 46,* 1522 (1975).

70. R. Isaacson, *J. Sci. Instr., 1,* 1137 (1968).

71. T. L. Bohan and H. J. Stapleton, *Rev. Sci. Instrum., 39,* 1707 (1968).

72. D. M. Marquardt, *J. Soc. Indust. Appl. Math., 11,* 431 (1963); also cf. Y. Bard, *Nonlinear Parameter Estimation,* Academic Press, New York, 1974, pp. 94-96.

73. K. J. Standley and R. A. Vaughan, *Phys. Rev., 139,* 1275 (1965).

74. R. L. Kyhl and B. D. Nageswara-Rao, *Phys. Rev., 158,* 284 (1967).

75. F. Bloch, *Phys. Rev., 102,* 104 (1956).

76. H. Shimizu, *J. Chem. Phys., 42,* 3603 (1965).

77. S. Schlick and L. Kevan, *J. Magn. Reson., 22,* 171 (1976).

78. M. Bowman, L. Kevan, and R. N. Schwartz, *Chem. Phys. Lett., 30,* 208 (1975).

79. W. Snipes and W. Bernhard, *J. Chem. Phys., 43,* 2921 (1965).

80. W. Kohnlein and J. H. Venable, Jr., *Nature, 215,* 618 (1967).

81. W. Maret, I. Andersson, H. Dietrich, H. Schneider-Bernlöhr, R. Einarsson, and M. Zeppezauer, *Eur. J. Biochem., 98,* 501 (1979).

82. W. B. Mims and J. Peisach, *J. Chem. Phys., 69,* 4921 (1978).

83. S. A. Dikanov, A. A. Shubin, and V. N. Parmon, *J. Magn. Reson., 42,* 474 (1981).

84. R. Aasa, R. Brändén, J. Deinum, B. G. Malmström, B. Reinhammar, and T. Vanngård, *Biochem. Biophys. Res. Commun., 70,* 1204 (1976).

85. R. Aasa, R. Brändén, J. Deinum, B. G. Malmström, B. Reinhammar, and T. Vänngård, *FEBS Lett., 61,* 115 (1976).

86. H. Beinert and W. H. Orme-Johnson, in *Magnetic Resonance in Biological Systems* (A. Ehrenberg, B. G. Malmström, and T. Vänngård, eds.), Pergamon, Oxford, 1967, pp. 221-247.

87. M. Sahlin, A. Gräslund, and A. Ehrenberg, *J. Magn. Reson., 67,* 135 (1986).

88. H. Rupp, K. K. Rao, D. O. Hall, and R. Cammack, *Biochim. Biophys. Acta, 537,* 255 (1978).

89. H. Rupp, A. de la Torre, and D. O. Hall, Biochim. Biophys. Acta, 548, 522 (1979).
90. P. Bertrand, J. P. Gayda, and K. K. Rao, J. Chem. Phys., 76, 4715 (1982).
91. H. Rupp and A. L. Moore, Biochim. Biophys. Acta, 548, 16 (1979).
92. V. V. Tarassov, J. Am. Chem. Soc., 80, 5052 (1958).
93. J. P. Gayda, P. Bertrand, A. Deville, C. More, G. Roger, J. F. Gibson, and R. Cammack, Biochim. Biophys. Acta, 581, 15 (1979).
94. L. Hederstedt, J. J. Maguire, A. J. Waring, and T. Ohnishi, J. Biol. Chem., 260, 5554 (1985).
95. H. Blum, J. R. Bowyer, M. A. Cusanovich, A. J. Waring, and T. Ohnishi, Biochim. Biophys. Acta, 748, 418 (1983).
96. S. I. Chan, G. W. Brudvig, C. T. Martin, and T. H. Stevens, in Electron Transport and Oxygen Utilization (C. Ho, ed.), Elsevier Biomedical, New York, 1982, pp. 171-177.
97. D. F. Blair, C. T. Martin, J. Gelles, H. Wang, G. W. Brudvig, T. H. Stevens, and S. I. Chan, Chem. Scr., 21, 43 (1983).
98. H. L. van Camp, Y. H. Wei, C. P. Scholes, and T. E. King, Biochim. Biophys. Acta, 537, 238 (1978).
99. T. H. Stevens, G. W. Brudvig, D. F. Bocian, and S. I. Chan, Proc. Natl. Acad. Sci. USA, 76, 3320 (1979).
100. G. W. Brudvig, T. H. Stevens, and S. I. Chan, Biochemistry, 19, 5275 (1980).
101. G. W. Brudvig, T. H. Stevens, R. H. Morse, and S. I. Chan, Biochemistry, 20, 3912 (1981).
102. C. P. Scholes, R. Janakiraman, H. Taylor, and T. E. King, Biophys. J., 45, 1027 (1984).
103. H. Uenoyama, T. Iizuka, H. Morimoto, and M. Kotani, Biochim. Biophys. Acta, 160, 159 (1968).
104. R. C. Herrick and H. J. Stapleton, J. Chem. Phys., 65, 4786 (1976).
105. R. Aasa, T. Vänngård, and H. B. Dunford, Biochim. Biophys. Acta, 391, 259 (1975).
106. T. Harami, Y. Maeda, Y. Morita, A. Trautwein, and U. Gonser, J. Chem. Phys., 67, 1164 (1977).
107. T. H. Moss, A. J. Bearden, R. G. Bartsch, and M. A. Cusanovich, Biochemistry, 7, 1583 (1968).
108. J. T. Colvin, R. Rutter, H. J. Stapleton, and L. P. Hager, Biophys. J., 41, 105 (1983).
109. C. E. Schulz, P. W. Devaney, H. Winkler, P. G. Debrunner, N. Doan, R. Chiang, R. Rutter, and L. P. Hager, FEBS Lett., 103, 102 (1979).

110. K. Morigaki, J. Phys. Soc. Japan, 19, 2064 (1964).
111. T. L. Estle and M. DeWit, Bull. Am. Phys. Soc., 6, 445 (1961).
112. J. N. McElearney, S. Merchant, G. E. Shankle, and R. L. Carlin, J. Chem. Phys., 66, 450 (1977).
113. G. Schneider, H. Eklund, C. Cedergren-Zeppezauer, and M. Zeppezauer, Proc. Natl. Acad. Sci. USA, 80, 5289 (1983).
114. E. Cedergren-Zeppezauer, J. P. Samama, and H. Eklund, Biochemistry, 21, 4895 (1982).
115. R. Aasa, M. Hanson, and S. Lindskog, Biochim. Biophys. Acta, 453, 211 (1976).
116. K. K. Kannan, I. Vaara, B. Notstrand, S. Lövgren, A. Borell, K. Fridborg, and M. Petef, in Drug Action at the Molecular Level (G. C. K. Roberts, ed.), University Park Press, Baltimore, 1979, pp. 73-79.
117. G. M. Zverev and N. G. Petelina, Soviet Physics-JETP (Engl. Trans.), 15, 820 (1962).
118. J. F. Johnson, R. E. Dietz, and H. J. Guggenheim, Appl. Phys. Lett., 5, 21 (1964).
119. W. Low, Phys. Rev., 109, 256 (1958).
120. M. W. Makinen, L. C. Kuo, J. J. Dymowski, and S. Jaffer, J. Biol. Chem., 254, 356 (1979).
121. K. Hardman and W. N. Lipscomb, J. Am. Chem. Soc., 106, 463 (1984).
122. D. C. Rees, M. Lewis, and W. N. Lipscomb, J. Mol. Biol., 168, 367 (1983).
123. C. I. Brändén, J. Jörnvall, H. Eklund, and B. Furugren, in The Enzymes (P. D. Boyer, ed.), Vol. 11, 3rd ed., Academic, New York, 1975, pp. 103-190.
124. J. Kvassman and G. Petterson, Eur. J. Biochem., 103, 565 (1980).
125. R. T. Dworschak and B. V. Plapp, Biochemistry, 16, 2716 (1977).
126. P. F. Cook and W. W. Cleland, Biochemistry, 20, 1805 (1981).
127. E. Bignetti, G. L. Rossi, and E. Zeppezauer, FEBS Lett., 100, 17 (1976).
128. M. W. Makinen, M. B. Yim, G. B. Wells, and W. Maret, manuscript in preparation.
129. N. Bloembergen, Physica, 15, 386 (1949).
130. A. Abragam, Phys. Rev., 98, 1729 (1955).
131. J. P. Korb and J. Maruani, J. Chem. Phys., 74, 1504 (1981).
132. A. G. Redfield, Phys. Rev., 98, 1787 (1955).

133. J. S. Hyde and R. V. S. Rao, *J. Magn. Reson.*, *29*, 509 (1978).
134. C. Mailer, T. Sarna, H. M. Swartz, and J. S. Hyde, *J. Magn. Reson.*, *25*, 205 (1977).
135. W. E. Antholine, J. S. Hyde, and H. M. Swartz, *J. Magn. Reson.*, *29*, 517 (1978).
136. J. S. Hyde and T. Sarna, *J. Chem. Phys.*, *68*, 4439 (1978).
137. T. Sarna, J. S. Hyde, and H. M. Swartz, *Science*, *192*, 1132 (1976).
138. G. D. Case, T. Ohnishi, and J. S. Leigh, Jr., *Biochem. J.*, *160*, 785 (1976).
139. A. V. Kulikov and G. I. Likhtenstein, *Biofizika*, *19*, 420 (1974).
140. A. V. Kulikov and G. I. Likhtenstein, *Adv. Mol. Relaxation and Interaction Processes*, *10*, 47 (1977).
141. T. Ohnishi, R. LoBrutto, J. C. Salerno, R. C. Bruckner, and T. G. Frey, *J. Biol. Chem.*, *257*, 14821 (1982).
142. T. Ohnishi, H. J. Harman, and A. J. Waring, *Biochem. Soc. Trans.*, *13*, 607 (1985).
143. M. Erecinska, J. K. Blasie, and D. F. Wilson, *FEBS Lett.*, *76*, 235 (1977).
144. C. Barlow and M. Erecinska, *FEBS Lett.*, *98*, 9 (1979).
145. J. K. Blasie, J. M. Pachense, A. Tavornina, P. L. Dutton, J. Stamatoff, P. Eisenberger, and G. Brown, *Biochim. Biophys. Acta*, *679*, 188 (1982).
146. G. W. Brudvig, D. F. Blair, and S. I. Chan, *J. Biol. Chem.*, *259*, 11001 (1984).
147. G. Goodman and J. S. Leigh, Jr., *Biochemistry*, *24*, 2310 (1985).
148. C. Mailer and C. P. Taylor, *Biochim. Biophys. Acta*, *322*, 195 (1973).
149. H. J. Stapleton, J. P. Allen, C. P. Flynn, D. G. Stinson, and S. R. Kurtz, *Phys. Rev. Lett.*, *45*, 1456 (1980).
150. H. J. Stapleton, *Comments on Molecular and Cellular Biophysics*, Vol. 3, 321 (1986).
151. C. Kittel, *Introduction to Solid State Physics*, 5th ed., Wiley, New York, 1976, pp. 131-135.
152. B. B. Mandelbrot, *Fractals, Form, Chance, and Dimension*, W. H. Freeman, San Francisco, 1977.
153. B. B. Mandelbrot, *The Fractal Geometry of Nature*, W. H. Freeman, New York, 1983.
154. S. Alexander and R. Orbach, *J. Phys. Lett.*, *43*, L625 (1982).
155. R. Rammal and G. Toulouse, *J. Phys. Lett.*, *44*, L13 (1983).

156. J. S. Helman, A. Coniglio, and C. Tsallis, *Phys. Rev. Lett.*, *53*, 1195 (1984).
157. H. Kon, *J. Biol. Chem.*, *243*, 4350 (1968).
158. P. J. Muench and H. J. Stapleton, *J. Chem. Phys.*, *82*, 2828 (1985).
159. N. Go, T. Noguti, and T. Nishikawa, *Proc. Natl. Acad. Sci. USA*, *80*, 3696 (1983).
160. J. P. Allen, Ph.D. thesis, University of Illinois-Urbana, 1982.

5

Electron Spin Echo: Applications to Biological Systems

Yuri D. Tsvetkov and Sergei A. Dikanov
Institute of Chemical Kinetics and Combustion
Novosibirsk 630090, USSR

1.	INTRODUCTION	207
2.	ESEEM SPECTROSCOPY	213
	2.1. ESEEM Theory	213
	2.2. Peculiarities of ESEEM Frequency Transformation	221
	2.3. ESEEM in Single Crystals	224
	2.4. ESEEM in Disordered Systems	227
3.	BIOLOGICAL APPLICATIONS OF ESEEM	240
	3.1. ^{1}H and ^{2}D Modulations	240
	3.2. ^{13}C Modulation	247
	3.3. ^{14}N and ^{15}N Modulations	248
	3.4. ^{31}P Modulation	256
4.	CONCLUSION	257
	ABBREVIATIONS	258
	REFERENCES	259

1. INTRODUCTION

Pulsed methods of ESR spectroscopy open up new possibilities for studying the structure and properties of paramagnetic particles, including ions and radicals in biological systems. As with NMR spectroscopy, echo methods prove to be extremely useful. The present chapter is devoted to applications of the ESE method to solving structural problems in biologically important systems. This method, based on an analysis of the ESE envelope modulation, supplements the continuous-wave ESR spectroscopy and the CW ENDOR in such investigations.

The ESE method is a special pulse version of ESR, which acts on a spin system in a constant magnetic field $\vec{H}_0$ by a series of either two or three microwave pulses of a magnetic field $\vec{H}_1(t) \perp \vec{H}_0$. This field oscillates with the magnetic resonance frequency $\nu_0 = (\gamma_e/2\pi)H_0$ (γ_e is the electron gyromagnetic ratio). One studies a spontaneous emission of the system, i.e., its echo signal [1,2].

A simple and pictoral description of a ESE signal generation is provided by a vector model based on analyzing the motion of the sample magnetization vector under the action of a variable magnetic field pulse.

In an external magnetic field $\vec{H}_0$ a sample with paramagnetic particles acquires a magnetic moment $\vec{M}_0$ which is parallel to $\vec{H}_0$ under thermal equilibrium. The magnetization evolution in a spin system is convenient to analyze in coordinates rotating about $\vec{H}_0 \parallel z$ with a frequency ν_0 (Fig. 1). Let the x axis of the rotating coordinates coincide with $\vec{H}_1$. Hence, in the field $\vec{H}_1$ the magnetization $\vec{M}_0$ rotates with an angular velocity $2\pi\nu_1 = \gamma_e H_1$ about the x axis on the plane yz and turns through the angle:

$$\theta = 2\pi\nu_1 t_p = \gamma_e H_1 t_p \tag{1}$$

within the time of the microwave pulse duration t_p. The field pulse is a 90° pulse if $\theta = \pi/2$. Under the action of a 90° pulse the magnetization $\vec{M}_0$, primarily orientated along the z axis, becomes directed along the y axis. The ESR line is usually inhomogeneously broadened due to various magnetic interactions. Its shape obeys the function $g(\nu_z)$ characterizing the intensity distribution in the ESR spectrum. After the action of a microwave pulse, every vector $\vec{M}(\nu_z) = \vec{M}_0 g(\nu_z)$ oscillates within a time τ on the plane xy with an angular velocity $2\pi \Delta\nu$ ($\Delta\nu = \nu_z - \nu_0$) in the rotating coordinates, acquires a phase

ELECTRON SPIN ECHO: APPLICATIONS TO BIOLOGICAL SYSTEMS

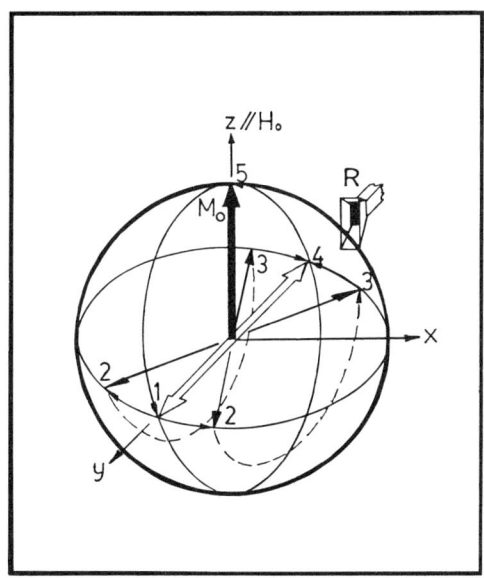

FIG. 1. A vector model of an ESE signal formation in rotating coordinates. A 90° pulse turns the magnetization vector $\vec{M}_0$ in the xy plane (1); $\vec{M}_0$ decomposes into elementary vectors with a rate T_2^{-1} (2); a 180° pulse turns the elementary vectors (3); $\vec{M}_0$ is restored along the y axis (4); an ESE signal forms; $\vec{M}_0$ is restored along the z axis due to spin-lattice relaxation with the rate T_1^{-1} (5).

$\rho_z = 2\pi \, \Delta\nu \, \tau$, and spreads into a "fan" of elementary vectors on the xy plane. This spread shows up as a fall of the magnetization M_y after the pulse at a time $T_2^* \sim 1/\Delta\nu_z^*$ [$\Delta\nu_z^*$ is a function of $g(\nu_z)$ width]. This magnetization change induces the so-called free induction decay signal. Consequently, at the time T_2^* the resultant magnetization vector spreads all over the xy plane. If now at an instant τ a 180° pulse acts on the system, every elementary vector turns through 180° about the x axis. In this case, the vector $M(\nu_z)$, on the xy plane, acquires an additional phase $(180° - 2\rho_z)$. After this second pulse every elementary vector goes on moving with the initial angular velocity and acquires the phase ρ_z again within the time τ. As a result, by the time 2τ the vector $\vec{M}(\nu_z)$ will have the total phase $\rho_z + (180° - 2\rho_z) + \rho_z = 180°$ irrespective of ν_z.

This means that the total magnetization M_0 is restored along the y axis and can be detected as a two-pulse or primary echo signal with a maximum at 2τ (Figs. 1 and 2).

Apart from the two-pulse method, ESE experiments often employ the three-pulse technique, which acts on a spin system with three pulses at times 0, τ, and $\tau + T$ (Fig. 2). In this case, the echo signal at $2\tau + T$ is formed by all the three pulses and termed a three-pulsed or stimulated echo.

If a spin system in a constant magnetic field H_0 is disturbed by a 90° pulse, there arise relaxation processes, which bring the system back to the thermodynamic equilibrium, when $\vec{M}_0 \parallel \vec{H}_0$. One of these processes, the spin-lattice relaxation, restores the longitudinal magnetization along $\vec{H}_0$ up to the primary value with the characteristic T_1 time. Another process, spin-spin or phase relaxation, induces random changes in the spin precession frequency about $\vec{H}_0$ and thus irreversibly dephases the transverse magnetization components at the time T_2. Due to the relaxation processes, the PE and SE signal envelopes, detected at different τ and T, decrease monotonously with increasing decay times. The echo signal decay kinetics are informative of various time stochastic processes inducing magnetic relaxation in spin systems [1-5].

An echo signal can arise due to the spin motion dephasing resulting from a static inhomogeneous broadening of the ESR lines within the time $T_2^* \sim 1/\Delta\nu_z^*$. This is a reversible process. Hence, the reversible dephasing effect makes it possible to use the spin-echo method to remove the disguising influence of inhomogeneous line broadening in the ESR spectrum and to investigate directly the irreversible spin magnetization relaxation associated with the homogeneous ESR line broadening.

In most cases, ESE experiments are performed in the X band, i.e., at a frequency of about 9.5 GHz ($\lambda \sim 3.2$ cm). For paramagnetic centers with $g \sim 2$ the resonance conditions arise in a constant magnetic field H_0 approximately equal to 3,300 Oe.

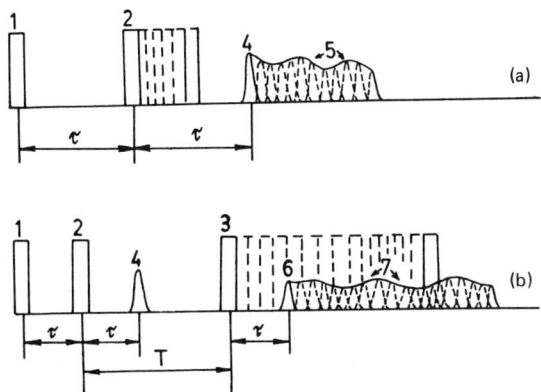

FIG. 2. The sequence of pulses and the ESE generation in the two (a) and three (b) pulse techniques: (1,2,3) microwave pulses; (4) PE signal; (5) PEEM; (6) SE signal; (7) SEEM.

At present about 30 ESE spectrometers are either at work or being set up at various laboratories in the world [6]. Normally, these are assembled of separate modules in every particular case.

The basic modules of any ESE spectrometer are a pulsed microwave source, a microwave receiver of ESE signals with a detector and a recording system, a sample cavity, a programmable pulse generator, an electromagnet with a power supply, and a module for the field stabilization. Particular arrangements of ESE spectrometers employed in biological experiments are described in the literature [7-13].

Important characteristics of ESE spectrometers are their sensitivity and time resolution. The sensitivity of an ideal coherent ESE spectrometer has been calculated [2] to be 10^{10}-10^{11} spins per sample under complete excitation of the ESR spectrum, at times $T_2 = 3 \cdot 10^{-7}$ sec, $T_1 = 3 \cdot 10^{-6}$-10^{-4} sec, and a signal detection time of about 1 sec. This corresponds to the sensitivity of modern cw ESR spectrometers. However, the sensitivity of real ESE spectrometers is lower for several reasons, which have been considered in detail [2]. The sensitivity of a ESE spectrometer is reduced in experiments with paramagnetic centers having either short relaxation times or an essential inhomogeneous line broadening of the ESR spectrum. In the latter case, the sensitivity losses are due to a finite microwave pulse amplitude equal to several oersteds. This gives only partial excitation of the ESR line. The sensitivity may be increased by prolonging the detection time by way of accumulating the ESE signal.

As in the case of any time domain spectroscopy, the ESE has an important parameter which is its time resolution. This parameter means the minimum magnetic relaxation time that can be detected by a spectrometer. The time resolution of an ideal spectrometer is restricted by the exciting microwave pulse duration normally equal to several tens of nanoseconds.

In the case of a real spectrometer, it is necessary also to take into account the cavity ringing, which is a reflected signal with an exponential decay. After switching the exciting pulse off, the ringing is usually so great that the receiver is overloaded for a certain dead time and cannot receive a weak useful signal. The dead time of the spectrometer can amount to some 150-200 nsec. Taking into account the microwave pulse duration and the dead time, the actual resolution of a ESE spectrometer is about 200-300 nsec.

Usually, magnetically diluted solid solutions of paramagnetic ions or of radicals in water or in hydrocarbons at 4-77 K show the signal decay of several microseconds for a PE and an order of magnitude higher for a SE.

As a rule, the ESE signal decay due to relaxation is accompanied by periodic falls and rises in the echo signal amplitude (Fig. 2), which have been termed modulation effects, or ESEEM. It has been found [1-3] that the main physical source of these effects are electron-nuclear and electron-electron couplings in a spin system. In most cases, the structural applications of the ESE method are associated with an analysis of the ESEEM induced by the coupling of unpaired electrons of paramagnetic centers with surrounding magnetic nuclei [8,10,11,13-20]. Therefore, we consider below the theory of such ESEEM and also some aspects of its analysis that might be useful for investigating paramagnetic centers in biological systems.

2. ESEEM SPECTROSCOPY

2.1. ESEEM Theory

2.1.1. Qualitative ESEEM Model

A simple qualitative explanation of ESEEM due to electron-nuclear couplings can be provided by taking into account the nuclear spin precession in the close vicinity of electrons.

Let an electron and a nucleus, separated by a distance r, be in a constant magnetic field $\vec{H}_0$. In this case, the electron is in a local field $\vec{H}_S = \vec{H}_0 + \vec{H}_{IS}$ and the nucleus is in a local field $\vec{H}_I = \vec{H}_0 + \vec{H}_{SI}$, where $\vec{H}_{SI}$ is the magnetic field induced by the electron at the location of the nucleus and $\vec{H}_{IS}$ is that induced by the nucleus at the location of the electron. At a distance of several ångströms $\vec{H}_{SI}$ equals several hundreds of oersteds, and $\vec{H}_{IS}$ is approximately three orders of magnitude lower. In this case, in an external magnetic field $H_0 \sim 3{,}000$ Oe, in the X band, the quantization axis of the electron spin practically coincides with $\vec{H}_0$, whereas the nuclear spin quantization axis may differ essentially from $\vec{H}_0$ (Fig. 3). At the location of the nucleus the local electron field $\vec{H}_{SI}$ depends on the sign of the electron spin projection onto the quantization axis; $m_S = \pm 1/2$. A sudden change in the electron spin state due to a microwave pulse $\vec{H}_1$ causes a simultaneous change in the local field direction of the nucleus from $\vec{H}_I(+)$ to $\vec{H}_I(-)$. If the pulse duration is much less than the nuclear Larmor precession period in the local field, the field changes suddenly, $\vec{H}_{SI}(m_S = +1/2) \to \vec{H}_{SI}(m_S = -1/2)$. The nuclei have no time to adiabatically follow such sudden changes of the resultant magnetic field direction $\vec{H}_I(-)$. Hence, as the electron spin changes its orientation under a microwave pulse, the nuclear spin orientation does not coincide with the new direction of $\vec{H}_I(-)$; the nucleus starts to precess about this direction (Fig. 3) and thus induces local variable fields $\vec{H}_\parallel(t) \parallel \vec{H}_0$ and $\vec{H}_\perp(t) \perp \vec{H}_0$ at the location of the electron. Because of the fast oscillation of the electron magnetic moment (as $\mu_e \gg \mu_I$), its interaction with the local field component $H_\perp(t)$ can be neglected. After the action of the pulse,

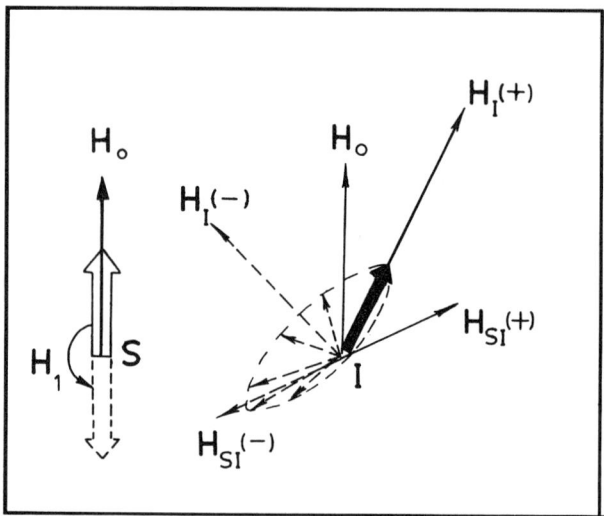

FIG. 3. Nuclear spin precession at different electron spin orientations.

the electron is thus in the oscillating local magnetic field $H_0 + H_{\parallel}(t)$. The value of $H_{\parallel}(t)$, e.g., for a $I = 1/2$ spin, can be either positive or negative depending on the initial electron spin state. Therefore, a portion of electron spins precess in the field $H_0 + |H_{\parallel}(t)|$, and the other spins precess in the field $H_0 - |H_{\parallel}(t)|$. A ESE signal arises if the spin dephasings in the local magnetic fields occur with the same rate after the first and the second microwave pulses. In oscillating local fields this requirement is met only if the interpulse intervals τ are divisible by the nuclear precession periods in the local fields. Failing this, the echo signal amplitude decreases due to incomplete electron spin dephasing on the fields $H_0 + |H_{\parallel}(t)|$ and $H_0 - |H(T)|$. The qualitative analysis has thus shown that the ESE signal envelope is modulated by the surrounding nuclear oscillation frequencies in the local fields induced by the external magnetic field and the electron magnetic moments.

2.1.2. Primary ESEEM

The ESEEM can be calculated exactly in terms of the density matrix formalism developed in the general form [21,22]. When an electron spin S = 1/2 is coupled with a nuclear spin I, the ESEEM results from hyperfine, nuclear Zeeman, and nuclear quadrupole interactions (no nqi arises for I = 1/2). The function that describes the primary ESEEM has the form [22]:

$$V(\tau) = k_0 + \sum_{i<j} k_{ij}^{(\alpha)} \cos 2\pi \nu_{ij}^{(\alpha)} \tau + \sum_{l<n} k_{ln}^{(\beta)} \cos 2\pi \nu_{ln}^{(\beta)} \tau$$
$$+ \sum_{i<j} \sum_{l<n} k_{ij,ln}^{(\alpha,\beta)} [\cos 2\pi(\nu_{ij}^{(\alpha)} + \nu_{ln}^{(\beta)})\tau + \cos 2\pi(\nu_{ij}^{(\alpha)} - \nu_{ln}^{(\beta)})\tau]$$
(2)

Hence, the ESEEM frequencies are frequencies of the nuclear transitions $\nu_{ij}^{(\alpha)}$ and $\nu_{ln}^{(\beta)}$ between energy levels with the same m_S (+1/2 for i, j and -1/2 for l, n) as well as the sums and the differences $\nu_{ij}^{(\alpha)} \pm \nu_{ln}^{(\beta)}$ of the nuclear frequencies for differing m_S. The coefficients, which set the modulation harmonic amplitudes, depend on the matrix elements characterizing the probability of electron-nuclear transitions between levels with different m_S. For an arbitrary relative value of the above-mentioned interactions, the expressions for modulation frequencies and amplitudes cannot be derived analytically except for a nucleus with I = 1/2.

The system of S = 1/2 and I = 1/2 has four energy levels. In the general case, the splitting of two levels with the same m_S = 1/2, equal to $h\nu_\alpha$, differs from that of the levels with m_S = -1/2, which is $h\nu_\beta$ (Fig. 4). Taking into account the hyperfine and the nuclear Zeeman couplings, the frequency expression is

$$\nu_{\alpha,\beta} = \left[\left(\nu_I \pm \frac{A}{2}\right)^2 + \frac{B^2}{4}\right]^{1/2} \tag{3}$$

Here $A = T_{zz}$, $B = (T_{zx}^2 + T_{zy}^2)^{1/2}$, T_{zi} are the elements of the hfi tensor in the laboratory coordinates, with the external magnetic field $\vec{H}_0$ directed along the z axis.

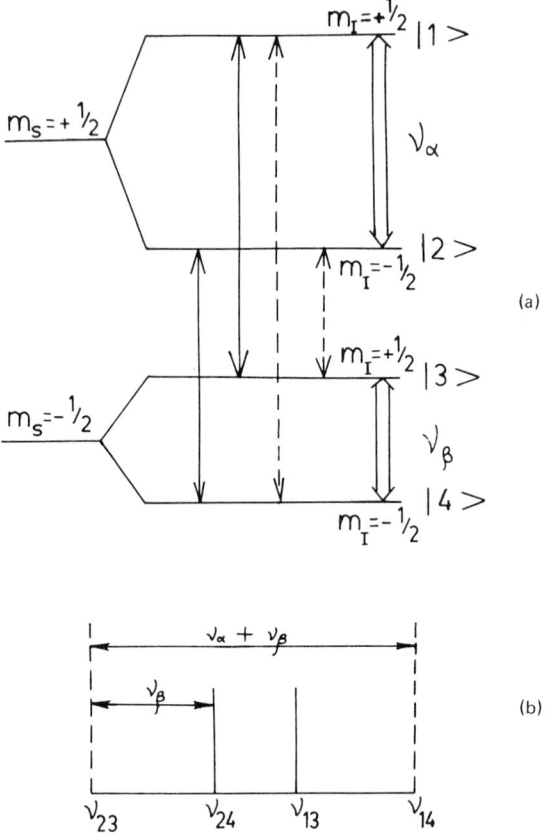

FIG. 4. Energy levels and different resonance transitions (a), and the ESR spectrum (b) for an electron spin $S = 1/2$ coupled with a nuclear spin $I = 1/2$.

Between the upper and the lower doublets there may occur four ESR transitions induced by microwave pulses at a resonance frequency. The probability of these transitions is determined by the values of corresponding matrix elements. Two of these transitions (Fig. 4, solid lines) occur without changing the nuclear spin projection and usually are called allowed transitions; the other two transitions (Fig. 4, dashed lines) are accompanied by a simultaneous change in the nuclear spin projection and are called forbidden transitions.

ELECTRON SPIN ECHO: APPLICATIONS TO BIOLOGICAL SYSTEMS 217

Besides, a radiofrequency field can here induce two more transitions (double lines), with frequencies ν_α and ν_β, between levels with the same m_S. These are the transitions observed in ENDOR experiments, and therefore the frequencies ν_α and ν_β are termed ENDOR frequencies.

The expression of the primary ESEEM derived for $S = 1/2$ and $I = 1/2$ [21,23,24] can be written in the explicit form as

$$V(\tau) = 1 - \frac{1}{2}k + \frac{1}{2}k[\cos 2\pi\nu_\alpha\tau + \cos 2\pi\nu_\beta\tau - \frac{1}{2}\cos 2\pi(\nu_\alpha + \nu_\beta)\tau$$

$$- \frac{1}{2}\cos 2\pi(\nu_\alpha - \nu_\beta)\tau] \quad (4)$$

The coefficient k in Eq. (4), determining the modulation frequency amplitude, is

$$k = \left(\frac{\nu_I B}{\nu_\alpha \nu_\beta}\right)^2$$

and proportional to the produce of allowed and forbidden transition probabilities in a spin system [21]. Hence, the necessary ESEEM condition is a simultaneous existence of allowed and forbidden transitions in the spin system.

The coefficient k involves the parameter B which depends on the values of nondiagonal elements of the hfi tensor. This suggests another important conclusion, namely, that the ESEEM can only occur under an anisotropic hyperfine interaction between an electron and a nucleus, i.e., in solids.

Among other theoretical results, it is important to mention the relations for ESEEM for the coupling of $S = 1/2$ with a nucleus of an arbitrary spin I derived through the formulas for $I = 1/2$ [25]. This result is approximate in that it has been obtained neglecting the nqi typical of nuclei with $I \geq 1$. The ESEEM theory can take into account nqi for a nucleus of an arbitrary spin only in the limiting case of the nuclear Zeeman interaction exceeding essentially hfi and nqi [26] (see Sec. 2.4.3), and also for $I = 1$, when nqi is much less than hfi [21,27].

For $S > 1/2$ only the theory of modulation of an electron triplet coupled with $I = 1/2$ has been analyzed [28].

2.1.3. Stimulated ESEEM

In the general form the function for the SEEM in the system $S = 1/2$ and I is [22]:

$$V(\tau,T) = \frac{1}{2}\left[V_\alpha(\tau,T) + V_\beta(\tau,T)\right] = \frac{k_0}{2} + \frac{1}{2}\sum_{i<j} k_{ij}^{(\alpha)} \cos 2\pi\nu_{ij}^{(\alpha)}\tau$$

$$+ \sum_{i<j} \cos 2\pi\nu_{ij}^{(\alpha)}(\tau + T)\left[\frac{1}{2}k_{ij}^{(\alpha)} + \sum_{1<n} k_{ij,1n}^{(\alpha,\beta)} \cos 2\pi\nu_{1n}^{(\beta)}\tau\right]$$

$$+ \frac{k_0}{2} + \frac{1}{2}\sum_{1<n} k_{1n}^{(\beta)} \cos 2\pi\nu_{1n}^{(\beta)}\tau + \sum_{1<n} \cos 2\pi\nu_{1n}^{(\beta)}(\tau + T)$$

$$\left[\frac{1}{2}k_{1n}^{(\beta)} + \sum_{i<j} k_{ij,1n}^{(\alpha,\beta)} \cos 2\pi\nu_{ij}^{(\alpha)}\tau\right] \tag{6}$$

In the case of $I = 1/2$, which can be calculated exactly, relation (6) takes the form [21]:

$$V(\tau,T) = \frac{1}{2}\left[V_\alpha(\tau,T) + V_\beta(\tau,T)\right]$$

$$= \frac{1}{2}\left\{1 - \frac{1}{2}k\left[(1 - \cos 2\pi\nu_\alpha\tau)\left(1 - \cos 2\pi\nu_\beta(\tau + T)\right)\right]\right\}$$

$$+ \frac{1}{2}\left\{1 - \frac{1}{2}k\left[\left(1 - \cos 2\pi\nu_\alpha(\tau + T)\right)(1 - \cos 2\pi\nu_\beta\tau)\right]\right\} \tag{7}$$

Note that for $T \to 0$ Eqs. (6) and (7) turn into the corresponding Eqs. (2) and (4) for the PE.

Experimentally, a SE is usually detected as a function of the time T at a constant τ. In this case, the structure of Eqs. (6) and (7) shows that the SEEM frequencies are only frequencies of nuclear transitions at $m_s = \pm 1/2$ without their sums and differences, as was the PE case. Moreover, the presence of an extra measurable time parameter τ in the SE technique results in the dependence of the modulation frequency amplitude and also in an interesting frequency suppression effect. For example, if $I = 1/2$, Eq. (7) shows the $\nu_{\alpha(\beta)}$ amplitude to be at maximum at

$$\tau_{max} = \frac{n + 1/2}{\nu_{\beta(\alpha)}} \tag{8}$$

and to fall to zero at

$$\tau_{min} = \frac{n}{\nu_{\beta(\alpha)}} \tag{9}$$

This means that if τ is set equal to an integer of ν_β periods, the term involving $\cos 2\pi\nu_\alpha(\tau + T)$ in Eq. (7) will vanish from the envelope, and the modulation will have only the frequency ν_β. If τ is set equal to an integer of ν_α periods, the SE will be modulated with the frequency ν_α. This regularity holds also for $I > 1/2$, although the quenching is not complete. The suppression effect is often used to discriminate modulation contributions from different nuclei and also to attribute frequencies to definite sets with different m_s.

2.1.4. Interaction with Several Nuclei

In solids unpaired electrons of paramagnetic centers interact often with a great number of magnetic nuclei of various types. If an electron spin interacts with N magnetic nuclei of an arbitrary spin I_i, the modulation effects are described by the formula [21]:

$$V_N(\tau) = \prod_{i=1}^{N} V_{I_i}(\tau) \tag{10}$$

The ESEEM under the interaction with several nuclei has been derived in an explicit form only for the PE [21]. However, experimental data are often interpreted using an equation of the type (10):

$$V_N(\tau,T) = \prod_{i=1}^{N} V_{I_i}(\tau,T) \tag{11}$$

which is inapplicable to SEEM calculation (e.g., see [29]). Relation (11) is also available in reviews [10,15].

A correct analysis by the method [21] has shown [25,30] that the SEEM under the interaction of an electron with several nuclei obeys a formula which differs from Eqs. (10) and (11) in structure and has the form:

$$V_N(\tau,T) = \frac{1}{2}\left[\prod_{i=1}^{N} V_{I_i\alpha}(\tau,T) + \prod_{i=1}^{N} V_{I_i\beta}(\tau,T)\right] \qquad (12)$$

Here $V_{I_i\alpha}(\tau,T)$ and $V_{I_i\beta}(\tau,T)$ are determined in Eq. (6).

2.1.5. *Partial Excitation*

The above results have been derived under the assumption that a spin system is under the action of microwave pulses with sufficiently great amplitude H_1 which excite the whole ESR spectrum and that all possible transitions are induced according to magnitudes proportional to their matrix elements. It is, however, possible that a certain transition is sufficiently far from resonance determined by the external magnetic field strength and, depending on the microwave field amplitude, is either completely unexcited or excited only partially. As a result, the modulation frequency amplitude changes. This may result, at the limit, in complete absence of some frequencies in the ESEEM. Let us exemplify this situation by various extreme cases of exciting the ESR spectrum components (shown in Fig. 4) using results reported [24]. For example, if the two edge lines, corresponding to ν_{23} and ν_{24} transitions, are only excited, the PEEM is [24]

$$V(\tau) = 1 - \frac{k}{2}(1 - \cos 2\pi\nu_\beta\tau) \qquad (13)$$

and for the SEEM

$$V(\tau,T) = 1 - \frac{k}{2}\left[1 - \frac{1}{2}\left(\cos 2\pi\nu_\beta\tau + \cos 2\pi\nu_\beta(\tau+T)\right)\right] \qquad (14)$$

That is, the ESEEM has only the modulation frequency equal to the frequency difference of the excited transitions $\nu_{24} - \nu_{23} = \nu_\beta$. Similarly, it is possible to see that the modulation does not show up under the excitation of either one or two lines providing that the corresponding transitions have no common energy level. These modulation rules for a partial excitation of a spectrum are general and applicable to other systems, where the number of possible transitions increases compared with the case of $I = 1/2$. In particular,

the partial excitation of several spectral components results in
modulation only if the corresponding transitions have a joint energy
level. Such a level is associated with a set of modulation frequencies, which are frequency differences of transitions from this
joint under microwave pulses. Experimental evidences for the above
rules are available [31,32].

So far the transition, excited at $\Delta\nu = \nu - \nu_0$ apart from resonance by a microwave pulse with duration t_p and strength $\nu_1 = (\gamma_e/2\pi)H_1$ has not been analyzed in detail. Recently the partial excitation effect was investigated [33] in a SEEM. For that purpose a general expression for the SEEM was derived and modified for the case of partial spectral excitation. That expression allowed the following conclusions [33]: When the SE is detected by a standard method with the time interval τ between the first and the second pulses fixed, the ESEEM frequency spectrum shows only the nuclear transition frequencies that vary their amplitude under partial excitation. Under partial excitation the two-dimensional frequency spectrum (see Sec. 2.2) obtained by detecting the SE at various τ shows some additional peaks, which are combinations of three nuclear transition frequencies. Two of them belong to a set with one m_s, and the third frequency to a set with an opposite m_s. These theoretical conclusions have been confirmed by experiments with F centers (electrons trapped at anion vacancies) in KCl single crystals.

Moreover, numerical calculations [33] demonstrated the influence of ν_1 on the modulation amplitude, which corroborated earlier qualitative assumptions [2,8] on effective excitation of transitions with frequencies in the range $(\nu_0 - \nu_1, \nu_0 + \nu_1)$ by a microwave pulse with ν_1 amplitude. In other words, a complete excitation of the spectrum requires ν_1 to exceed the maximum sum of the nuclear frequencies corresponding to different m_s.

2.2. Peculiarities of ESEEM Frequency Transformation

An experimental ESEEM is a linear combination of a constant component and decaying oscillations, which in the general case have different

initial decay amplitudes and decay functions. Taking into account
the total relaxation decay, the ESE envelope can be written as

$$V(t) = G_{rel}(t)[a_0 + \sum_{i=1}^{K} a_K g_K(t) \cos 2\pi\nu_K t] \qquad t = \tau, T \qquad (15)$$

In this case, the modulation frequencies can be determined, at first
sight, by the cosine Fourier transformation. However, in practice
the frequency analysis meets with some difficulties. The principal
difficulty stems from the fact that the experimentally detected
ESEEM is a cut part of the functions V(t) determined within the
interval from the starting value, set by the spectrometer's dead
time t_d, to the final time t_f. The FT analysis of the cut signal
decay without any preliminary processing does the following:

1. It gives a peak with side lobes near zero frequency (the
   amplitude of this peak usually exceeds essentially that
   of the resonance line associated with ν_K frequencies).
2. It underestimates the amplitude of ESEEM harmonics, thus
   increasing the signal-to-noise ratio and distorting the
   real relative amplitudes of different harmonics.
3. The side lobes of the intensity peak near zero frequency
   and the most intensive ESEEM spectral components may dis-
   guise lines corresponding to real ESEEM frequencies with
   small amplitudes.

For the first time the above factors have been mentioned [7]
as violating the modulation spectra; also it has been proposed that
the detected decay function should be approximated by the polynomial
P(t) using the method of least squares, and then the modulus FT of
the difference V(t) - P(t) follows. This procedure allows one practically to remove the lowest frequency part of the spectrum and to
obtain the ESEEM spectra convenient for analysis.

The side lobes of the real spectral components, which arise
after the cosine FT of the ESEEM detected in a limited time interval,
can be readily interpreted [34] if the detection interval is represented as

$$V_{real} = h(t)V(t) \tag{16}$$

where $h(t)$ is a rectangular window determined as

$$h(t) = \begin{cases} 1, & \text{at } t_d \leq t \leq t_f \\ 0, & \text{at any other } t \end{cases}$$

The transformation of the incomplete function $V_{real}(t)$ is accompanied by a simultaneous transformation of $h(t)$, which results in false spectral components.

The modulus FT does not lead to the appearance of side lobes. However, the lines of this spectrum are wider and have long wings. Moreover, the nonlinearity of the modulus FT leads to distortions of the relative spectral intensity at ranges with overlapping lines.

To suppress the side lobes, it is possible to use the so-called apodization technique, i.e., prior to FT the experimental ESE signal envelope is multiplied by functions which weaken the echo signal at the ends of the detection time interval and thus ensure smooth passages to zero line [34]. As apodization functions of the simplest type one may use a triangle, a cosine, a trapezoid, and a gaussian function. A negative property of the apodization technique is that while increasing the side lobes decay rate, it also increases the basic maxima widths, i.e., affects the spectrum resolution. The apodization technique has been successfully employed [34] to analyze the SEEM for Cu^{2+} in the Cs_2ZnCl_4 single crystal, which accompanied the ESE signal decay during several hundreds of microseconds.

The apodization technique proves to be less effective when applied to polyorientated samples, where the modulation decays at best within 10-20 μsec, and very often even within several microseconds. In this case, a smoothing factor weakens the initial part of the ESEEM, thereby reducing its efficiently detected part.

It has been proposed to analyze a ESE modulation with a short decay (when every frequency is represented by just several oscillations) by the cosine FT supplemented by the restoration of data into the dead time region. For that purpose a special algorithm has been developed [35] exemplified by its applications to the modulation analysis in biological systems.

Parametric methods of spectral estimation based on the autoregressive model have recently been employed to analyze ESEEM [36, 37]. These methods use the experimental time dependence V(t) and autocorrelation to extend the available data to both sides in infinite time limits. This procedure gives estimates of the spectrum. The spectrum obtained does not have distortions caused by the smoothing factors, and besides the extrapolation of data outside the detection interval leads to its high resolution.

An important requirement imposed by the AR analysis on the signal is its stationarity, i.e., all its frequency components must have constant amplitudes within the detection time. An ESEEM involves many different frequency components with different decay rates, particularly when an electron interacts with many nuclei. In these conditions they cannot be always made approximately stationary by multiplying them by some increasing function. Therefore, the above-mentioned methods do not yield adequate results for different spectral parts.

In conclusion, it should be mentioned that the FT and AR methods applied to a ESEEM analysis have advantages as well as limitations. Therefore, these are often employed simultaneously, which increases the reliability of the spectral information obtained. Some examples of this analyses are shown in Figures 7, 9, and 10 (see Secs. 2.4.4 and 3.3).

2.3. ESEEM in Single Crystals

Since ESEEM frequencies, which are those of nuclear transitions, depend on hfi parameters, studies on the angular dependence of modulation frequencies in single crystals solve the same problems as an analysis of angular dependencies of ESR and ENDOR spectra, i.e., to determine the hfi and nqi tensor components and the orientation of their principal axes.

ELECTRON SPIN ECHO: APPLICATIONS TO BIOLOGICAL SYSTEMS   225

The ESE method is particularly useful for determining the parameters which cannot be readily obtained from ESR spectra, e.g., parameters for hfi with nuclei having low magnetic moments and modulation frequencies of several tens of kilohertz to 1 MHz. This frequency range cannot be ENDOR-investigated.

ESEEM in single crystals is studied usually for paramagnetic ions (as an admixture), radicals generated under irradiation, and photoexcited triplet states of admixture molecules. The most complete review on ESEEM investigations of paramagnetic centers in single crystals has been made recently [20].

A good example of up-to-date applications of the ESE method to studies on weak hyperfine couplings in single crystals has been furnished [38] by ESEEM investigations of the lowest photoexcited triplet state of free-base porphin in an n-octane single crystal. The experiments were performed at T = 1.4 K with per-deuterated n-octane single crystals involving free-base porphin deuterated at the peripheral hydrogen positions. The deuteration was aimed at improving the ESE excitation conditions by narrowing the ESR lines, and also at increasing the phase relaxation time.

The ESEEM analysis allowed full sets of hfi and nqi parameters (principal values and directions of the principal axes of the tensors) for protonated and unprotonated ^{14}N in the triplet state of free-base porphin.

A further consideration of hfi and nqi data, as an example of an analysis of the ^{14}N coupling in $\pi\pi^*$ states of aza aromatic molecules, made it possible to infer [38] a number of important general quantum chemical and structural conclusions. The hfi data showed that the anisotropic part of the interaction is almost entirely determined by the one-centered interaction of magnetic dipoles. On the other hand, the isotropic part is strongly affected by multicenter interactions.

The knowledge of the nqi constants, which differ essentially in two pairs of nitrogens in a porphin molecule, made it possible to determine the populations of the valency orbitals and to compare these with theoretical calculations. The most important aspect of the investigations [38] is the determination of not only the above-mentioned values, which in principle can be obtained by nuclear quadrupole resonance, but also of the direction of the principal axes for the quadrupole tensors. This has confirmed the anticipated rotation of the field gradient axes in the hydrogen bond formation and revealed such a feature of porphin molecular structure in the excited state as the pyrrole ring nonplanarity.

As mentioned [38], the hfi and nqi with ^{14}N in triplet states of aza aromatic molecules in crystals have been investigated in the last 20 years by various physical methods, including ESR, ODMR in zero field, and ENDOR. However, these methods supply only incomplete information, if at all, on the interactions with nitrogen. Therefore, the example being considered is an excellent illustration of potentialities of the ESE method as applied to studies on interactions with nitrogen, which is present in many molecules playing an important role in biological systems.

Besides the ESEEM analysis, hyperfine couplings can be determined also by the SE technique combined with the pulsed action of a radiofrequency field on magnetic nuclei, i.e., the pulsed ENDOR technique [1,2]. The method is to provide an rf pulse between the second and the third ESE microwave pulses. The rf pulse induces transitions between nuclear sublevels, thereby decreasing the SE amplitude, if the frequency of this pulse coincides with that of a corresponding transition. The dependence of the ESE signal amplitude on the rf frequency gives a PENDOR spectrum. The first PENDOR experiments were performed by Mims [39]. This method was employed [39] to determine hfi parameters for Ce^{3+} ions with neighboring ^{183}W nuclei in the calcium wolframate lattice. The ESEEM and rf pumping effects for Ce^{3+} ions were investigated in detail in a ruby single crystal [40,41]. PENDOR spectral analysis allowed hfi and nqi parameters for the 34 nuclei ^{27}Al with $I = 5/2$ surrounding a chromium ion. The PEEM calculated with those parameters fairly well agreed with the experiment. The cw ENDOR method had been used earlier [42] to determine the coupling parameters corresponding only to 13 neighboring ^{27}Al nuclei in this system.

The PENDOR method was also used successfully to investigate the lowest triplet state of free-base porphin in an n-octane single crystal [43]. An analysis in the angular dependence of the PENDOR spectra allowed hfi tensors for the two interior protons, and thereby for the spin densities on nitrogen atoms and the locations of protons within the ring. The PENDOR technique was mentioned [43] to have an important advantage over the cw ENDOR. A cw ENDOR signal intensity is about 1% of that for an ESR signal. The time domain spectroscopy and the detection of nuclear transitions by the ESE signal amplitude reduction make it possible to raise the sensitivity by over an order of magnitude, which is of importance in studying species with low concentrations of paramagnetic centers.

The above points demonstrate that the ESE method, combined with a ESEEM frequency analysis, and also the PENDOR technique, provide valuable information on the hyperfine interaction in single crystals. These methods can be used with advantage in cases where the cw ESR and ENDOR techniques do not ensure the required sensitivity and resolution.

In real biological systems paramagnetic centers are usually orientated in a random manner. However, detailed and fundamental information on a single crystal can be gained by experiments with paramagnetic centers abstracted from either real biological systems or their synthetic analogues.

2.4. ESEEM in Disordered Systems

2.4.1. ESEEM Under Weak Anisotropic hfi

The anisotropy of hfi and their dependence on the spatial orientation of paramagnetic centers result in a smearing of energy levels in disordered spin systems. Eventually, this leads to additional losses of the monochromatic character of movement of the spin system during free precession periods. This gives an additional modulation frequency spread $\Delta\nu$ in paramagnetic particles, which are randomly orientated relative to the external magnetic field. In a time domain this spread results in the modulation decay within a time of about $1/\Delta\nu$.

As a rule, ESEEM in disordered systems are observed at times of about 1-10 μsec. This means that the hfi anisotropy is some 0.1-1 MHz. Since in disordered samples the linewidth usually does not exceed 10 MHz, the ESEEM results from the interaction disguised by inhomogeneous broadening in the ESR spectra of solids. This unique property of the ESE method is fully confirmed by experiment. At present a main line in ESE applications to physicochemical investigations is associated with studies on weak hfi and nqi in disordered solids [1,2,8,10,13-20].

A ESEEM analysis should consider first the widely met case of weak anisotropic hfi of an electron spin with matrix nuclei, I = 1/2, far remote from a paramagnetic center. In the point-dipole approximation the quantities A and B, characterizing the hfi with one nucleus, are simply related to the geometry of the relative electron-nucleus position and to the isotropic hfi constant,

$$A = T_\perp (3 \cos^2\theta - 1) + a$$
$$B = 3T \sin\theta \cos\theta$$
$$T_\perp = g_e g_I \beta_e \beta_I / hr^3 \tag{17}$$

Here r is the electron-nucleus separation, θ is the angle between the vectors of the external magnetic field $\vec{H}_0$ and $\vec{r}$, a is the isotropic hfi constant in NHz.

Under the assumption that initially a = 0, the anisotropic hfi, as compared with the nuclear Zeeman coupling, may be characterized by the parameter $T_\perp/\nu_I$, which follows from Eq. (17) to be

$$\frac{T_\perp}{\nu_I} = \frac{g_e \beta_e}{H_0} \frac{1}{r^3} \tag{18}$$

Relation (18) does not contain nuclear magnetic parameters and is determined solely by r in a set H_0. In X-band ESR experiments, when H ~ 3300 Oe, the ratio $T_\perp/\nu_I$ is about 0.1-0.2 for paramagnetic centers with g ~ 2 at r = 3-4 Å.

Thus, the Zeeman coupling of sufficiently remote matrix nuclei exceeds essentially the anisotropic hfi, that is the condition

ELECTRON SPIN ECHO: APPLICATIONS TO BIOLOGICAL SYSTEMS 229

$\nu_I \gg A, B$ holds. In the approximation of extremely weak hfi the frequencies $\nu_{\alpha,\beta}$ of Eq. (3) can be presented as $\nu_{\alpha,\beta} \simeq \nu_I \pm A/2$. Simple trigonometric transformations of Eq. (4) yield for the PEEM from a nucleus with $I = 1/2$:

$$V(\tau) = 1 - \frac{1}{2}\frac{B^2}{\nu_I^2}\left[1 - 2\cos 2\pi\nu_I\tau \cos 2\pi\left(\frac{A\tau}{2}\right) + \frac{1}{2}\cos 2\pi(2\nu_I\tau)\right.$$

$$\left. + \frac{1}{2}\cos 2\pi(A\tau)\right] \qquad (19)$$

Averaging Eq. (19) over all orientations of the magnetic field and taking into account Eq. (17), we obtain for a disordered system:

$$\langle V(\tau)\rangle = 1 - \frac{3}{10}\frac{g_e^2\beta_e^2}{H_0^2\nu^6}\left[2 - 4\cos 2\pi\nu_I\tau\, \phi(T_\perp\tau) + \cos 2\pi(2\nu_I\tau)\right.$$

$$\left. + \phi(2T_\perp\tau)\right] \qquad (20)$$

where

$$\phi(T_\perp t) = \frac{15}{4}\int_0^\pi \sin^2\theta \cos^2\theta \cos\left[\pi T_\perp t(3\cos^2\theta - 1)\right]\sin\theta d\theta$$

$$t = \tau, 2\tau$$

Numerical integration of the function $\phi(T_\perp\tau)$ shows it to decrease monotonously from unity to zero as the parameter $T_\perp t$ rises. If $\pi T_\perp t \simeq 3.6$, this function alters its sign and then does not exceed 0.05 by absolute value [14,20,44].

Relation (20) allows one to infer that if an electron and a nucleus with $I = 1/2$ are coupled by a weak (compared with the nuclear Zeeman) anisotropic hfi, the PEEM is of a comparatively simple shape in a disordered sample. The modulation harmonic of the initial part (small τ) coincides with the nuclear Zeeman frequency ν_I; as τ increases, it changes to $2\nu_I$. At sufficiently high τ the modulation harmonic is only $2\nu_I$ and its amplitude is constant. As the electron-nucleus separation r increases, the initial modulation amplitude decreases proportionally with $1/\nu^6$, and the decay time of the ν_I harmonic grows in proportion to $\tau_0 = 7.2h\nu^3/g_3 g_I \beta_e \beta_I$.

In the case of extremely weak hfi, the expression for the SEEM is analogous to Eq. (20) for the PEEM

$$<V(\tau,T)> = 1 - \frac{6}{5} \frac{g_e^2 \beta_e^2}{H_0^2 \nu^6} \sin^2 2\pi \left(\frac{\nu_I \tau}{2}\right) \left\{1 - \cos 2\pi\nu_I(\tau + T)\phi[T_\perp(\tau + T)]\right\} \quad (21)$$

That is, the SE does not have the $2\nu_I$ harmonic corresponding to the frequency $(\nu_\alpha + \nu_\beta)$, and the modulation is only the decaying harmonic ν_I.

When an electron interacts with N nuclei of the same type at distances ν_i, the modulations from different nuclei must be summed up, if account is taken only of the first-order terms with respect to B^2/ν_I^2. The modulation amplitude and the ν_I harmonic decay become dependent on the spatial distribution of the nuclei. The modulation amplitude is proportional to the function of the number of nuclei and the distances to them $\Sigma_i (1/\nu_i^6)$. The full decay time of the ν_I harmonic depends on the distance to the farthest nuclei whose contribution to modulation can be measured. The time dependence of the decreasing harmonic amplitude at ν_I is determined by the distance distribution of the nuclei around the electron.

These properties of ESEEM make it possible to gain information on the environments of paramagnetic centers by determining the number of nuclei and the distances to them. Figure 5 shows the PE and the SEEM for hydrogen atoms trapped at 77 K in the γ-irradiated glassy solution of 8 M H_2SO_4/H_2O. This matrix has only one type of magnetic nuclei, i.e., ^1H. The ESE signals are detected in an external magnetic field, H_0 = 3625 Oe. In this field the proton Zeeman frequency is ν_I = 15.43 MHz, which corresponds to a modulation period of 64.8 nsec. This period is observed at the initial part PE, then the modulation frequency is gradually doubled. Dashed lines show the ν_I harmonic decay. The SE has only the proton Zeeman frequency. This modulation has been analyzed quantitatively [45] in order to determine the surrounding structure of a trapped atom.

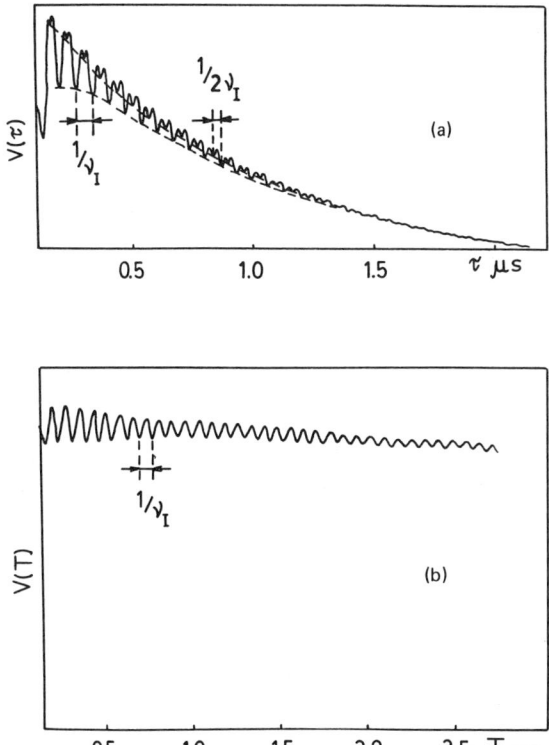

FIG. 5. The PE (a) and SEEM (b) for H atoms in 8 M H_2SO_4/H_2O.

2.4.2. Arbitrary hfi ESEEM Spectra

The modulation shape and the above simple relationship between modulation characteristics and hfi parameters may change essentially if the condition of extremely weak hfi does not hold. Therefore, in the general case, the ESEEM in disordered systems has to be analyzed with the exact expressions (4) and (6) for the PE and SEEM. These expressions are numerically integrated for various hfi models, and calculations are then compared with experiments in search of the best fit.

However, this is not the most efficient and convenient way, especially in the case of an unpaired electron interacting with several nuclei with different hfi parameters. In this case it is necessary to vary many independent parameters to make the time dependencies coincide.

In case of monocrystals, another way of analyzing the complex modulation is to transform the ESE envelope to the frequency domain and process the ESEEM spectra. The simplest method of gaining structural information is to determine spectral line maxima and to find their relationship with hfi parameters. This method is often employed to analyze ESEEM spectra of disordered systems.

In a disordered sample the shape and the position of a maximum of a spectral line is determined by the angular dependence of the amplitude factor k taken with the statistic weight of the orientation $\sin \theta$.

To derive the analytical expression for θ_{max} corresponding to a maximum $k \sin \theta$, let us consider the three extreme versions of the relation between $|T_\perp|$ and $|\nu_I \pm (a/2)|$ given in Table 1. Table 1 lists $k \sin \theta$ and the angles θ_{max} obtained from analysis of the derivatives of $k \sin \theta$, and also the values of $\nu_\alpha(\theta_{max})$ and $\nu_\beta(\theta_{max})$ corresponding to spectral line maxima in the above-mentioned extreme cases.

The lineshape in a modulation spectrum is also of interest. Therefore, the ideal PE and SEEM spectra were calculated [46] as the FT of Eqs. (4) and (6) neglecting the constant terms.

The data obtained [46] and also listed in Table 1 show the spectral components to be smooth, bell-shaped lines with one maximum, whose position depends on the relative values of $|T_\perp|$ and $|\nu_I \pm (a/2)|$.

Under extremely weak hfi $|T_\perp| \ll |\nu_I \pm (a/2)|$ the position of the maxima for ν_α and ν_β coincides practically with $|\nu_I \pm (a/2)|$, i.e., depends on isotropic hfi parameters. The anisotropic hfi effect shows up as decay of the harmonics in the time domain, and as changes in the lineshape and width (the latter increases with anisotropic hfi) in the frequency domain. The PEEM shows not only

TABLE 1

Values of $k\sin\theta$, Angles θ_{max}, and Maxima of the Modulation Spectral Lines ν_α and ν_β, Corresponding to These Angles, for Different Relations Between hfi Parameters

Approximation	$k\sin\theta$	θ_{max}	$\nu_\alpha(\theta_{max})$	$\nu_\beta(\theta_{max})$
$\|T_\perp\| \ll \|\nu_I \pm \frac{a}{2}\|$	$\dfrac{9\nu_I^2 T_\perp^2}{\left(\nu_I^2 - \dfrac{a^2}{4}\right)^2}\sin^3\theta\cos^2\theta$	51°	$\|\nu_I + \frac{a}{2}\|$	$\|\nu_I - \frac{a}{2}\|$
$\|\nu_I - \frac{a}{2}\| \ll \|T_\perp\| \ll \|\nu_I + \frac{a}{2}\|$	$\dfrac{36\nu_I^2 \sin^3\theta\cos^2\theta}{\left(\nu_I + \dfrac{a}{2}\right)^2 (3\cos^2\theta + 1)}$	59°	$\|\nu_I + \frac{a}{2}\|$	$\|0.67\,T_\perp + 0.14\left(\nu_I - \frac{a}{2}\right)\|$
$\|T_\perp\| \gg \|\nu_I + \frac{a}{2}\|$	$\dfrac{144\,\nu_I^2 \sin^3\theta\cos^2\theta}{T_\perp^2 (3\cos^2\theta + 1)^2}$	65°	$\|0.61\,T_\perp - 0.4\left(\nu_I + \frac{a}{2}\right)\|$	$\|0.61\,T_\perp - 0.4\left(\nu_I - \frac{a}{2}\right)\|$

Note: For calculation, the isotropic hfi constant is set positive; otherwise the second inequality should be inverted.

Source: Reproduced by permission from Ref. 46.

$|\nu_I \pm (a/2)|$ frequencies but also a narrow line at $\nu_\alpha + \nu_\beta \simeq 2\nu_I$. This frequency prevails in the ESEEM at great τ, since its decay depends on the terms of the next order of smallness as compared with the harmonics at ν_α and ν_β.

If the condition of extremely weak hfi does not hold, the spectral line maxima may shift essentially from the positions determined by the isotropic parameters a and ν_I. The size of this shift is informative of the value of anisotropic hfi. The high-frequency line maximum is less sensitive to anisotropic hfi than the low-frequency one. A substantial rise in $|T_\perp|$ results, eventually, in a strong line broadening. This reflects the absence of modulation from nuclei with high anisotropic hfi constants, as established experimentally in disordered samples.

In a disordered system the ESEEM spectra differ essentially from the ENDOR ones. The position of ENDOR line absorption maxima always differs from $|\nu_I \pm (a/2)|$; every line may correspond to two or three peaks in the spectra taken as the first derivative [46].

Summing up the results of the two previous sections, it should be said that a ESEEM and its frequency spectrum in a disordered sample are informative of isotropic and anisotropic hfi. A typical value of these interactions can be much lower than or equal to ν_I, which is of particular importance for nuclei with low magnetic moments or small hfi constants. The above conclusions are valid for nuclei with I = 1/2 involving such isotopes as ^1H, ^{13}C, ^{15}N, ^{19}F, and ^{31}P.

2.4.3. Effects of Weak nqi on ESEEM

Nuclei with I ≥ 1 possess a quadrupole moment, and therefore an ESEEM analysis of these nuclei must take into account electron-nuclear hfi as well as nqi. Since the problem of ESEEM involving nqi cannot be solved in the general form, the nqi effect was considered [26] for the case, when nqi and hfi were small compared with the nuclear Zeeman interaction. The relative size of hfi and nqi might be arbitrary. The applicability criterion for the weak nqi approximation is the fulfillment of $\nu_I \gg 3/2I\ K$ [26], where $K = e^2qQ/4h$, q is a field gradient, e^2qQ/h is a quadrupole coupling constant. The parameter $3/2I\ K$ determines the order of frequency variations of nuclear transitions due to nqi.

Structural studies on the environments of paramagnetic centers in matrices, when groups of hydrogen atoms belong to different molecules or molecular fragments, often require distinguishing of the modulation contribution only from a definite group of nuclei. For that purpose hydrogen is replaced by deuterium. A deuterium nucleus has a spin I = 1. Its magnetic moment is so that $\nu_I \simeq 2$ MHz in X-band experiments. As follows from an analysis of available data [47], the nqi constant does not exceed 0.4 MHz, that is, $(3/2I) \cdot K < 0.15$ MHz. Thus, for a deuterium nucleus in the X band the Zeeman coupling exceeds nqi substantially. Therefore, many investigators analyzed experimental modulation effects of deuterium nuclei in disordered systems using theoretical relations derived for I = 1 and neglecting nqi. However, estimates and numerical calculations [26,27,48] show that even a weak nqi leads to qualitatively new modulation effects in disordered systems. Therefore, experiments must be analyzed by taking nqi into account in order to avoid incorrect results.

In the above-mentioned approximation the PEEM for I = 1 follows the expression [26]:

$$V(\tau) = 1 - \frac{4}{3}\frac{B^2}{\nu_I^2}\left[1 - 2\cos 2\pi\nu_I\tau \cos 2\pi\left(\frac{A}{2}\tau\right) \cos \gamma\tau \right.$$

$$\left. + \cos 2\pi(2\nu_I\tau) \cos 2\gamma\tau + \cos 2\pi A\tau\right] \quad (22)$$

where $\gamma = (3/4)(eQV_{zz}/h)$; eQ is a nuclear quadrupole moment; V_{zz}, a component of the electrical field gradient tensor. Note that as nqi tends to zero ($\gamma \to 0$), Eq. (22) turns into one analogous to Eq. (19) for I = 1/2. The only difference is the numerical coefficient determined by the spin factor $I(I + 1)$. The nqi effect that readily manifests itself in disordered systems is the decay of the $2\nu_I$ frequency, whose amplitude must remain constant, if an electron and a nucleus are coupled only by hfi. This results from the time-dependent term $\cos 2\gamma\tau$ which appears at the $2\nu_I$ harmonic.

The ν_I harmonic decays now under the joint action of hfi and nqi and depends on the relative value of $T_\perp$ and 4K. Under anisotropic hfi the decay time increases in proportion to r^3. When 4K is small compared with $T_\perp$, the ν_I decay depends mainly on hfi. A fall in $T_\perp$ relative to 4K leads to a faster decay of ν_I than it has been expected from the dependence $\tau_0 \sim r^3$. On the contrary, if 4K exceeds $T_\perp$, the decay time is determined mainly by nqi. This conclusion is valid for both the PE and SE [27]. Thus, irrespective of the deuterium nuclei arrangement around a paramagnetic center in a disordered system, nqi affects the ν_I modulation harmonic decay, which occurs at a time $t_Q = yK/3\pi$ in any case, even if induced only by nqi. As follows from Eq. (22), this time is determined by averaging the function $\cos \gamma t$ ($t = \tau, 2\tau$), where $y = 2-3$ depending on the relative orientation of the hfi and nqi tensors.

For example, if $4K = 0.215$ MHz [49] (for D_2O ice), the $2\nu_I$ harmonic in the PEEM will decay within $\tau \simeq 2-3$ μsec, while the ν_I nqi-induced decay will take $\tau \simeq 4-6$ μsec. These qualitative estimates are in good agreement with experimental results on paramagnetic centers in heavy-water solutions [30,48,50].

The weak nqi effect on the ESEEM from deuterium nuclei can be illustrated as follows. Figure 6 shows the PEEM for deuterium atoms trapped in γ-irradiated 8 M D_2SO_4/D_2O at 77 K. In contrast to the above-mentioned case (see Fig. 5), the trapped atoms are surrounded not by protons but by deuterium nuclei, with the Zeeman frequency $\nu_I = 2.27$ MHz in an external field with $H_0 = 3467$ Oe. Hence, the modulation period is 440 nsec. Alongside changes in the modulation frequency, the substitution of deuterons for protons results also in an increase of the modulation amplitude at the initial part in correspondence with the contribution from the spin factor $I(I + 1)$. Moreover, a deuterated system does not show the transition from ν_I to $2\nu_I$ with increasing τ; the $2\nu_I$ harmonic appears only at $\tau \leq 2.5$ μsec and has a small amplitude.

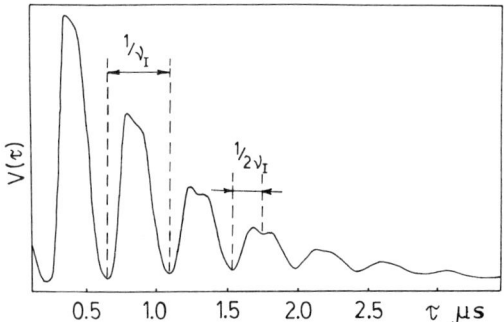

FIG. 6. The PEEM for D atoms in 8 M D_2SO_4/D_2O.

In hydrocarbons the quadrupole coupling constant of a deuterium nucleus bound to a carbon atom is 0.05-0.1 MHz [47], which is much lower than this constant in water. As a result, the nqi effect starts to show up at long times. However, even in this case nqi influences the ESEEM amplitude decay.

2.4.4. ESEEM under nqi Comparable with the Nuclear Zeeman Coupling

The requirement of a weak nqi compared with the Zeeman coupling is well fulfilled for deuterium but violated for ^{14}N, which also has a spin of I = 1. An analysis of published data demonstrates that 3/2I K for ^{14}N runs from 1.1 to 2.3 MHz [47,51], while the Zeeman frequency ν_I is 1.04 MHz in H_0 = 3390 Oe, i.e., the nqi for ^{14}N in the X band either is comparable with or even exceeds the nuclear Zeeman coupling.

In disordered systems ^{14}N nuclei induce various modulation frequencies differing essentially from the Zeeman frequency. Since the problem of ESEEM frequencies and their amplitudes under strong nqi has not been solved analytically, magnetic coupling data are usually obtained by an ESEEM frequency transformation.

The qualitative analysis [52,53,86] has shown that the ESEEM frequencies from ^{14}N that can show up in the SE of a disordered system depend on the ratio ν_{ef}/K, where $\nu_{ef} = |\nu_I \pm (a/2)|$. The two ν_{ef} values correspond to an electron spin orientation $m_s = \pm 1/2$ onto the external magnetic field direction.

If $\nu_{ef}/K < 1$, the nuclear transition modulation frequencies will be close to the NQR frequencies:

$$\nu_+ = K(3 + \eta) \qquad \nu_- = K(3 - \eta) \qquad \nu_0 = 2K\eta \qquad (23)$$

where η is the asymmetry parameter. Otherwise, when $\nu_{ef}/K \gtrsim 1$, there will arise only one modulation frequency:

$$\nu_m \simeq 2[\nu_{ef}^2 + K^2(3 + \eta^2)]^{1/2} \qquad (24)$$

corresponding to the $\Delta m_I = 2$ transition. The two frequencies corresponding to the $\Delta m_I = 1$ transitions will not be observed in the ESEEM spectra because of the strong orientation dependence.

A comparison of Eqs. (23) and (24) shows, in particular, that the frequencies (23) are independent of the external magnetic field, while the frequency (24) must increase with the field. When the ESR spectrum is sufficiently wide for its different parts to be excited by ESE pulses, this dependence is sometimes useful for determining the correspondence of the spectrum frequency to transitions (23) or (24). An example of the external field dependence of the frequency (24) was observed [54] in studying the complexes of VO^{2+}-acetylacetonate with nitrogen-containing donor bases.

Since the SEEM is a superposition of nuclear transitions for two electron spin projections, the maximum number of ESEEM frequencies from one nucleus in the SE is four. This number may increase in the PE or under interactions of an electron spin with several nuclei in the SE.

The reliability of Eqs. (23) and (24) can be verified by calculating the nuclear transition frequency spectrum for these parameters and by comparing this spectrum with an experimental one. In this case, the nuclear frequency spectrum is calculated as a ENDOR frequency spectrum neglecting the anisotropic nqi, which is assumed to be small and therefore to produce no effect on the frequencies [55].

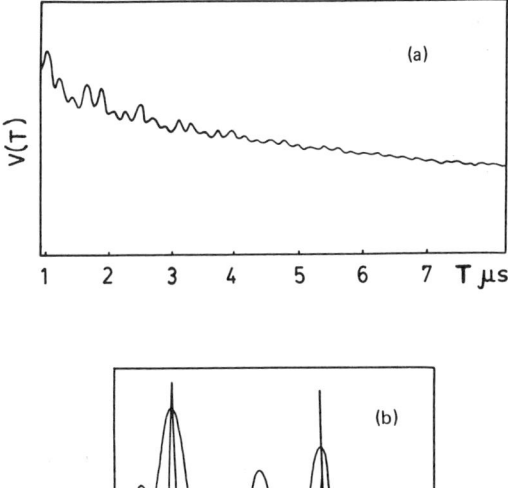

FIG. 7. The SEEM (a) for the radicals R at τ = 410 nsec, and the frequency spectra (b) for this envelope obtained by the FT (broad lines) and by the AR method (narrow lines).

Figure 7(a) shows an example of the SE envelope for the radical:

at τ = 410 nsec [52]. Figure 7(b) shows the frequency spectrum corresponding to the stimulated echo signal decay from Figure 7(a). The spectrum has three components with frequencies 1.3, 3.5, and 4.8 MHz. The three frequencies agree with those known for nitrogen NQR for the =NOH group of various compounds [51]. Besides, these frequencies show a property typical of NQR frequencies: the sum of two of them equals the third one. Hence, it is possible to say that the frequencies obtained by analyzing the radical SEEM are NQR frequencies for ^{14}N of the oximino group. As a result, Eq. (23) can readily yield K = 1.4 MHz and η = 0.47 [52].

If ^{15}N is substituted for ^{14}N, the modulation in a disordered system can be calculated numerically with good accuracy for various model situations. This makes it possible to determine the anisotropic hfi for the system under study, i.e., to gain a more detailed information on the structure of a paramagnetic center.

3. BIOLOGICAL APPLICATIONS OF ESEEM

In many biological systems paramagnetic centers are either present constantly or generated under an external action. Such systems are usually amorphous, with randomly orientated paramagnetic centers. This is a serious obstacle to structural investigations on these centers by the cw ESR spectroscopy. Definite advances have been made in this area using the ESE method, mainly due to pioneer works by Mims, Peisach, and their colleagues.

Structural data on paramagnetic centers in biological systems were obtained by analyzing ESEEM from various types of nuclei. Below these data will be discussed together with a brief characteristic of the analysis method employed.

3.1. ^1H and ^2D Modulations

A way of ESEEM analysis for obtaining structural data is to calculate the ESEEM numerically and to find the best fit with the experiment. This is performed usually in the approximation of uncorrelated nuclei, neglecting the real geometric arrangement of nuclei around a paramagnetic center and assuming their relative position

ELECTRON SPIN ECHO: APPLICATIONS TO BIOLOGICAL SYSTEMS   241

to be uncorrelated. In this case, for example, the PEEM from nuclei in a disordered sample is calculated for point dipoles as

$$V_N(\tau) = \prod_{i=1}^{N} <V_{I_i}(\tau,\theta)>_\theta \qquad (25)$$

For nuclei of the same type and at the same distance from a paramagnetic center, Eq. (25) assumes the form:

$$V_N(\tau) = <V_{I_i}(\tau,\theta)>_\theta \qquad (26)$$

Such calculations were performed [50,56] in numerical simulation of the PEEM for the complexes of Nd^{3+} and Cu^{2+} with H_2O and D_2O, and also for the complexes of copper with some other ligands in glassy water solutions. For example, in calculating the proton modulation for Nd^{3+}, the first coordination sphere was assumed to involve nine water molecules, as in the case of neodymium ethyl sulfate and neodymium bromate crystals. For that number of water molecules in the first coordination sphere the best fit between calculation and experiment was attained at the distance r = 3.0 ± 0.1 Å to 18 protons. It was found for the Cu^{2+}-aquo complex that four protons at a distance of 2.4 Å are the closest ones to the ion. Alongside the nuclei of the first sphere, the calculations also took into account the other matrix protons, whose position was set by a continuous distribution starting with a certain minimum distance (r_{min} = 4 Å for Nd^{3+} and r_{min} = 2.8 Å for Cu^{2+}): The distribution density was as that for water protons. There is one more example of modulation calculations for deuterium nuclei neglecting nqi [50]. The comparison of calculation and experiment showed [50] the presence of an additional ESEEM amplitude decay due to nqi in the experiment.

The structural data on Nd^{3+} and Cu^{2+} surroundings in water were then used to study the accessibility of paramagnetic centers in copper-containing proteins for water molecules in frozen water solutions. These proteins have two types of paramagnetic copper ions. The type I is characterized by a small $A_\parallel$ in the ESR spectra and by an intense optical absorption near 600 nm; the type II has $A_\parallel$ several times as high as that of the type I, and a low absorption intensity in the visible spectrum [57]. Type I Cu^{2+} centers are present in some low molecular weight electron-transporting proteins, including azurin, plastocyanin, and stellacyanin, as well as in copper-containing oxidases, such as laccase, ascorbate oxidase, and ceruloplasmin.

An attempt was made [58] to investigate type I Cu^{2+} centers in laccase, stellacyanin, and azurin for their accessibility for surrounding solvent molecules using the ESE method. For that purpose the ESEEM for Cu^{2+} was detected for proteins dissolved in H_2O and D_2O under similar conditions. The quotient of the two envelopes has the modulation from deuterium nuclei divided by the modulation from substituted protons. The remaining proton frequency and the decay due to the difference in the phase relaxation times for ions in the two species were removed by a high- and low-pass filtering. Thereafter the resulting normalized ESEEM curves had a characteristic modulation at the Zeeman frequency of deuterium nuclei in the external field applied. That proved the presence of deuterium nuclei near copper ions in proteins. To make quantitative conclusions on the nature of the deuterium modulation observed, its modulation and decay, characterizing the arrangement of nuclei around a paramagnetic center, were compared with the ESEEM parameters obtained in the same way for Cu^{2+}- and Nd^{3+}-aquo complexes in H_2O and in H_2O/D_2O mixtures with various relative concentrations of the components. The comparison was performed for the initial lengths of the normalized ESEEM curves characterized by low nqi effects.

The observed deuterium modulation amplitude was proportional to $\Sigma_i(1/\nu_i^6)$ at small τ. Estimates of this sum, made for various models of relative positions of an ion and surrounding deuterium nuclei, showed that a Cu^{2+} ion was not buried into a protein molecule but was either on its surface or inside a shallow crevice with an outlet to the surrounding medium, i.e., had free access to the solvent. It was also found that the closest deuterium nuclei of the solvent were at a distance of about 3-3.5 Å from an ion.

The above data obtained by the ESE method show that the metal centers in copper-containing proteins in solutions are more open and hence more accessible for water molecules than is indicated by x-ray crystallography of crystalline proteins [58].

Similar experiments were performed [59] to investigate water coordination to metmyoglobin. The metmyoglobin ESE is strongly modulated due to the interaction with ^{14}N nuclei of porphyrin and imidazole ligands to iron. Therefore, to evaluate the effect of the interactions with water, the SEEM was determined in two samples of metmyoglobin, i.e., in H_2O and D_2O. The second envelope was then divided by the first one. The FT of that normalized function gave a spectrum with pronounced lines induced by the interaction with ^2D.

The lines were a triplet, with the central frequency coinciding with the Zeeman frequency of deuterium and increasing with the external field from 3,384-3,985 Oe. At the same time, the line splitting did not vary and equaled 0.4 MHz. Therefore, it was concluded [59] that the central spectral line belonged to the deuterium nuclei of ambient water molecules which were not directly connected with the heme iron, and also to the exchangeable deuterons in the vicinity of paramagnetic centers. The other two lines were attributed to the frequencies $\nu_{\alpha,\beta} \simeq |\nu_I \pm (a/2)|$ from deuterium nuclei with an isotropic hfi constant a = 0.8 MHz. The latter nuclei belonged to water molecules bonded directly to the heme iron since the heme did not have any other close-distant ligands with exchangeable protons, which could produce such an effect on the ESEEM

As mentioned [59], using the ESE method it was shown for the first time that water near the heme in metmyoglobin is actually bonded to iron rather than located near it as was concluded from x-ray crystallography, ESR with $H_2^{17}O$, and proton relaxation data.

On the other hand, similar experiments with methemoglobin [59] showed only one line with deuterium frequency in the ESEEM spectrum of the normalized curve. This proved the absence of any spin density at the surrounding deuterium nuclei and hence a weaker coupling with water than in metmyoglobin.

Summing up the results [58,59], it should be said that the quotient method with isotopic substitution may prove to be useful to study the interaction of an unpaired electron with a definite group of nuclei when it interacts with a great number of different nuclei contributing to the ESEEM.

ESE investigations on coordination and interatomic distances in an enzyme-substrate system were carried out [60] in experiments on the binding of cytochrome P-450$_{scc}$ with cholesterol-22,22-d$_2$ deuterated in one position:

and also with its derivatives 2,2-hydroxycholesterol-22-d and 20-azacholesterol-22,22-d$_2$. An indication for complex formation was deuterium modulation that arose in the ESE signal envelope after adding the deuterated substrate to the system.

The deuterium modulation was analyzed quantitatively only for 22(R)-hydroxycholesterol-22-d, which has only one deuteron. Numerical SEEM calculations yielded the distance between an unpaired electron of P-450$_{scc}$ and a deuteron, 4 ± 1 Å. On the contrary, experiments with the "unnatural" deuterated isomer 22(S)-hydroxycholesterol-22-d showed no modulation from deuterium. This indicated a long distance (over 6 Å) between the heme and the ^{2}D nucleus in the complex. The deuterium modulation in cholesterol-22,22-d$_2$ and 20-azacholesterol-22,22-d$_2$ was not analyzed in detail. However, its presence in both cases proved the proximity of these molecules to the heme.

The ESE method was employed to study iron-sulfur proteins to gain information on the proton exchange near iron-sulfur centers depending on their environments [61,62]. There occur two common types of polynuclear iron-sulfur proteins [63]. First, ferredoxins which consist of molecules with either one 2Fe-2S* cluster, or one and two 4Fe-4S* clusters. Second, high potential iron-sulfur proteins (HIPIP) contain a 4Fe-4S* cluster whose structure is nearly identical to that of the 4Fe-4S* in ferredoxins. The proteins under study may undergo one-electron redox transformations resulting in either appearance or disappearance of ESR spectra.

The PEEM for 2Fe-2S* and 4Fe-4S* of ferredoxins and HIPIP have a modulation at a frequency coinciding with the proton Zeeman frequency in the applied external magnetic field, which points to the presence of protons near the clusters. The storage of ferredoxin 2Fe-2S* and 4Fe-4S* in D_2O prior to the one-electron reduction of an iron-sulfur cluster leads to deuterium along with proton modulation in the PEEM. This proves the H-D exchange in the vicinity of a paramagnetic center. The deuterium modulation amplitude depends on the storage time of the oxidized ferredoxin in D_2O before reduction. The exchange half-time is about a minute. If ferredoxin is stored in D_2O after its reduction in H_2O, the deuterium modulation arises after the same time interval but with much lesser amplitude [61,62].

On the other hand, a long storage of oxidized or reduced HIPIP in D_2O does not result in H-D exchange near a paramagnetic center. This agrees with x-ray crystallographic data showing that an iron-sulfur cluster is less accessible for solvent molecules in HIPIP than in 4Fe-4S* ferredoxin. The H-D exchange in HIPIP starts when a protein molecule has unfolded, thus making a cluster accessible for water molecules.

Nitroxide radicals are widely used in biological investigations as spin labels and probes. The ESE method can also supply valuable structural information on these systems.

The ESE method was applied [64-66] to study micellar systems, with x-doxylstearic acids:

$$HOOC(CH_2)_{x-2}\underset{\underset{\underset{}{}}{}}{C}(CH_2)_{17-x}CH_3$$

$$\text{with } O, N\text{-}O \text{ ring}, \quad x = 5, 7, 10, 12, 16$$

as spin probes. The ESE signal from the spin probes was deuterium modulated in frozen micellar solutions prepared with D_2O or with surfactants deuterated in their head group or counterions. The ESEEM amplitude was measured as a function of x, which made it possible to consider the arrangement of nitroxide and the spin probe conformation in micelles.

The nitroxide radical solvation was investigated [67-69] in frozen solutions of alcohols and water. Figure 8 shows the primary ESE signal envelopes for 2,2,6,6-tetramethylpiperidine-1-oxyl in CD_3OH and CH_3OD. The modulations are seen to differ substantially. The use of compounds with different deuterated fragments made it possible to observe the complexing of a radical with a solvent

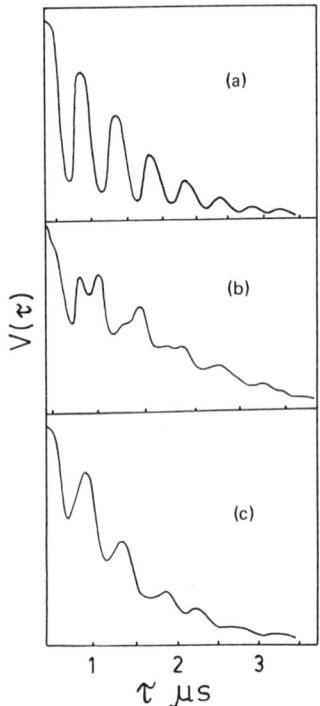

FIG. 8. The PEEM for 2,2,6,6-tetramethylpiperidine-1-oxyl radicals in CD_3OH (a), CH_3OD (b), and for 2,2,5,5-tetramethyl-4-phenyl-3-imidazoline-3-oxide-1-oxyl in CH_3OD (c).

molecule accompanied with the spin density transfer to the deuterium nucleus of the OD group. The isotropic hfi constant for that nucleus and the distance to it were calculated numerically. It was found that the distance to the nucleus increased with the geometric size of a solvent molecule in the series: water, methanol, ethanol. The complexing of nitroxide with hydroxyl-containing molecules was shown to depend on their structure. For example, in contrast to piperidine radicals, imidazoline nitroxide radicals did not complex with solvent molecules (Fig. 8).

3.2. ^{13}C Modulation

The use of ^{13}C-labeled compounds in structural ESE studies has already been described [70]. The affinity of transferrin to metal ions is regulated by bicarbonate and other anions, oxalate included. However, it has been uncertain whether anions are directly bound to metal ions or to some other protein molecule fragments sufficiently remote from the ions. Experiments were performed with samples of transferrin, in which Cu^{2+} was substituted for Fe^{3+}, which is usually bound to the protein under physiological conditions. To detect the oxalate coordination of copper ions, the PEEM for Cu^{2+}-transferrin was compared in the presence of ^{12}C and ^{13}C oxalate in the sample. The division of the PEEM by one another yielded a curve with a modulation period similar to that observed for the PE of model complexes of Cu^{2+} with ^{13}C-oxalate in a water solution. These experiments proved oxalate to bind copper ions in the protein. A similar method was employed to show oxalate to be directly coordinated to metal ions also in Cu^{2+}-conalbumin-oxalate complexes [71]. On the other hand, the ESE for Cu^{2+}-conalbumin complexes prepared with ^{13}C-carbonate did not show modulation from ^{13}C, which points to a weak hfi with that nucleus.

3.3. ^{14}N and ^{15}N Modulations

An example of an analysis of ESEEM from ^{14}N and ^{15}N is furnished by studies on Cu^{2+}-diethylenetriamine-imidazole and Cu^{2+}-(imidazole)$_4$ complexes in water-glycerol glasses (pH 8.3) [55]. The modulation effects result from the interaction of ion electron spins with remote nitrogen nuclei at the third position of the imidazoline cycle. The coordinating nitrogen atoms have comparatively high hfi constants and thus give no modulation.

For both complexes containing ^{15}N-imidazole the SE shows two modulation frequencies, 0.25 and 2.9 MHz, in the magnetic field H_0 = 3330 Oe. Such frequencies mean that the isotropic constant of these nuclei is close to $2\nu_I$. Numerical PEEM calculations agree with the experiment at a_{15_N} = 2.5 MHz and r = 2.9 Å.

The SEEM for the complexes with ^{14}N-imidazole shows three frequencies, 0.7, 1.4, and 4 MHz. The spectrum with these frequencies was calculated with the isotropic hfi constant a_{14_N} = 1.75 MHz taken from ^{15}N-imidazole experiments, and with the quadrupole parameters K = 0.356 MHz and η = 0.98 obtained by the NQR method [72] for a protonated nitrogen atom at the third position of imidazole.

The two low frequencies, 0.7 and 1.4 MHz, were attributed [55] to NQR frequencies which coincided, within experimental accuracy, with the data published for protonated nitrogen in imidazole 0.698, 0.719, and 1.417 MHz (the two frequencies near 0.7 MHz were not resolved in the spectrum). The frequency 4 MHz belonged to the transition with Δm_I = 2 [Eq. (24)] in the nuclear sublevels, with m_s opposite to that of the NQR frequencies.

The above described investigations have shown that the parameters for hfi with the nucleus of uncoordinating nitrogens of imidazole are the same in copper complexes with one and with four imidazole molecules. For pH 8.3 this nitrogen atom is protonated in the complexes. The experiments with Cu^{2+}-diethylenetriamine-imidazole and Cu^{2+}-(imidazole)$_4$ complexes were considered as models. The results were then repeatedly used by Mims, Peisach, et al. in their experiments with proteins involving the types I and II copper.

The type I copper was investigated in stellacyanine [73,74], laccase from *Rhus vernicifera* [75,76], ascorbate oxidase from *Cucurbita pepo medullosa* [76]. The copper ions were shown to be imidazole-coordinated at these centers. This conclusion was based on the modulation frequencies 0.7, 1.4, and 4 MHz observed in ESE experiments with the type I copper ions. These frequencies are typical for the protonated nitrogen at the third position of imidazole.

The imidazole coordination of the type II copper ions was also observed in laccase and ascorbate oxidase [76]. The only difference was the splitting of the 0.7-MHz line into two spectral components with a splitting of 0.2 MHz. This splitting pointed to a slight perturbation of the ^{14}N quadrupole field. That perturbation was believed [76] to result from a weak H bond between the remote imidazole nitrogen and the protein structure. A similar splitting was also detected for the type II copper galactose oxidase [77] and in holo superoxide dismutase [78]. However, no splitting of this kind was observed in superoxide dismutase without Zn^{2+}, normally bound to an imidazole nitrogen atom at the third position.

The coordination with the imidazole fragment was discovered in the complexes of Cu^{2+} and Co^{2+} with bleomycin [79] and of Cu^{2+} with glycyl-L-histidyl-L-lysine [80] and with conalbumine [71]. The complex of Fe^{3+} with bleomycin [79] and the iron-sulfur protein in succinate-ubiquinone oxidoreductase [81] were found to involve nitrogen-containing ligands; however, these were not reliably attributed to particular molecular groups.

The ESEEM in a series of low-spin heme compounds were investigated [82] in glassy solutions. The experiments were carried out in the model complexes of heme-mercaptoethanol-nitrogen-containing bases (^{14}N, ^{15}N-imidazole, pyridine, propylamine) and heme proteins involving cytochrome c, some myoglobin derivatives, and cytochrome P-450. The FT analysis of the SEEM showed the isotropic hfi constants of a Fe^{3+} electron spin with porphyrin nitrogen nuclei to be almost the same in complexes of both groups and to vary within 4-5 MHz. The ESE for the model heme complexes, with the bases as axial ligands, showed NQR frequencies from the base nitrogen participating in the coordination, which allowed the isotropic hfi constant for the nitrogen nucleus to be estimated as 2 MHz. The ESEEM analysis demonstrated the presence of nitrogen-containing ligands in Fe^{3+} of heme proteins [82]; yet the hfi and nqi parameters for nitrogen nuclei differed from those in the model complexes. The nitrogen modulation was also analyzed in hemoglobin azide [83].

Two hydrogenases with different properties can be purified from *Methanobacterium thermoautotrophicum*. In vitro, one hydrogenase (FH_2ase) catalyzes the reduction of the physiological two-electron acceptor 7,8-didemethyl-8-hydroxy-5-deazaflavin (F_{420}) by hydrogen, while the other (MVH_2ase) reduces only the artificial one-electron acceptor methylviologen [84]. Both enzymes involve paramagnetic nickel. ESE studies on these centers reveal differences between the two hydrogenases. Near the paramagnetic nickel in FH_2ase there is ^{14}N, which is absent in MVH_2ase. The SE for FH_2ase is modulated with an essential amplitude, which is not the case for the MVH_2ase ESE.

The FT of the SEEM for FH_2ase has three narrow lines, 0.4, 1.2, and 1.6 MHz, and a broad line at 4 MHz. A spectrum similar to that observed was calculated for K = 0.42 MHz, η = 0.48, and a = 1.8 MHz. By calculations, the anisotropic hfi with a nitrogen nucleus was found to be small, and $r_{Ni-N} \geq 3$ Å. The hfi constant obtained has suggested [84] that a nitrogen atom does not participate in the direct coordination with nickel but is either a part of the aromatic heterocyclic ligand or forms a strong H bond with the ligand. Yet the comparison of the calculated nqi parameters with those available in the literature has not proved helpful for its attribution to a definite, nitrogen-containing, molecular group among those constituting FH_2ase.

The ESEEM from ^{14}N and ^{15}N for radical cations of chlorophyll a and bacteriochlorophyll a was investigated and compared [85-89] with the ESEEM for P-700$^+$ and P-860$^+$ centers, primary electron donors in the plant photosystem I and in the purple bacteria photosystem in order to gain information on the structure of the reaction center.

Figures 9 and 10 show the PE and SEEM for BChl a$^+$ with ^{14}N and ^{15}N nuclei. Since the isotopic substitution of nitrogen atoms influences the modulation effects observed, these effects seem to be induced by the interaction of an unpaired electron in the radical cation with nitrogen nuclei. The SE taken at various τ (Figs. 9 and 10) demonstrates the frequency suppression effect. Figures 9 and 10 show also the ESEEM frequency spectra obtained by the FT and AR analyses. These spectra, as well as those taken for other τ, were used [89] to reliably find a great number of modulation frequencies in the ESE for BChl a$^+$ with ^{14}N and ^{15}N (Tables 2 and 3). The ESEEM frequencies for P-860$^+$ with ^{14}N and ^{15}N were found [88] in the same way (see Tables 2 and 3).

A comparison of the ESEEM frequencies for BChl a$^+$ and P-860$^+$ with ^{14}N and ^{15}N shows the absence of the proportionality corresponding to the values of magnetic moments. Hence, the ESEEM frequencies of these paramagnetic centers with ^{14}N are essentially nqi-dependent.

As in the case of Chl a$^+$ investigated earlier [87], the ESEEM frequencies for BCHl a$^+$ and P-860$^+$ with ^{14}N can be arranged into two sets for which the NQR frequency property holds, i.e., the sum of two of them equals the third one. Table 2 lists sets of these frequencies, and K and η obtained by Eq. (23) and attributed firstly in [87] and then in [88,89] to two types of pyrrole nitrogen atoms.

In contrast, the ESEEM for P-700$^+$ with ^{14}N showed only two frequencies differing from one another by a factor of two (Table 2).

The observation of the sets of modulation frequencies corresponding to Eq. (23) is somewhat unexpected, as the ENDOR data (see below) prove a substantial hfi. Therefore, these results and their tentative interpretation require further theoretical consideration and, perhaps, numerical calculations of the spectra.

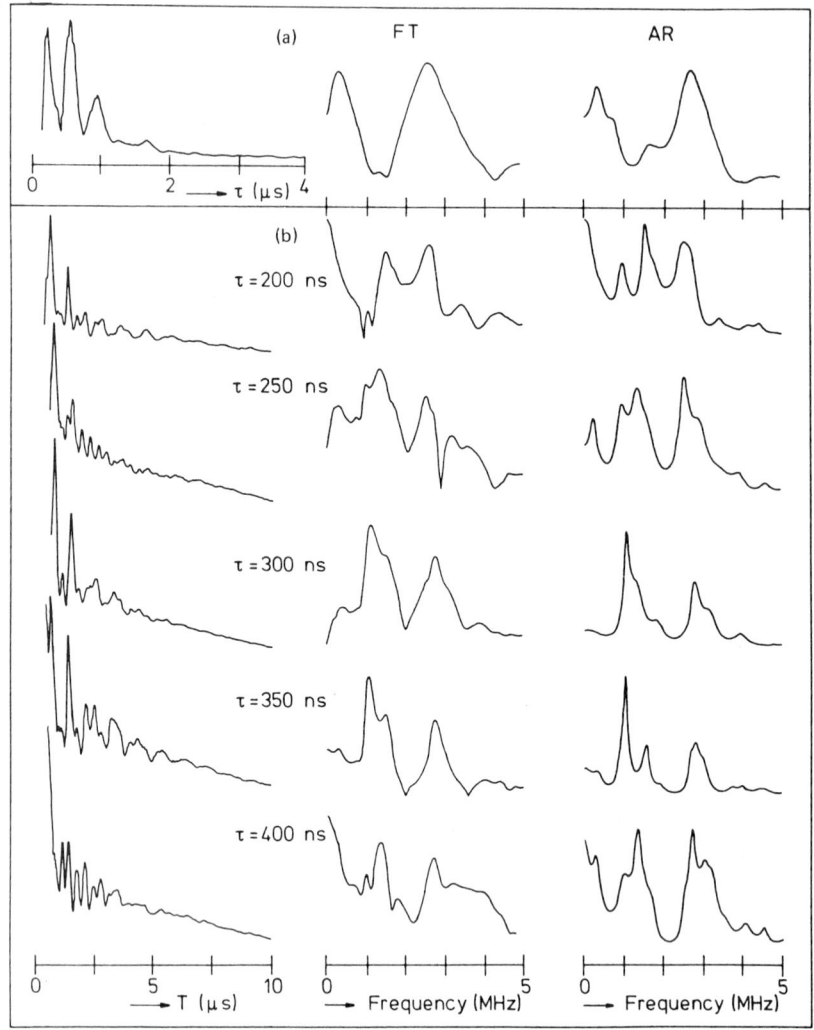

FIG. 9. The PE (a) and SEEM (b) of BChl a$^+$ with ^{14}N for values of τ as indicated, and frequency spectra obtained by the FT and AR analysis. (Reproduced by permission from Ref. 89.)

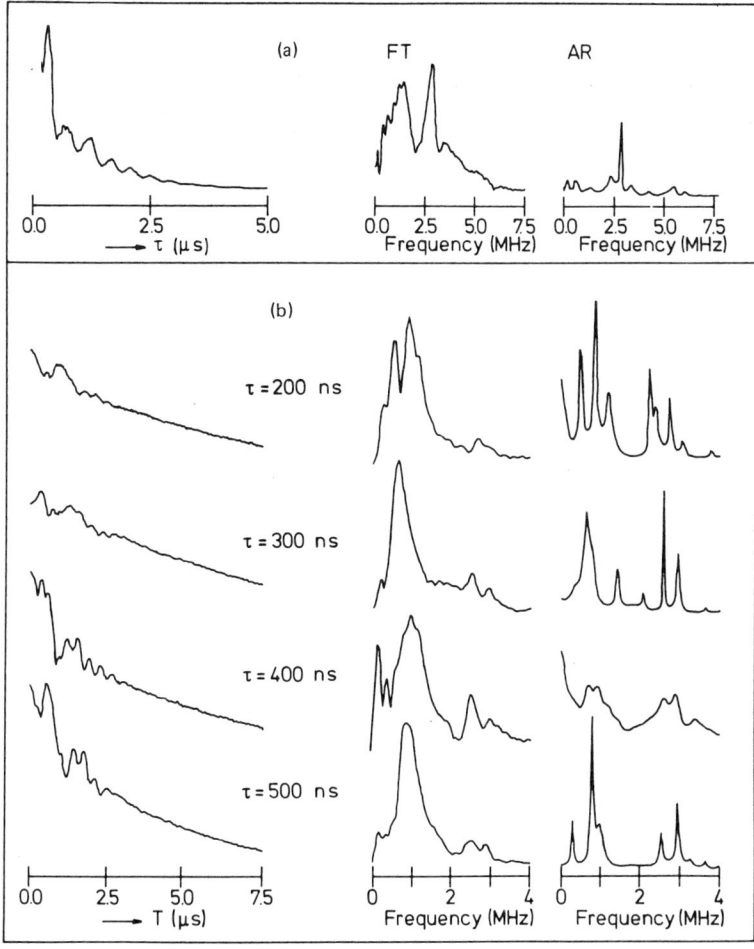

FIG. 10. The same as in Fig. 9 for BChl a$^+$ with ^{15}N. (Reproduced by permission from Ref. 89.)

TABLE 2

Stimulated ESEEM Frequencies and ^{14}N nqi Parameters for
ChI a$^+$, P-700$^+$, BChI a$^+$, and P-860$^+$

Center	Modulation frequencies (MHz)			K_a MHz	η_a	K_b MHz	η_b	Ref.
ChI a$^+$	0.8	(a)	(b)					
	1.6		(b)					
	2.0	(a)		0.8	0.5	0.67	0.6	87
	2.4		(b)					
	2.8	(a)						
P-700$^+$	1.3	(a)		0.65	1			87
	2.6	(a)						
BChI a$^+$	0.95	(a)						
	1.1		(b)					
	1.5	(a)	(b)	0.68	0.7	0.7	0.79	89
	2.55	(a)						
	2.7		(b)					
P-860$^+$	0.5	(a)						
	0.9	(a)	(b)	0.42	0.68	0.68	0.66	88
	1.6	(a)	(b)					
	2.5		(b)					

TABLE 3

Stimulated ESEEM Frequencies for BChI a$^+$, P-860$^+$,
ChI a$^+$, and P-700$^+$ with ^{15}N

Center	Modulation frequencies (MHz)	Ref.	ENDOR data (MHz) [90]		
			$A_\perp{}^a$	$A_\parallel{}^a$	A^b
BChI a$^+$	0.5, 1.0, 2.5, 2.95	89	2.1, 3.0	5.6, 7.2	3.15, 3.25, 4.05, 4.4
P-860$^+$	0.49, 0.71, 1.38, 2.6	88	1.9, 2.3	2.9, 3.1	1.05, 1.61, 2.22, 2.6
ChI a$^+$	0.35, 0.55, 0.85, 1.15, 1.45, 1.7, 2.15, 2.6, 2.75	c			
P-700$^+$	0.3, 0.5, 1.1, 1.4, 2.5	c			

^{a}Experimental data for frozen samples.
^{b}Experimental data for liquid solutions.
^{c}Our data.

BChl a^+ and P-860$^+$ with ^{14}N and ^{15}N have been intensively investigated also by ENDOR at room and low temperatures [90]. The ENDOR absorption lines, both for liquids and solids, can only be detected at $\gtrsim$1.5 MHz. Consequently, the low-frequency spectral part, which might be useful for interpreting the spectrum (especially in solids), turns out to be lost.

However, the ENDOR spectra were successfully used [90] to determine the isotropic hfi constants for nitrogen nuclei in BChl a^+ and P-860$^+$ at room temperature (Table 3). The ENDOR spectrum for solid BChl a^+ with ^{15}N showed two intensive lines at 2.1 and 3.0 MHz, and two weak lines at 4.2 and 5.0 MHz, ascribed to nitrogen nuclear $A_\perp$ and $A_\parallel$ of the two groups. The intensive and weak lines for P-860$^+$ were slightly displaced toward low frequencies, 1.9, 2.3, and 2.9, 3.1 MHz, respectively. These data were used [90] to infer a dimeric model of P-860$^+$. The model implies the reaction center to include two chlorophyll molecules with delocalized spin density.

As compared to ENDOR spectra, the ESEEM ones make it possible to cover the range of low frequencies with intensive peaks. When analyzed, these spectra prove to be asymmetric relative to the ^{15}N Zeeman frequency ν_I. The frequencies observed do not correspond to the relation $\nu_{\alpha,\beta} \simeq |\nu_I \pm (a/2)|$ which can be calculated using the hfi constants published [90]. This agrees with the solid ENDOR spectra shape pointing to comparable anisotropic and isotropic hfi in these paramagnetic centers. Moreover, an analysis of the modulation frequencies for BChl a^+ and P-860$^+$ demonstrated [88,89] the position of some maxima to correspond to $A_\perp$ observed in the ENDOR spectra [90]. Yet the general structure of the ESEEM spectrum and the origin of the lines observed were not explained.

The ESEEM spectra from ^{15}N were taken for Chl a^+ and P-700$^+$. They were asymmetric relative to the Zeeman frequency of ^{15}N too and had frequencies listed in Table 3. At present there are no ENDOR data available for these particles, and therefore we cannot compare results obtained by different methods.

Another paramagnetic form of P-860 is its triplet state (^{3}P-860). It is produced by the recombination of a photoinduced radical pair P-860$^+$I$^-$, where I$^-$ is the intermediate electron acceptor bacteriopheophytin. The short lifetime of ^{3}P-860 (5 µsec at room temperature and about 110 µsec at cryogenic temperatures) hampers gaining the cw ENDOR information on hfi. This lifetime, however, is sufficient for the ESE method employed [91] to investigate the hfi and nqi with ^{14}N and ^{15}N in ^{3}P-860 and ^{3}BChl a.

The data obtained for ^{3}P-860 [91] made it possible to infer that on a magnetic resonance time scale of 10^{-6}-10^{-7} sec the triplet state ^{3}P-860 is localized on two BChl a molecules, while on an optical time scale it is 10^{-13} sec on one molecule [92].

At present a great body of experimental data has been gained on the chlorophyll cation radicals P-700 and P-860 using the ESE method. This information has been analyzed only qualitatively because of the complex structure of the centers involving several magnetically nonequivalent nuclei. Therefore, the important hfi and nqi data for nitrogen nuclei require a more detailed quantitative analysis.

3.4. ^{31}P Modulation

The Nd^{3+} modulation by adenosine triphosphate was investigated [93] in glassy water-glycerol solutions. The addition of ATP to the solutions led to ^{31}P modulation of the Nd^{3+} ESE. The modulation amplitude increased with the ATP concentration up to the metal to nucleotide ratio of 1:3 which corresponded to saturation of the ion coordination sites. Nearly the same metal to nucleotide ratio was required to detect the maximum effect in the optical spectrum at room temperature. The proximity of optical to ESE data made it possible to conclude [93] that an ESEEM analysis might be applied to titration, when the optical titration is impossible, as in cases with Ce^{3+}-ATP complexes [16,93]. The ^{31}P modulation effects observed in the Nd^{3+} ESE were compared with numerical calculations with the NMR data [94]. Two alternative spatial ^{31}P distributions were assumed around a Nd^{3+} ion coordinated by two ATP molecules. First phosphorus nuclei were

ELECTRON SPIN ECHO: APPLICATIONS TO BIOLOGICAL SYSTEMS   257

at distances of 3.2, 3.2, and 6.1 Å from the ion; second, they were
at 3.3, 3.5, and 5.1 Å. The ESEEM calculations proved the former
set distances.

4. CONCLUSION

As was already mentioned above, this chapter is devoted to the ESE
applications to studies on biological systems based on ESEEM analysis.
The material discussed has demonstrated the ESE method to be useful
for gaining qualitative and quantitative information on the structure
around a paramagnetic center in a biological system. This is the
property that has attracted much attention of investigators to the
ESE method as applied to biological systems.

However, the potentialities of the ESE method are much wider:
as a pulsed version of ESR spectroscopy, it can be successfully combined with various physical actions on a system. For example, some
kinetic ESE versions are applied to studies on fast processes, with
paramagnetic particles produced by the light of a powerful lamp [95],
a laser [96], or an electron pulse [97]. In these experiments the
ESE method is employed as a time-resolved spectroscopy enabling one
to detect unstable paramagnetic particles under photolysis, radiolysis,
or during their conversions. An example of such experiments with biological systems is an investigation on paramagnetic particles generated in photosynthetic systems under photoinduced electron transfer.
These experiments have been reviewed in detail by Hoff [11].

The ESE method, in combination with a pulsed action of an electrical field, was used to measure the linear electrical field effect
and to study the binding and the symmetry of local environments of
paramagnetic centers in proteins [8,16,81].

Among other potentialities of the ESE method which seem to be
useful for biological studies but have not been employed yet are the
following. The ESE method is a direct and informative technique of
determining the spatial paramagnetic center distributions in solids.
It can be used to detect whether or not the particle distribution is

homogeneous; to measure the local concentrations; to restore the distribution function and its variations during diffusion or reaction in the case of a pair distribution of paramagnetic centers; and to determine the number of spins per group in the case of a group distribution. The techniques of such measurements, their theory and physicochemical applications are available [4,5,98]. The microscopic distribution of paramagnetic particles in a sample has been recently determined [99] by the ESE tomography.

The cw ESR spectroscopy is insensitive to rotational mobility times above 10^{-7} sec. The methods of ESE pulsed saturation [100] and of ESE with a magnetic field jump [101] make it possible to measure the rotational mobility of paramagnetic particles or their fragments in single crystals and amorphous substances with times above 10^{-7} sec.

It is possible to expect from the foregoing that the ESE applications to biological studies will grow in importance and eventually hold a firm place among other magnetic resonance methods.

ABBREVIATIONS

AR	autoregressive
ATP	adenosine 5'-triphosphate
BChl	bacteriochlorophyll
Chl	chlorophyll
cw	continuous wave
ENDOR	electron nuclear double resonance
ESE	electron spin echo
ESEEM	electron spin echo envelope modulation
ESR	electron spin resonance
FT	Fourier transformation
hfi	hyperfine interaction
HIPIP	high potential iron-sulfur protein
nqi	nuclear quadrupole interaction
NQR	nuclear quadrupole resonance

ODMR	optically detected magnetic resonance
PE	primary echo
PEEM	primary echo envelope modulation
PENDOR	pulsed ENDOR
rf	radiofrequency
SE	stimulated echo
SEEM	stimulated echo envelope modulation

REFERENCES

1. W. B. Mims, in *Electron Paramagnetic Resonance* (S. Geschwind, ed.), Plenum Press, New York, 1972, p. 263.
2. K. M. Salikhov, A. G. Semenov, and Yu. D. Tsvetkov, *Electron Spin Echo and its Applications,* Science Novosibirsk, 1976.
3. *Time Domain Electron Spin Resonance* (L. Kevan and R. N. Schwartz, eds.), Wiley-Interscience, New York, 1979.
4. Yu. D. Tsvetkov, *Uspekhi khim.*, *52,* 1514 (1983).
5. A. M. Raitsimring and K. M. Salikhov, *Bull. Magn. Reson.*, *7,* N3 (1985).
6. W. B. Mims and Yu. D. Tsvetkov, *International List of ESE Spectrometers,* Preprint of Bell Laboratories, New Jersey, Murray Hill, 1982.
7. W. E. Blumberg, W. B. Mims, and D. Zuckerman, *Rev. Sci. Instrum.*, *44,* 546 (1973).
8. W. B. Mims and J. Peisach, in *Biological Magnetic Resonance,* Vol. 3 (L. J. Berliner and J. Reuben, eds.), Plenum Press, New York, 1981, p. 213.
9. A. G. Semenov, M. D. Schirov, V. D. Zhidkov, V. E. Khmelinsky, and E. V. Dvornikov, Preprint N3, Institute of Chemical Kinetics and Combustion, USSR Academy of Science, Novosibirsk, 1980.
10. P. A. Narayana and L. Kevan, *Magn. Reson. Rev.*, *1,* 234 (1983).
11. A. J. Hoff, *Quart. Rev. Bioph.*, *17,* 153 (1984).
12. S. L. Tan, J. S. Waugh, and W. H. Orme-Johnson, *J. Chem. Phys.*, *81,* 576 (1984).
13. J. R. Norris, M. D. Thurnauer, and M. K. Bowman, in *Advances in Biological and Medical Physics,* Vol. 17 (J. H. Lawrence, J. W. Gofman, and T. L. Hayes, eds.), Academic Press, New York, 1980, p. 365.
14. S. A. Dikanov, V. F. Yudanov, and Yu. D. Tsvetkov, *Zh. Strukt. Khim.*, *18,* 460 (1977) (*J. Struct. Chem.*, *18,* 370 (1977)).

15. L. Kevan, in *Time Domain Electron Spin Resonance* (L. Kevan and R. N. Schwartz, eds.), Wiley-Interscience, New York, 1979, p. 279.
16. W. B. Mims and J. Peisach, in *Biological Applications of Magnetic Resonance* (R. G. Shulman, ed.), Academic Press, New York, 1979, p. 221.
17. L. Kevan, *J. Phys. Chem.*, *85*, 1628 (1981).
18. W. B. Mims, in *Fourier, Hadamard and Hilbert Transforms in Chemistry* (A. G. Marshall, ed.), Plenum Press, New York, 1982.
19. T.-S. Lin, *Chem. Rev.*, *84*, 1 (1984).
20. S. A. Dikanov and Yu. D. Tsvetkov, *Zh. Strukt. Khim.*, *26*, N5, 136 (1985).
21. W. B. Mims, *Phys. Rev. B.*, *5*, 2409 (1972).
22. W. B. Mims, *Phys. Rev. B.*, *6*, 3543 (1972).
23. L. G. Rowan, E. L. Hahn, and W. B. Mims, *Phys. Rev. A.*, *137*, 61 (1965).
24. G. M. Zhidomirov and K. M. Salikhov, *Teor. Eksp. Khim.*, *4*, 514 (1968) (*Theor. Exp. Chem.*, *4*, 332 (1968)).
25. S. A. Dikanov, A. A. Shubin, and V. N. Parmon, *J. Magn. Reson.*, *42*, 474 (1981).
26. A. A. Shubin and S. A. Dikanov, *J. Magn. Reson.*, *53*, 1 (1983).
27. A. A. Shubin and S. A. Dikanov, *J. Magn. Reson.*, *63*, 000 (1985).
28. D. J. Sloop, H.-L. Yu, T.-S. Lin, and S. I. Weissman, *J. Chem. Phys.*, *75*, 3746 (1981).
29. M. Romanelli, M. Narayana, and L. Kevan, *J. Chem. Phys.*, *80*, 4044 (1984).
30. S. A. Dikanov, V. F. Yudanov, and Yu. D. Tsvetkov, *J. Magn. Reson.*, *34*, 631 (1979).
31. V. F. Yudanov, A. M. Raitsimring, and Yu. D. Tsvetkov, *Teor. Eksp. Khim.*, *4*, 520 (1968).
32. V. F. Yudanov, V. P. Soldatov, and Yu. D. Tsvetkov, *Zh. Strukt. Khim.*, *15*, 600 (1974) (*J. Struct. Chem.*, *15*, 510 (1974)).
33. H. Barkhuijsen, R. de Beer, B. J. Pronk, and D. van Ormondt, *J. Magn. Reson.*, *61*, 284 (1985).
34. R. P. J. Merks and R. de Beer, *J. Magn. Reson.*, *37*, 305 (1980).
35. W. B. Mims, *J. Magn. Reson.*, *59*, 291 (1984).
36. S. M. Key and S. L. Marple, *Proc. IEEE*, *69*, 1380 (1981).
37. D. van Ormondt and K. Nederveen, *Chem. Phys. Lett.*, *82*, 443 (1981).
38. D. J. Singel, W. A. J. A. van der Poel, J. Schmidt, and J. H. van der Waals, *J. Chem. Phys.*, *81*, 5453 (1984).

39. W. B. Mims, *Proc. Roy. Soc. (London)*, *283*, 452 (1965).
40. D. Grischkowsky and S. R. Hartman, *Phys. Rev.*, *B*, *2*, 60 (1970).
41. P. F. Liao and S. R. Hartman, *Phys. Rev. B*, *8*, 69 (1973).
42. N. Laurence, E. C. McIrvine, and J. Lambe, *J. Phys. Chem. Solids*, *23*, 515 (1962).
43. W. A. J. A. van der Poel, D. J. Singel, J. Schmidt, and J. H. van der Waals, *Mol. Phys.*, *49*, 1017 (1983).
44. P. A. Narayana, M. K. Bowman, L. Kevan, V. F. Yudanov, and Yu. D. Tsvetkov, *J. Chem. Phys.*, *63*, 3365 (1975).
45. S. A. Dikanov, Yu. D. Tsvetkov, and A. V. Astashkin, *Chem. Phys. Lett.*, *91*, 515 (1982).
46. A. V. Astashkin, S. A. Dikanov, and Yu. D. Tsvetkov, *Chem. Phys. Lett.*, *122*, 259 (1985).
47. G. K. Semin, T. A. Babushkina, and G. G. Yakobson, *Applications of Nuclear Quadrupole Resonance in Chemistry*, Chemistry, Leningrad Branch, 1972.
48. S. A. Dikanov, Yu. D. Tsvetkov, A. V. Astashkin, and A. A. Shubin, *J. Chem. Phys.*, *79*, 5785 (1983).
49. P. Waldstein, S. B. Rabideau, and J. A. Jackson, *J. Chem. Phys.*, *41*, 3407 (1964).
50. W. B. Mims, J. Peisach, and J. L. Davis, *J. Chem. Phys.*, *66*, 5536 (1977).
51. I. A. Safin and D. Ya. Osokin, *Nuclear Quadrupole Resonance in Compounds of Nitrogen*, Science, Moscow, 1977.
52. A. V. Astashkin, S. A. Dikanov, and Yu. D. Tsvetkov, *Zh. Strukt. Khim.*, *25*, N. 1, 53 (1984).
53. S. A. Dikanov, A. V. Astashkin, and Yu. D. Tsvetkov, *Zh. Strukt. Khim.*, *25*, N. 2, 35 (1984).
54. A. V. Astashkin, S. A. Dikanov, and Yu. D. Tsvetkov, *Zh. Strukt. Khim.*, *26*, N. 3, 53 (1985).
55. W. B. Mims and J. Peisach, *J. Chem. Phys.*, *69*, 4921 (1978).
56. W. B. Mims and J. L. Davis, *J. Chem. Phys.*, *64*, 4836 (1976).
57. R. Malkin, in *Inorganic Biochemistry* (G. B. Eichhorn, ed.), Elsevier, Amsterdam, 1973, Chap. 21.
58. W. B. Mims, J. L. Davis, and J. Peisach, *Biophys. J.*, *45*, 755 (1984).
59. J. Peisach and W. B. Mims, *J. Biol. Chem.*, *259*, 2704 (1984).
60. S. E. Groh, A. Nagahisa, S. L. Tan, and W. H. Orme-Johnson, *J. Am. Chem. Soc.*, *105*, 7445 (1983).
61. J. Peisach, N. R. Orme-Johnson, W. B. Mims, and W. H. Orme-Johnson, *J. Biol. Chem.*, *252*, 5643 (1977).

62. N. R. Orme-Johnson, W. B. Mims, W. H. Orme-Johnson, R. G. Bartsch, M. A. Cusanovich, and J. Peisach, *Bioch. Bioph. Acta. 748*, 68 (1983).

63. W. H. Orme-Johnson, in *Inorganic Biochemistry* (G. B. Eichhorn, ed.), Elsevier, Amsterdam, 1973, Chap. 22.

64. E. Szaidzinska-Pietek, R. Maldonado, L. Kevan, and R. R. M. Jones, *J. Am. Chem. Soc., 106*, 4675 (1984).

65. E. Szaidzinska-Pietek, R. Maldonado, L. Kevan, R. R. M. Jones, and M. J. Coleman, *J. Am. Chem. Soc., 107*, 784 (1985).

66. E. Szaidzinska-Pietek, R. Maldonado, and L. Kevan, *J. Phys. Chem., 89*, 1547 (1985).

67. S. A. Dikanov and Yu. D. Tsvetkov, *Zh. Strukt. Khim., 20*, 824 (1979) (*J. Struct. Chem., 20*, 699 (1979)).

68. S. A. Dikanov, A. V. Astashkin, and Yu. D. Tsvetkov, *Zh. Strukt. Khim., 23*, N. 3, 11 (1982) (*J. Struct. Chem., 23*, 333 (1982)).

69. S. A. Dikanov and Yu. D. Tsvetkov, *Zh. Strukt. Khim., 20*, 934 (1979) (*J. Struct. Chem., 20*, 800 (1979)).

70. J. Zweier, P. Aisen, J. Peisach, and W. B. Mims, *J. Biol. Chem., 254*, 3512 (1979).

71. J. Zweier, J. Peisach, and W. B. Mims, *J. Biol. Chem., 257*, 10314 (1982).

72. M. J. Hunt, A. L. Mackay, and T. Edmonds, *Chem. Phys. Lett., 34*, 473 (1975).

73. W. B. Mims and J. Peisach, *Biochemistry, 15*, 3863 (1976).

74. J. Peisach and W. B. Mims, *J. Biol. Chem., 254*, 4321 (1979).

75. B. Mondovi, M. T. Graziani, W. B. Mims, R. Oltzik, and J. Peisach, *Biochemistry, 16*, 4198 (1977).

76. L. Avigliano, J. L. Davis, M. T. Graziani, A. Marchesini, W. B. Mims, B. Mondovi, and J. Peisach, *FEBS Lett., 136*, 80 (1981).

77. D. J. Kosman, J. Peisach, and W. B. Mims, *Biochemistry, 19*, 1304 (1980).

78. J. A. Fee, J. Peisach, and W. B. Mims, *J. Biol. Chem., 256*, 1910 (1981).

79. R. M. Burger, A. D. Adler, S. B. Horwitz, W. B. Mims, and J. Peisach, *Biochemistry, 20*, 1701 (1981).

80. J. H. Freedman, L. Pickart, B. Weinstein, W. B. Mims, and J. Peisach, *Biochemistry, 21*, 4540 (1982).

81. B. A. C. Ackrel, E. B. Kearney, W. B. Mims, J. Peisach, and H. Beinert, *J. Biol. Chem., 259*, 4015 (1984).

82. J. Peisach, W. B. Mims, and J. L. Davis, *J. Biol. Chem., 254*, 12379 (1979).

83. Y. V. S. Ramakrishna and P. A. Narayana, *J. Chem. Phys.*, *75*, 1123 (1981).
84. S. L. Tan, J. A. Fox, N. Kojima, S. T. Walsh, and W. H. Orme-Johnson, *J. Am. Chem. Soc.*, *106*, 3064 (1984).
85. M. K. Bowman, J. R. Norris, M. C. Thurnauer, J. Warden, S. A. Dikanov, and Yu. D. Tsvetkov, *Chem. Phys. Lett.*, *55*, 570 (1978).
86. S. A. Dikanov, Yu. D. Tsvetkov, M. K. Bowman, and A. V. Astashkin, *Chem. Phys. Lett.*, *90*, 149 (1982).
87. S. A. Dikanov, A. V. Astashkin, Yu. D. Tsvetkov, and M. G. Goldfeld, *Chem. Phys. Lett.*, *101*, 206 (1983).
88. A. de Groot, A. J. Hoff, R. de Beer, and H. Scheer, *Chem. Phys. Lett.*, *113*, 286 (1985).
89. A. J. Hoff, A. de Groot, S. A. Dikanov, A. V. Astashkin, and Yu. D. Tsvetkov, *Chem. Phys. Lett.*, *118*, 40 (1985).
90. W. Lubitz, R. A. Isaakson, E. C. Abresch, and G. Feher, *Proc. Natl. Acad. Sci. USA*, *81*, 7792 (1984).
91. A. de Groot, R. Evelo, A. J. Hoff, R. de Beer, and H. Scheer, *Chem. Phys. Lett.*, *118*, 48 (1985).
92. H. J. den Blanken and A. J. Hoff, *Biochim. Biophys. Acta*, *681*, 365 (1982).
93. T. Shimizu, W. B. Mims, J. Peisach, and J. L. Davis, *J. Chem. Phys.*, *70*, 2249 (1979).
94. P. Tanswell, J. Thornton, A. V. Korda, and R. J. P. Williams, *Eur. J. Biochem.*, *57*, 135 (1975).
95. A. D. Milov, M. D. Schirov, and V. E. Khmelinsky, *Dokl. Akad. Nauk SSSR*, *218*, 878 (1974).
96. A. D. Trifunac and J. R. Norris, *Chem. Phys. Lett.*, *59*, 140 (1978).
97. A. D. Trifunac, J. R. Norris, and R. G. Lowler, *J. Chem. Phys.*, *71*, 4380 (1979).
98. A. D. Milov, A. B. Ponomarev, and Yu. D. Tsvetkov, *Chem. Phys. Lett.*, *110*, 67 (1984).
99. A. D. Milov, A. Yu. Pusep, S. A. Dzuba, and Yu. D. Tsvetkov, *Chem. Phys. Lett.*, *119*, 421 (1985).
100. S. A. Dzuba, K. M. Salikhov, and Yu. D. Tsvetkov, *Chem. Phys. Lett.*, *79*, 568 (1981).
101. S. A. Dzuba, A. G. Maryasov, K. M. Salikhov, and Yu. D. Tsvetkov, *J. Magn. Reson.*, *58*, 95 (1984).

Author Index

Numbers in parentheses are reference numbers and indicate that an author's work is referred to although his name may not be cited in the text. Underlined numbers give the page on which the complete reference is listed.

A

Aasa, R., 152(65), 153(65), 165(84,85), 172(105), 179(115), 202, 203, 204
Abragam, A., 84(9), 85(9), 86(9), 101; 131(1), 134(1), 136(1), 185(130), 190(1), 199, 204
Abresch, E. C., 254(90), 255(90), 263
Ackrel, B. A. C., 249(81), 257(81), 262
Adams, M. W., 73(105), 80
Addison, A. W., 106(15), 115(15), 126
Adler, A. D., 249(79), 262
Adman, E. T., 29(46), 78
Agostinelli, E., 25(49), 31(49), 32(49), 33(49), 78
Aisen, P., 25(60), 39(60), 78; 247(70), 262
Akhvldiani, I. G., 148(50), 150(50), 201
Aktas, B., 84(8), 101
Alcock, R. M., 106(6), 126
Alegria, A. E., 106(7,10,11), 126
Alexander, S., 195(154), 205
Allen, J. P., 151(58,60), 154(58,60), 167(58,60), 190(58,149), 191(58), 192(58,60), 193(58), 196(58,60,149), 197(160), 198(60), 201, 205, 206
Allendoerfer, R. D., 106(5), 126
Anderson, J. R., 22(31), 23(31), 24(31), 77

Andersson, I., 163(81), 202
Angelos, S. G., 21(28), 77
Antholine, W. E., 24(37), 26(37), 77; 186(135), 205
Antonsson, B., 90(28), 102
Artman, J. O., 139(27), 140(27), 143(27), 200
Ash, D. E., 84(10), 90(29), 91(29), 98(47), 101, 102, 103
Astashkin, A. V., 230(45), 232(46), 233(46), 235(48), 236(48), 237(52,53,86), 238(54), 239(52), 240(52), 246(68), 250(86,87,89), 251(87,89), 252(89), 253(89), 254(87,89), 255(89), 261, 262, 263
Avigliano, L., 249(76), 262

B

Babushkina, T. A., 235(47), 237(47), 261
Baker, G. J., 25(49), 31(49), 32(49), 33(49), 78
Baldwin, T. O., 59(92), 67(92), 80
Banerjee, R., 45(76), 79
Banks, R. D., 98(50), 103
Baram, A., 86(20), 101
Barkhuijsen, H., 221(33), 260
Barlow, C., 188(144), 205
Bartlett, N., 45(83), 64(83), 65(83), 67(83,94), 68(83), 69(94), 79, 80
Bartsch, R. G., 173(107), 203; 244(62), 245(62), 262

265

Bearden, A. J., 173(107), 203
Beinert, H., 21(27), 35(55), 36(55), 37(57), 77, 78; 166(86, 87), 167(86), 202; 249(81), 257(81), 262
Bekauri, P. I., 143(37), 200
Bemski, G., 136(23), 151(57), 200, 201
Benecky, M. J., 73(105), 80
Bennett, J. E., 45(73), 79
Bernhard, W., 162(79), 202
Bertrand, P., 167(90,93), 168(90), 190(90,93), 196(90), 197(90), 203
Berulava, B. G., 143(37), 200
Bignetti, E., 183(127), 204
Blair, D. F., 169(97), 188(146), 203, 205
Blake, C. C., 98(49,50), 103
Blasie, J. K., 188(143,145), 205
Bleaney, B., 84(9), 85(9), 86(9), 101; 131(1), 134(1), 136(1), 190(1), 199
Block, F., 160(75), 202
Bloembergen, N., 139(27), 140(27), 141(28), 142(28), 143(27), 144(28), 154(28), 185(129), 200, 204
Blum, H., 146(49), 147(49), 149(49), 150(49), 167(95), 201, 203
Blum, R., 114(44), 127
Blumberg, W. E., 30(47), 46(86), 78, 80; 106(14), 115(14), 126; 211(7), 222(7), 259
Boas, J. F., 20(26), 21(26), 28(26), 33(26), 77
Bocian, D. F., 169(99), 203
Bohan, T. L., 155(71), 202
Borell, A., 179(116), 204
Boscaino, R., 143(38), 200
Bowers, C. P., 106(12), 108(12), 126
Bowman, M. K., 132(12), 148(51), 150(51), 162(78), 199, 201, 202; 211(13), 212(13), 228(13), 229(44), 237(86), 250(85,86), 259, 261, 263
Bowyer, J. R., 167(95), 203
Box, H. C., 2(9), 15(22), 76, 77

Brai, M., 143(38), 200
Brändén, C. I., 182(123), 204
Brändén, R., 165(85), 202
Brill, A. S., 151(61), 154(61), 171(61), 196(61), 201
Brown, G., 188(145), 205
Brown, T. G., 22(31), 23(31,35), 24(31,41), 25(60), 28(41), 29(41), 30(41), 39(60), 71(98), 77, 78, 80
Bruckner, R. C., 187(141), 189(141), 205
Brudvig, G. W., 25(56), 36(56), 78; 169(96,97,99-101), 188(146), 203, 205
Bryant, T. N., 98(51), 103
Burger, R. M., 249(79), 262
Burgess, C., 106(6), 126
Burlamacchi, L., 86(16,17), 101; 106(9), 126

C

Calvo, R., 15(23), 77; 151(57), 201
Cammack, R., 166(88), 167(88,93), 190(93), 202, 203
Campbell, S. J., 151(64), 152(64), 201
Carlin, R. L., 179(112), 204
Case, G. D., 186(138), 205
Caspary, W. J., 106(13), 126
Cassoly, R., 45(77), 79
Castner, T. G., 141(29), 142(29), 148(29), 150(29), 188(29), 200
Cedergren-Zeppezauer, E., 176(113), 179(113,114), 182(113), 204
Chan, S. I., 25(56), 36(56), 70(97), 78, 80; 169(96,97,99-101), 188(146), 203, 205
Chance, B., 30(47), 78
Chiang, R., 173(109), 203
Chien, J. C. W., 40(62,63), 41(62), 42(62,63), 46(87), 49(87), 51(87), 55(87), 56(87), 57(87), 59(87), 61(87), 79, 80
Christahl, M., 56(90), 80
Ciccarello, I., 143(38), 200

AUTHOR INDEX

Cleland, W. W., 82(3-5), 101; 183(126), 204
Cline, J. F., 24(43), 25(42, 43), 29(42,43), 30(43), 31(43), 32(43), 33(42,43), 37(42), 73(105), 78, 80
Clough, S., 141(30), 142(30), 143(30), 163(30), 200
Cohn, M., 82(1), 90(30), 94(33, 35), 96(38), 100(1), 101, 102
Coleman, M. J., 245(65), 262
Colman, P. M., 29(45), 78
Colvin, J. T., 151(58-60), 154(58-60), 167(58-60), 173(108), 190(58), 191(58), 192(58,60), 193(58), 194(59), 195(59), 196(58-60,108), 198(59,60), 201, 203
Conception, R., 106(7), 126
Coniglio, A., 195(156), 206
Connolly, B. A., 90(23), 91(23), 97(23), 102
Conroy, S. C., 98(51), 103
Cook, P. F., 183(126), 204
Cook, R. J., 11(17), 12(18), 77
Cramer, S. P., 37(57), 78
Culvahouse, J. W., 133(14), 199
Cusanovich, M., 167(95), 173(107), 203; 244(62), 245(62), 245(62), 262

D

Davis, J. L., 236(50), 241(50, 56), 242(58), 243(58), 244(58), 249(76,82), 250(82), 256(93), 261, 262, 263
Deal, R. M., 2(3), 76
de Beer, 221(33), 222(34), 223(34), 250(88), 251(88), 254(88), 255(88), 256(91), 260, 263
Debrunner, P. G., 173(109), 203
de Groot, A., 250(88,89), 251(88,89), 252(89), 253(89), 254(88,89), 255(88,89), 256(91), 263
Deinum, J., 165(84,85), 202
Delaney, M. F., 110(37), 127
de la Torre, A., 166(89), 167(89), 203

den Blanken, H. J., 256(92), 263
Desideri, A., 25(49), 31(49), 32(49), 33(49), 78
Desmeules, P., 94(34), 102
Devaney, P. W., 173(109), 203
Deville, A., 167(93), 190(93), 203
DeWit, M., 179(111), 204
Dickinson, L. C., 3(11), 40(62, 63), 41(62), 42(62,63), 46(87), 49(87), 51(87), 55(87), 56(87), 57(87), 59(87), 61(87), 69(11), 76, 79, 80
Dietrich, H., 163(81), 202
Dietz, R. E., 179(118), 204
Dikanov, S. A., 165(83), 202; 212(14,20), 217(25-27), 219(25,30), 225(20), 228(14,20), 229(14,20), 230(45), 232(46), 233(46), 234(26), 235(26,27, 48), 236(27,30,48), 237(52,53, 86), 238(54), 239(52), 240(52), 246(67-69), 250(85-87,89), 251(87,89), 252(89), 253(89), 254(87,89), 255(89), 259, 260, 261, 262, 263
Doan, N., 173(109), 203
Dobson, M. J., 98(51), 103
Dodgen, H. W., 88(21), 91(21), 101
Doetschman, D. C., 45(80,81), 79
Dousmanis, C., 54(88), 80
Dunaway-Mariano, D., 82(3), 101
Dunford, H. B., 172(105), 203
Dutton, P. L., 188(145), 205
Dvornikov, E. V., 211(9), 259
Dworschak, R. T., 183(125), 204
Dymowski, J. J., 176(120), 182(120), 204
Dzuba, S. A., 258(99-101), 263

E

Eccleston, J. F., 90(29), 91(29), 102
Eckstein, F., 82(2), 100(2), 101
Edmonds, T., 248(72), 262
Einarsson, R., 163(81), 202
Eisenberger, P., 2(2), 76; 188(145), 205

Eklund, H., 176(113), 179(113, 114), 182(113,123), 204
Elzinga, M. E., 96(42), 102
England, T. S., 144(45), 201
Erecinska, M., 188(143,144), 205
Eschenfelder, A. H., 144(46), 201
Estle, T. L., 179(111), 204
Evans, P. R., 98(50), 103
Evelo, R., 256(91), 263

F

Fairhurst, S. A., 42(67), 79
Farach, H. A., 131(4), 199
Farver, O., 32(51), 78
Fee, J. A., 21(30), 24(38), 25(58), 27(38), 37(58), 77, 78; 249(78), 262
Feher, G., 2(1), 8(15), 44(1), 70(15), 76; 136(24), 137(24), 143(36), 144(47), 171(24), 172(24), 200, 201; 254(90), 255(90), 263
Feng, D. F., 144(40), 200
Fiamingo, F. G., 151(61,62), 154(61,62), 155(62), 171(61,62), 196(61), 201
Finn, C. B. P., 135(17), 199
Fleischer, E. W., 23(34), 77
Flynn, C. P., 151(58), 154(58), 167(58), 190(58,149), 191(58), 192(58), 193(58), 196(58,149), 201, 205
Forrer, J., 19(24), 77
Fothergill, L. A., 98(51), 103
Fox, J. A., 250(84), 263
Freedman, J. H., 249(80), 262
Freeman, H., 24(43), 25(43), 29(43), 30(43), 31(43), 32(43), 33(43), 78
Freeman, H. C., 29(45), 78
Freund, H. G., 15(22), 77
Frey, P. A., 94(36), 95(36), 102
Frey, T. G., 187(141), 189(141), 205
Fridborg, K., 179(116), 204
Froncisz, W., 35(55), 36(55), 78
Fujii, A., 26(36), 77
Fukuyama, J. M., 136(21), 151(21), 152(21), 174(21), 175

[Fukuyama, J. M.], (21), 176(21), 177(21), 179(21), 200
Furugren, B., 182(123), 204

G

Gaber, B. P., 21(30), 77
Gampp, H., 108(25,26), 109(31-33), 110(31-33), 111(33,39,41, 42), 112(25), 113(25), 114(25), 115(25), 116(25), 117(25), 118(25,41), 119(41), 120(26,46), 121(26), 122(26), 123(26), 124(25,26,39,41,42), 127
Gans, P., 107(24), 110(24), 127
Gayda, J. P., 167(90,93), 168(90), 190(90,93), 196(90), 197(90), 203
Gelles, J., 169(97), 203
George, P., 45(73), 79
Gere, E. A., 143(36), 200
Gersonde, K., 56(90), 80
Gewirth, A. A., 32(50), 78
Gibson, J. F., 167(93), 190(93), 203
Gibson, Q. H., 44(71), 79
Gill, J. C., 133(15), 199
Go, N., 198(159), 206
Goldfeld, M. G., 250(87), 251(87), 254(87), 263
Golub, G. H., 110(36), 127
Gonser, U., 172(106), 203
Goodhart, P. J., 94(37), 95(37), 102
Goodman, B. A., 105(1), 106(1), 114(1), 115(1), 126
Goodman, G., 188(147), 205
Goody, R. S., 85(14,15), 86(14), 87(15), 90(14,15,23,27), 91(14, 15,23,27), 92(15), 93(15), 96(27), 97(23), 98(43,45), 101, 102, 103
Gordon, G., 115(45), 127
Gould, D. C., 106(16), 126
Gray, H. B., 24(43), 25(43), 29(43), 30(43), 31(43), 32(43), 33(43), 37(57), 78
Graziani, M. T., 33(52), 78; 249(75,76), 262
Griffiths, J. S., 44(72), 45(73), 79

AUTHOR INDEX

Grischkowsky, D., 226(40), 261
Grishan, C. M., 109(27), 127
Groh, S. E., 244(60), 261
Gross, K.-H., 96(41), 102
Guggenheim, H. J., 179(118), 204
Günthard, H. H., 19(24,25), 32 (32), 40(61), 41(61), 42(61), 77, 79
Guss, J. M., 29(45), 78

H

Hager, L. P., 72(102,103), 80; 173(108,109), 196(108), 203
Hahn, E. L., 217(23), 260
Hall, D. O., 166(88,89), 167(88, 89), 202, 203
Hampton, D. A., 151(61), 154 (61), 171(61), 196(61), 201
Hänisch, G., 109(29), 127
Hanson, M., 179(115), 204
Harami, T., 172(106), 203
Hardman, K., 176(121), 182(121), 204
Hardy, G. W., 98(50), 103
Harman, H. J., 187(142), 205
Harrington, W. F., 98(44), 103
Hartley, F. R., 106(6), 126
Hartman, S. R., 226(40,41), 261
Hartzell, C. D., 21(27), 77
Hase, H., 148(51), 150(51), 201
Haser, R., 98(50), 103
Haspra, D., 120(46), 127
Hederstedt, L., 167(94), 203
Helman, J. S., 136(23), 195(156), 200, 206
Hendrich, M., 106(52), 109(61), 110(61), 118, 119
Henry, Y., 45(76), 79
Herbert, I. R., 151(64), 152(64), 201
Herrick, R. C., 171(104), 172 (104), 190(104), 196(104), 203
Hoff, A. J., 211(11), 212(11), 250(88,89), 251(88,89), 252 (89), 253(89), 254(88,89), 255 (88,89), 256(91,92), 257(11), 259, 263
Hoffman, B. M., 22(31), 23(31, 35), 24(31,41,43), 25(42,43, 54,60), 28(41), 29(41-43), 30

[Hoffman, B. M.], (41,43), 31 (43), 32(43), 33(42,43), 35 (54), 37(42), 39(60), 71(98, 99), 72(102,103), 73(105-107), 77, 78, 80
Höhn, M., 41(66), 46(87), 49(87), 51(87), 55(87), 56(87,89), 57 (87), 59(87), 61(87,89), 79, 80
Höhn-Berlage, M., 45(83), 64(83), 65(83), 67(83), 68(83), 79
Holmes, K. C., 98(45), 103
Hori, H., 40(64,65), 42(64), 45 (78), 54(78), 71(100), 79, 80; 84(7), 101
Horwitz, S. B., 249(79), 262
Huisjen, M., 155(68), 202
Hunt, J. P., 88(21), 91(21), 101
Hunt, M. J., 248(72), 262
Hüttermann, J., 41(66), 45(83, 84), 46(87), 49(87), 51(87), 55(87), 56(87), 57(87), 59(87), 61(87), 64(83,84), 65(83,84), 67(83), 68(83), 79, 80
Huynh, B. H., 73(105), 80
Hwang, K. J., 45(74), 79
Hyde, J. S., 10(16), 23(33), 24 (37), 26(37), 28(40), 35(55), 36(55), 76, 77, 78; 146(48), 151(54), 155(67-69), 157(67), 163(54), 185(54), 186(54,133-137), 187(54), 188(54,133), 201, 202, 205

I

Iitaka, Y., 26(36), 77
Iizuka, T., 171(103), 172(103), 203
Ikeda-Saito, M., 40(64,65), 42 (64), 45(78), 54(78), 79; 84 (7), 101
Imamura, T., 59(92), 67(92), 80
Ingram, D. J. E., 2(3), 45(73), 76, 79
Inubushi, T., 8(15), 70(15), 76
Isaacson, R. A., 2(1), 8(15), 15(21,23), 44(1), 70(15), 76, 77; 136(24), 137(24), 155(70), 156(70), 171(24), 172(24), 200, 202; 254(90), 255(90), 263
Iwasaki, M., 4(12), 76

J

Jackson, J. A., 236(49), 261
Jaffe, E. K., 82(1), 100(1), 101
Jaffer, S., 176(120), 182(120), 204
Janakiraman, R., 169(102), 170(102), 203
Jardetzky, O., 98(46), 103
Jeffries, C. D., 133(16), 199
Jensen, L. H., 29(46), 78
Jensen, P., 25(42), 29(42), 33(42), 37(42), 78
Johnson, J. F., 179(118), 204
Jones, R. R. M., 245(64,65), 262
Jörin, E., 40(61), 41(61), 42(61), 79
Jörnvall, J., 182(123), 204

K

Kaden, T. A., 109(29,30), 112(30,43), 113(30,43), 114(30,43), 127
Kalbitzer, H. R., 85(14,15), 86(14), 87(14), 90(14,15,23,28), 91(14,15,23), 92(15), 93(15), 96(40), 97(23), 101, 102
Kalinowski, H., 136(23), 200
Kang, C. H., 71(98,99), 80
Kannan, K. K., 179(116), 204
Kaplan, J. I., 96(39), 102
Kappl, R., 45(83), 64(83), 65(83), 67(83), 68(83), 79
Kaziro, Y., 90(25), 102
Kearney, E. B., 249(81), 257(81), 262
Kennedy, W. R., 106(12), 108(12), 126
Kenyon, G. L., 94(31,34,37), 95(37), 102
Kevan, L., 2(5), 76; 132(12), 144(40,41), 148(51), 150(51), 160(77), 161(77), 162(77,78), 199, 200, 201, 202; 211(10), 212(10,15,17), 219(10,15,29), 228(10,15,17), 229(44), 245(64-66), 259, 260, 261, 262
Key, S. M., 224(36), 260
Khakhanashvili, O. G., 143(37), 200

Khmelinsky, V. E., 211(9), 257(95), 259, 263
Kiel, A., 143(39), 200
Kim, J. E., 136(21), 151(21), 152(21), 174(21), 175(21), 176(21), 177(21), 179(21), 200
Kim, Y. W., 70(96), 80
King, T. E., 21(27), 25(53), 35(53,55), 36(53,55), 46(85), 48(85), 52(85), 54(85), 61(85), 62(85), 63(85), 64(85), 77, 78, 80; 169(98,102), 170(102), 203
Kingsman, A. J., 98(51), 103
Kingsman, S. M., 98(51), 103
Kispert, L. D., 2(5), 76
Kittel, C., 193(151), 205
Kittl, W. S., 107(18), 108(18), 113(18), 115(18), 126
Kohnlein, W., 162(80), 202
Kojima, N., 250(84), 263
Kokoszka, G. F., 115(45), 127
Kon, H., 196(157), 206
Kopylov, V. V., 148(50), 150(50), 201
Korb, J. P., 143(42-45), 185(131), 200, 204
Korda, A. V., 256(94), 263
Kormaz, M., 84(8), 101
Kosman, D. J., 2(8), 76; 249(77), 262
Kotani, M., 152(66), 171(103), 172(103), 202, 203
Kulikov, A. V., 187(139,140), 205
Kuo, L. C., 136(20-22), 144(42,43), 145(22), 150(22), 151(20-22,42,43,56), 152(21,22), 162(42,43), 163(42,43), 164(43), 165(42,43), 171(22), 172(22), 174(20,21), 175(21), 176(21,22,120), 177(21), 179(20,21,43), 180(56), 181(42,56), 182(42,120), 183(43), 184(43), 200, 201, 204
Kurtz, S. R., 190(149), 196(149), 205
Kvassman, J., 183(124), 204
Kwiram, A., 7(14), 76
Kyhl, R. L., 159(74), 202

L

Lambe, J., 226(42), 261
Lapidot, A., 8(15), 70(15), 76
Larson, G. H., 133(16), 199
Laurence, N., 226(42), 261
Lawton, W. H., 110(35), 127
Leibl, W., 45(84), 64(84), 65 (84), 79
Leigh, J. S., Jr., 40(65), 79; 86(19), 94(35), 101, 102; 186 (138), 188(147), 205
Lerch, K., 72(101), 80
Leupold, C. M., 90(27), 91(27), 96(27), 102
Levin, P. D., 151(61), 154(61), 171(61), 196(61), 201
Lewis, M., 182(122), 204
Leyh, T. S., 89(22), 94(22,36,37), 95(36,37), 98(47), 102, 103
Liao, P. F., 226(41), 261
Likhtenstein, G. I., 187(139, 140), 205
Lilga, K. T., 15(22), 77
Lin, T.-S., 212(19), 217(28), 228(19), 260
Lindberg, P. N. T., 109(28), 127
Lindskog, S., 179(115), 204
Lipscomb, W. N., 176(121), 182 (121,122), 204
Lloyd, J. P., 148(52), 150(52), 201
Lo, G. Y.-S., 88(21), 91(21), 101
LoBrutto, R., 46(85), 48(85), 52 (85), 54(85), 61(85), 62(85), 63(85), 64(85), 80; 187(141), 189(141), 205
Lövgren, S., 179(116), 204
Low, W., 179(119), 204
Lowler, R. G., 257(97), 263
Lozos, G. P., 22(31), 23(31), 24 (31), 77
Lubitz, W., 254(90), 255(90), 263
Lulich, C., 15(23), 77
Lum, V., 24(43), 25(43), 29(43), 30(43), 31(43), 32(43), 33 (43), 78
Luz, Z., 86(18,20), 101

M

Mackay, A., 248(72), 262
Maeda, Y., 172(106), 203
Maeder, M., 109(31-33), 110(31-33), 111(33,39,41,42), 118(41), 119(41), 120(46), 124(39,41,42), 127
Maggio, M. S., 110(35), 127
Maguire, J. J., 167(94), 203
Mailer, C., 146(48), 186(134), 190(148), 201, 205
Makinen, M. W., 136(19-22), 144 (42-44), 145(22), 150(22), 151 (19-22,42-44,55,56), 152(21,22), 154(44), 162(42-44), 163 (42-44), 164(15,43,55), 165(42-44), 171(22), 172(22), 174(19-21), 175(21), 176(21,22,55,120), 177(21), 179(19-21,43,55), 180(56), 181(42,56), 182 (42,120), 183(19,43,44,55), 184(43,55,128), 200, 201, 204
Maldonado, R., 245(64-66), 262
Malkin, R., 242(57), 261
Malmström, B. G., 152(65), 153 (65), 165(84,85), 202
Mandelbrot, B. B., 194(152,153), 205
Manenkov, A. A., 135(18), 200
Mannherz, H. G., 98(43), 102
Marchesini, A., 249(76), 262
Maret, W., 136(20), 151(20,55), 163(81), 164(55), 174(20), 179 (20,55), 183(55), 184(55,128), 200, 201, 202, 204
Margenau, H., 109(34), 127
Margerum, D. W., 106(12), 108 (12), 126
Margoliash, E., 25(54), 35(54), 71(98,99), 78, 80
Markham, G. H., 83(6), 84(6), 85 (6,12), 86(12), 101
Marple, S. L., 224(36), 260
Marquardt, D., 111(40), 127; 158 (72), 202
Marquetant, R., 90(23), 91(23), 96(40), 97(23), 102
Martin, C. T., 25(56), 36(56), 70(97), 78, 80; 169(96,97), 203

Martini, G., 86(17), 101; 106(9), 126
Maruani, J., 143(32-35), 150(53), 185(131), 200, 201, 204
Marx, F., 96(42), 102
Maryasov, A. G., 258(101), 263
Mascarenhas, R., 8(15), 46(85), 48(85), 52(85), 54(85), 61(85), 62(85), 63(85), 64(85), 70(15), 76, 80
Mason, H. S., 106(16), 126
Mattuck, R. D., 131(9), 199
McCain, D. C., 106(8), 126
McElearney, J. N., 179(112), 204
McIrvine, E. C., 226(42), 261
McLaughlin, A. C., 94(35), 102
Meirovitch, E., 85(13), 101
Merchant, S., 179(112), 204
Merett, M., 98(50), 103
Merks, R. P. J., 222(34), 223(34), 260
Meyer, C. J., 109(33), 110(33), 111(33,39,41,42), 118(41), 119(41), 124(39,41,42), 127
Mildvan, A. S., 82(4), 94(34), 101, 102
Miller, C. K., 23(34), 77
Miller, D. L., 90(24), 102
Milov, A. D., 257(95), 258(98,99), 263
Mims, W. B., 72(101), 80; 165(82), 202; 208(1), 210(1), 211(6-8), 212(1,8,16,18), 215(21,22), 217(21,23), 218(21,22), 219(21), 221(8), 222(7), 224(35), 226(1,39), 228(1,8,16,18), 236(50), 238(55), 241(50,56), 242(58), 243(58,59), 244(58,59,61,62), 245(61,62), 247(70,71), 248(55), 249(73-82), 250(82), 256(16,93), 257(8,16,81), 259, 260, 261, 262, 263
Mondovi, B., 249(75,76), 262
Moore, A. L., 167(91), 203
Moore, J. M., 98(52), 99(52), 103
Morata, M., 29(45), 78
More, C., 167(93), 190(93), 203
Morigaki, K., 179(110), 204
Morimoto, H., 171(103), 172(103), 203

Morita, Y., 172(106), 203
Morpurgo, L., 25(49), 31(49), 32(49), 33(49,52), 78
Morris, V. A., 29(45), 78
Morse, R. H., 169(101), 203
Mortenson, L. E., 73(105), 80
Moss, T. H., 21(27), 77; 173(107), 203
Muench, P. J., 196(158), 206
Münck, E., 73(105), 80
Muraoka, Y., 26(36), 77
Muromtsev, V. I., 148(50), 150(50), 201
Murphy, G. M., 109(34), 127
Muto, H., 4(12), 76
Myers, R. J., 106(8), 126

N

Nagahisa, A., 244(60), 261
Nagel, R. L., 2(1), 44(1), 76
Nageswara Rao, B. D., 85(12), 86(12), 94(33), 96(38,39), 101, 102; 159(74), 202
Nakamura, H., 26(36), 77
Nakatani, T., 26(36), 77
Narayana, P. A., 144(41), 200; 211(10), 212(10), 219(10,29), 228(10), 229(44), 250(83), 259, 260, 261, 263
Nederveen, K., 224(37), 260
Nelson, M., 73(106), 80
Newman, B. D., 85(11), 101
Nguyen, A. C., 94(37), 95(37), 102
Nishikawa, T., 198(159), 206
Nitschke, W., 45(79,84), 64(84), 65(84), 79
Noack, M., 115(45), 127
Noda, L., 96(38), 102
Noguti, T., 198(159), 206
Norris, J. R., 211(13), 212(13), 228(13), 250(85), 257(95,96), 259, 263
Notstrand, B., 179(116), 204

O

O'Connor, S. E., 109(27), 127
Ohba, Y., 25(48), 31(48), 78

AUTHOR INDEX

Ohnishi, T., 146(49), 147(49), 149(49), 150(49), 167(94,95), 186(138), 187(141,142), 189(141), 201, 203, 205
Oltzik, R., 249(75), 262
Orbach, R., 131(5,10,11), 135(11,17), 136(5,10,11), 190(5), 195(154), 199, 205
Orme-Johnson, N. R., 244(61,62), 245(61,62), 261, 262
Orme-Johnson, W. H., 73(106, 107), 80; 166(86), 167(86), 202; 211(12), 244(60-63), 245(61,62), 250(84), 259, 261, 262, 263
Orsega, E. F., 33(52), 78
Orton, J. W., 133(2), 199
Oseroff, S. B., 15(23), 77
Osokin, D. Ya., 237(51), 239(51), 261
Otsuka, J., 152(66), 202
Ottavani, M. F., 86(17), 101

P

Pachense, J. M., 188(145), 205
Pake, G. E., 148(52), 150(52), 201
Palmer, G., 42(68), 79
Pappenhagen, T. L., 106(12), 108(12), 126
Parello, J., 98(46), 103
Parmeggiani, A., 90(26), 102
Parmon, V. N., 165(83), 202; 217(25), 219(25), 260
Pattison, M. R., 70(96), 80
Peake, B. M., 109(28), 127
Pearson, J. E., 86(19), 101
Pecht, I., 32(50), 78
Peisach, J., 24(41,43), 25(43), 28(41), 29(41,43), 30(41,43, 47), 31(43), 32(43), 33(43), 46(86), 72(101), 78, 80; 106(14), 115(14), 126; 165(82), 202; 211(8), 212(8,16), 221(8), 228(8,16), 236(50), 238(55), 241(50), 242(58), 243(58,59), 244(58,59,61,62), 245(61,62), 247(70,71), 248(55), 249(73-82), 250(82),
[Peisach, J.], 256(16,93), 257(8, 16,81), 259, 260, 261, 262, 263
Penfield, K. W., 32(50), 78
Percival, P. W., 155(69), 202
Pereyra, V., 110(36), 127
Perkins, R. E., 98(51), 103
Pershan, P. S., 2(2), 76; 139(27), 140(27), 243(27), 200
Perutz, M. F., 44(70), 79
Petef, M., 179(116), 204
Petelina, N. G., 179(117), 204
Petering, D. H., 24(37), 26(37), 77
Petersen, J. L., 22(31), 23(31), 24(31), 77
Peterson, R. L., 45(82), 64(82), 79
Petterson, G., 183(124), 204
Philipps, A. W., 98(50), 103
Phillips, S. E. V., 43(69), 65(93), 79, 80
Pickart, L., 249(80), 262
Plapp, B. V., 183(125), 204
Ponomarev, A. B., 258(98), 263
Poole, C. P., Jr., 14(19), 77; 131(4), 199
Portis, A. M., 139(25,26), 140(25), 141(25,26), 142(25), 147(25), 150(25), 200
Pound, R. V., 141(28), 142(28), 144(28), 154(28), 200
Poupko, R., 85(13), 86(18), 101
Powers, L., 30(47), 78
Prokhorov, A. M., 135(18), 200
Pronk, B. J., 221(33), 260
Provotorov, B. N., 142(31), 200
Purcell, E. M., 141(28), 142(28), 144(28), 154(28), 200
Pusep, A. Yu., 258(99), 263

R

Rabideau, S. B., 236(49), 261
Rainer, M. J. A., 107(21,23), 115(21,23), 126, 127
Raitsimring, A. M., 210(5), 221(31), 258(5), 259, 260
Ramakrishna, Y. V. S., 250(83), 263
Rammal, R., 195(155), 205

Ramshaw, J. A. M., 29(45), 78
Rao, K. K., 166(88), 167(88,90), 168(90), 190(90), 196(90), 197(90), 202, 203
Rao, R. V. S., 186(133), 188(133), 205
Raynor, J. B., 25(49), 31(49), 32(49), 33(49), 78; 105(1), 106(1), 114(1), 115(1), 126
Redfield, A. G., 186(132), 204
Reed, G. H., 83(6), 84(6,7), 85(6,12), 86(12,19), 89(22), 90(29), 91(29), 94(22,31,36,37), 95(36,37), 98(47,52), 99(52), 101, 102, 103
Rees, D. C., 182(122), 204
Reiman, C. W., 29(44), 78
Rein, H., 45(75), 79
Reinhammar, B., 24(43), 25(42,43), 29(42,43), 30(43), 31(43), 32(43), 33(42,43), 37(42), 78; 165(84,85), 202
Ribeiro, A., 98(46), 103
Rice, D. W., 98(49,50), 103
Richardson, D. C., 21(29), 38(29), 77
Richardson, J. S., 21(29), 38(29), 77
Riggs, A., 59(92), 67(92), 80
Rigo, A., 33(52), 78
Rist, G. H., 10(16), 23(33), 28(40), 76, 77, 78
Ristau, O., 45(75), 79
Rizos, A. K., 45(80,81), 79
Roberts, J. E., 24(41,43), 25(43,54,60), 28(41), 29(41,43), 30(41,43), 31(43), 32(43), 33(43), 35(54), 39(60), 71(98,99), 72(102,103), 73(106,107), 78, 80
Rode, B. M., 107(18-23), 108(18), 113(18), 115(18-23), 126, 127
Rodgers, M. E., 98(44), 103
Roger, G., 167(93), 190(93), 203
Romanelli, M., 86(17), 101; 219(29), 260
Rösch, P., 85(14), 86(14), 90(14), 91(14), 96(40,41), 101, 102
Rosevaer, P. R., 94(34), 102
Rossi, G. L., 183(127), 204

Rotilio, G., 33(52), 78
Roughton, F. J. W., 44(71), 79
Rowan, L. G., 217(23), 260
Rubin, B. H., 21(29), 38(29), 77
Rubinstein, M., 86(20), 101
Rupp, H., 166(88,89), 167(88,89,91), 202, 203
Rutishauser, H., 110(38), 127
Rutter, R., 72(102,103), 80; 173(108,109), 196(108), 203

S

Safin, I. A., 237(51), 239(51), 261
Safronov, S. N., 148(50), 150(50), 201
Salerno, J. C., 187(141), 189(141), 205
Salikhov, K. M., 208(2), 210(2,5), 211(2), 212(2), 217(24), 220(24), 221(2), 226(2), 228(2), 258(5,100,101), 259, 260, 263
Samama, J. P., 179(114), 204
Sammons, R. D., 94(36), 95(36), 102
Sanadze, T. I., 143(37), 200
Sander, S., 90(26), 102
Sands, R. H., 24(38), 25(58), 27(38), 37(58), 72(104), 77, 78, 80
Sarna, T., 146(48), 186(134,136,137), 201, 205
Scheer, H., 250(88), 251(88), 254(88), 255(88), 256(91), 263
Scheler, W., 45(75), 79
Schirmer, H., 96(40,42), 102
Schirov, M. D., 211(9), 257(95), 259, 263
Schlick, S., 160(77), 161(77), 162(77), 202
Schmidt, J., 225(38), 226(38), 227(43), 260, 261
Schneider, E. E., 144(45), 201
Schneider, G., 176(113), 179(113), 182(113), 204
Schneider-Bernlöhr, H., 163(81), 202
Schoenborn, P. B., 65(93), 80
Scholes, C. P., 2(1), 3(10), 4(10), 8(15), 15(21), 25(53,56),

AUTHOR INDEX

[Scholes, C. P.], 35(53,55), 36 (53,55,56), 44(1,10), 46(85), 48(85), 52(85), 54(85), 61(85), 62(85), 63(85), 64(85), 70(10, 15,97), 76, 77, 78, 80; 136 (24), 137(24), 169(98,102), 170 (102), 171(24), 172(24), 200, 203

Schonland, D. S., 7(13), 76

Schulz, G. E., 96(42), 102; 173 (109), 203

Schwartz, R. N., 162(78), 202

Schwarz, H. R., 110(38), 127

Schweiger, A., 2(4), 4(4), 7(4), 8(4), 10(4), 12(4), 19(4,24, 25), 22(4), 23(4,32), 40(4,61), 41(61), 42(61), 44(4), 70(4), 72(4), 76, 77, 79

Scott, C. A., 141(30), 142(30), 143(30), 163(30), 200

Scott, R. A., 37(57), 78

Scovil, H. E. D., 144(47), 201

Sealy, R. C., 24(37), 26(37), 77

Seidel, W., 15(20), 77

Seiter, C. H. A., 21(28), 77

Semenov, A. G., 208(2), 210(2), 211(2,9), 212(2), 221(2), 226 (2), 228(2), 259

Semin, G. K., 235(47), 237(47), 261

Shankle, G. E., 179(112), 204

Shapiro, E., 21(27), 77

Shapiro, S., 139(27), 140(27), 143(27), 200

Sharnoff, M., 29(44), 78

Shaw, P. J., 98(51), 103

Shaw, R. W., 35(55), 36(55), 37 (57), 78

Shiga, T., 45(74), 79

Shimizu, H., 160(76), 202

Shimizu, T., 256(93), 263

Shubin, A. A., 165(83), 202; 217 (25-27), 219(25), 234(26), 235 (26,27,48), 236(27,48), 260, 261

Sieker, L. C., 29(46), 78

Singel, D. J., 225(38), 226(38), 227(43), 260, 261

Slappendel, S., 152(65), 153 (65), 202

Sloop, D. J., 217(28), 260

Snipes, W., 162(79), 202

Soldatov, V. P., 221(32), 260

Solomon, E. I., 32(50), 78

Speck, S. H., 25(54), 35(54), 78

Spraggins, T. A., 109(27), 127

Srinivasan, R., 2(3), 76

Stamatoff, J., 188(145), 205

Standley, J., 131(3), 136(3), 139 (3), 150(3), 159(73), 160(3), 190(3), 199, 202

Stapleton, H. J., 131(5), 151(58-60), 154(58-60), 155(71), 167 (58-60), 171(104), 172(104), 173(108), 190(5,58,104,149, 150), 191(58), 192(58,60,150), 193(58), 194(59), 195(59), 196 (58-60,104,108,149,150,158), 198(59,60), 199, 201, 202, 203, 205, 206

Stein, G., 243(36), 269

Stenkamp, R. E., 29(46), 78

Stevens, K. W. H., 131(6), 136 (6), 139(6), 169(96,97,99-101), 199, 203

Stevens, T. H., 25(56), 36(56), 78

Stevenson, G. R., 106(7,10,11), 126

Stiefel, E., 110(38), 127

Stinson, D. G., 151(58), 154(58), 167(58), 190(58,149), 191(58), 192(58), 193(58), 196(58,149), 201, 205

Strandberg, M. W., 131(9), 199

Sullivan, P. D., 106(4), 126

Sutcliffe, L. H., 42(67), 79

Swanson, M., 25(54), 35(54), 78

Swartz, H. M., 146(48), 186(134, 135,137), 201, 205

Sylvestre, E. A., 110(35), 127

Symons, M. C. R., 3(11), 45(82, 83), 64(82,83), 65(83), 67(83, 94), 68(83), 69(11,94), 76, 79, 80

Szajdzinska-Pietek, E., 245(64-66), 262

Szumowski, J., 45(80,81), 79

T

Takabe, T., 25(48), 31(48), 78
Takita, T., 26(36), 77
Tan, S. L., 211(12), 244(60), 250(84), 259, 261, 263
Tanswell, P., 256(94), 263
Tarassov, V. V., 167(92), 203
Tasaki, A., 152(66), 202
Tavornina, A., 188(145), 205
Taylor, C. P. S., 215(74), 216(74), 219(74), 227
Taylor, H., 169(102), 170(102), 190(148), 203, 205
Thomas, K. A., 21(29), 38(29), 77
Thorkildsen, R., 151(61,63), 154(61,63), 155(63), 156(63), 171(61,63), 196(61), 201
Thornton, J., 256(94), 263
Thurnauer, M. D., 211(13), 212(13), 228(13), 250(85), 259, 263
Tiezzi, E., 106(9), 126
Toulouse, G., 195(155), 205
Trautwein, A., 172(106), 203
Trientham, D. R., 98(47), 103
Trifunac, A. D., 257(96,97), 263
Tsallis, C., 195(156), 206
Tsvetkov, Yu. D., 208(2), 210(2, 4), 211(2,6), 212(2,14,20), 219(30), 221(2,31,32), 225(20), 226(2), 228(2,14,20), 229(14, 20,44), 230(45), 232(46), 233(46), 235(48), 236(30,48), 237(52,53), 238(54), 239(52), 240(52), 246(67-69), 250(85-87, 89), 251(87,89), 252(89), 253(89), 254(87,88), 255(89), 258(4,98-101), 259, 260, 261, 262, 263
Tuite, M. F., 98(51), 103
Twilfer, H., 56(90), 80
Tyuma, I., 45(74), 79

U

Uenoyama, H., 171(103), 172(103), 203
Umezawa, H., 26(36), 77

Unruh, W. P., 133(14), 199
Urban, W., 85(11), 101
Urbancich, J., 52(5), 82
Utterback, S. G., 45(80), 79

V

Vaara, I., 179(116), 204
Van Camp, H. L., 15(21), 24(38), 25(53,58), 27(38), 35(53), 36(53), 37(58), 77, 78; 169(98), 203
Vandeberg, J. L., 98(48), 103
van der Poel, W. A. J. A., 225(38), 226(38), 227(43), 260, 261
van der Waals, J. H., 225(38), 226(38), 227(43), 260, 261
Vänngård, T., 28(40), 39(59), 78; 165(84,85), 172(105), 202, 203
van Ormondt, D., 221(33), 224(37), 260
van Vleck, J. H., 131(7,8), 199
Vasavada, K. V., 96(39), 102
Vaughan, R. A., 131(3), 136(3), 139(3), 150(3), 159(73), 160(3), 190(3), 199, 202
Veldink, G. A., 152(65), 153(65), 202
Venable, J. H., Jr., 162(80), 202
Venkatappa, M. P., 29(45), 78
Venters, R., 25(42), 29(42), 33(42), 37(42), 73(106), 78, 80
Viglino, P., 33(52), 78
Villafranca, J. J., 84(10), 101
Vliegenthart, J. F. G., 152(65), 153(65), 202
Vollmer, R. T., 106(13), 126

W

Wagner, E., 57(95), 67(95), 69(95), 80
Wagner, G. C., 151(60), 154(60), 167(60), 192(60), 196(60), 198(60), 201
Wajnberg, E., 136(23), 200
Waldstein, P., 236(49), 261
Walker, J. B., 94(32), 102

AUTHOR INDEX

Walker, N. P., 98(51), 103
Walsh, S. T., 250(84), 263
Wang, G., 73(105), 80
Wang, H., 25(56), 36(56), 78; 169(97), 203
Warden, J., 250(85), 263
Waring, A. J., 167(94,95), 187(142), 203, 205
Warwick, C. B., 151(64), 152(64), 201
Watson, H. C., 58(91), 64(91), 66(91), 67(91), 69(91), 80; 98(51), 103
Waugh, J. S., 211(12), 259
Webb, L. E., 23(34), 77
Webb, M. R., 90(29), 91(29), 98(47), 102, 103
Wedler, F. C., 84(10), 101
Wei, Y. H., 25(53), 35(53,55), 36(53,55), 46(85), 48(85), 52(85), 54(85), 61(85), 62(85), 63(85), 64(85), 78, 80; 169(98), 203
Weidner, R. T., 144(46), 201
Weinstein, B., 249(80), 262
Weissbach, H., 90(24), 102
Weissfloch, C. F., 133(13), 199
Weissman, S. I., 217(28), 260
Wells, G. B., 136(20,21), 144(43,44), 151(20,21,43,44), 152(21), 159(44), 162(43,44), 163(43,44), 164(43), 165(43,44), 174(20,21), 175(21), 176(21), 177(21), 179(21,43), 183(43,44), 184(43,128), 200, 201, 204
Wendell, P. L., 98(51), 103
Werner, E. R., 107(19,20,22), 115(19,20,22), 126, 127
Whiffen, D. H., 12(18), 77
Williams, R. J. P., 256(94), 263
Wilson, D. F., 188(143), 205
Wilson, G. E., 90(30), 102
Winkler, H., 173(109), 203
Wittinghofer, A., 85(14,15), 86(14), 87(15), 90(14,15,27,28), 91(14,15,27), 92(15), 93(15), 96(27), 101, 102
Wolf, W. P., 135(17), 199
Woodgate, J. H., 151(64), 152(64), 201

Y

Yakobson, G. G., 235(47), 237(47), 261
Yim, M. B., 136(19-22), 144(43,44), 145(22), 150(22), 151(19-22,43,44,55), 152(21,22), 159(44), 162(43,44), 163(43,44), 164(19,43,55), 165(43,44), 171(22), 172(22), 174(19-21), 175(21), 176(21,22,55), 177(21), 179(19-21,43,55), 183(19,43,44,55), 184(43,55,128), 200, 201, 204
Yokoi, H., 24(39), 25(48), 27(39), 28(39), 31(48), 77, 78
Yonetani, T., 40(64,65), 42(64), 45(78), 54(78), 71(100), 79, 80; 84(7), 101
Yoshida, H., 144(40), 200
Yu, H.-L., 217(28), 260
Yudanov, V. F., 212(14), 219(30), 220(31,32), 228(14), 229(14,44), 236(30), 259, 260, 261

Z

Zeppezauer, E., 183(127), 204
Zeppezauer, M., 163(81), 176(113), 179(113), 182(113), 202, 204
Zetter, M. S., 88(21), 91(21), 101
Zhidkov, O. P., 148(50), 150(50), 201
Zhidkov, V. D., 211(9), 259
Zhidomirov, G. M., 217(24), 220(24), 260
Zuberbühler, A. D., 109(29,30,32,33), 110(32,33), 111(33,39,41,42), 112(30,43), 113(30,43), 114(30,43), 118(41), 119(41), 120(46), 124(39,41,42), 127
Zuckerman, D., 211(7), 222(7), 259
Zverev, G., 179(117), 204
Zweier, J., 247(70,71), 262

Subject Index

A

Absorption bands and spectra (see also Spectrophotometry), 242, 256
  polarized, 188
Acetate, as ligand, 178
Acetylacetonate, 238
Acidity constants, 113, 114
Active site (see also Coordination spheres), 182, 184
Adenosine 5'-diphosphate, see 5'-ADP
Adenosine 5'-monophosphate, see 5'-AMP
Adenosine 5'-triphosphate, see 5'-ATP
Adenylate kinase, 96, 97
Adiabatic process, 138
5'-ADP, 93-98
  diastereoisomers, 94, 95
  ^{17}O-labeled, 94, 95, 98
Affinity constants, see Stability constants
Albumin, con-, 247, 249
Alcohols (see also individual names), 246
Alcohol dehydrogenase, liver, 163, 164, 174, 176, 182-184
Aldehyde(s), 184
  oxidase, 166
Algorithms, 107, 108, 118, 124, 223
  least squares (see also Least squares), 158
  Marquardt, 158
  Newton-Gauss-Marquardt, 111
Alkaline earth ions, see individual names
Allosteric effectors, 45, 49
Aluminum(III), ^{27}Al, 226
Amides, 3,7-diazanonanedioic acid di-, 112-120, 124
Amines (see also Amino groups and individual names)
  propyl-, 249
  4,7,10-triazatridecane-1,13-di-, 120-124
Amino acids, see individual names
Aminoacyl-tRNA, 90
Amino groups, 26
Ammonia, 27
5'-AMP, 96, 97
Anisotropic
  g tensor (see also Tensor), 19
  hyperfine coupling (or interaction), 7, 10, 11, 34, 217, 227-232, 234, 236, 240, 250, 255
  terms, 114, 122
Antibiotics (see also individual names), 90
Antiferromagnetic coupling, 21, 34
Arthropodes, 19
Ascorbate oxidase, 242, 249
Aspartic acid (and residues), 57, 59, 98
Association constants, see Stability constants
5'-ATP, 93-98, 107, 256
  ^{17}O-labeled, 89, 96, 97
  thio analogs, 97
Azide, 21, 37, 69, 190, 191, 196, 250
Azotobacter vinelandii, 73
Azurin, 20, 24, 28-32, 242

B

Bacillus subtilis, 167

[*Bacillus*]
  stearothermophilus, 90, 167, 168
Bacteria (*see also* individual names), 250-256
Bacteriopheophytin, 256
Barbituric acid, 162
Beans, 24
  soy-, *see* Soybean
Bicarbonate, *see* Hydrogen carbonate
Binding constants, *see* Stability constants
Binuclear complexes (*see also* Cluster), 34
Bleomycin, 22, 24-28, 249
Blood
  bovine, 38
  cells, 38
  human, 38
  serum, 38
Bohr
  effect, 44
  magneton, *see* Magnetic susceptibility measurements
Bonding geometry (*see also* Coordination spheres), 12, 45
Bovine
  blood cells, 38
  heart, 34, 35
  pancreatic trypsin inhibitor, 198
  superoxide dismutase, 21, 37, 38

C

Cadmium(II), 82
  acetate, 160, 161
Calcium acetate, 160, 161
Camphor, $\underline{D}$(+)-, 171, 172
Carbon, ^{13}C, 39, 163, 234, 247
Carbonate(s), 39
  hydrogen, 39, 247
Carbonic anhydrase, 178
Carboxypeptidase A, 163, 174, 176, 178, 180-182
Catalases, 42, 71
Catalytic sites, *see* Active site

Cells, 90
  blood, 38
Cerium(III), 226, 256
Ceruloplasmin, 20
Cesium ion, 29
Chelating ligands, *see* individual names
Chlorine
  ^{35}Cl, 4
  ^{37}Cl, 4
Chlorophyll, 250, 254
  bacterio-, 250-256
Cholesterol, 244
  20-aza-, 244
  2,2-hydroxy-, 244
Chromatium D, 166
Chromium (oxidation state not defined), 226
Chromium(III), 82, 159, 163
Clostridium pasteurianum, 72, 73
Cluster(s)
  Fe-S, *see* Ferredoxins and Iron
  multi-metal, 166-171
Cobalamin, 40
Cobalt (oxidation state not defined)
  ^{59}Co, 164
  ^{60}Co, 162
Cobalt(II), 21, 133, 249
  coordination numbers, 136, 178, 179, 183
  crystal field, 175
  substituted proteins, 40-42, 65, 163, 164, 173-184
Cobalt(III), 82
Colorimetry, *see* Absorption bands and spectra and Spectrophotometry
Computer
  desk, 113, 124
  programs, 109, 111, 158
Conformation(al)
  of proteins, 198
  transitions, 45
Cooperativity, 44
Coordination spheres (*see also* Active site), 1-75, 81-100, 115, 123, 124, 136, 163, 174, 176, 178, 179, 182, 183, 241
Copper (oxidation state not defined), 64, 165

SUBJECT INDEX

[Copper]
^{63}Cu, 24, 25, 29, 160, 161
^{65}Cu, 29, 39
type I, 20, 241, 242, 249
type II, 20, 242, 249
type III, 20, 37
Copper(I), 35-37
Copper(II), 2, 21-26, 106, 107, 112-124, 133, 162, 166, 169-171, 196, 223, 241, 242, 247-249
bis(oxychinolate), 23
bis(salicylaldoxine), 23
bleomycin, 24, 26, 27
proteins, 19-40, 241-243
substituted proteins, 19-40
tetraphenylporphyrin, 22
Correlation time, 85, 142, 143
Corrins, 40, 41
Coupling
antiferromagnetic, 21, 34
constant, 17, 23, 40, 84, 89, 91, 94, 97, 100, 106
dipolar, see Dipolar coupling
electron nuclear, see Electron-nuclear coupling
hyperfine, see Hyperfine coupling
isotropic proton, 35
^{14}N, 29, 40
orbit-lattice, 132
small proton, 19
spin-orbit, 132, 174, 175
tensor, 89, 99
Creatine, 93-96
kinase, 93-96
phosphate, 93-96
Cryogenic
studies, 154
temperatures, 15, 151, 256
Cryostat, 15-17, 151, 152
Crystal(s)
field (see also Ligand field), 171, 173-175
single, see Single crystals
Crystal structures, see X-ray
Cucurbita pepo medulosa, 249
Curie law, 152, 153
Cyanate, 69
thio-, 94
Cyanide, 21, 25, 37, 38, 44, 69
^{13}C, 163

Cysteine (and residues), 29-32, 36, 37, 64, 72, 184
radical, 169
Cytochrome(s), 42
a, 70
c, 169, 190, 249
c, ferri-, 152, 153, 190, 191
c, peroxidase compound ES, 71, 72
oxidases, see Oxidases
P-450, 190, 191, 249
P-450$_{cam}$, 171
P-450$_{scc}$, 244
Cytochrome c oxidase, 19-21, 25, 34-37, 46, 54
nitrosyl, 61-64
yeast, 70

D

Debye
limit, 135
temperature, 190, 197
Dehydrogenases
alcohol, see Alcohol dehydrogenase
succinate, 167
Deprotonation constants, see Acidity constants
Detergents, see Surfactants
Deuterium, 26, 31, 36, 235-237, 240-247
oxide, 33-35, 62, 242, 243, 246
Diamagnetic metal ions, see individual names
1,2-Diaminoethane, see Ethylenediamine
2,6-Diaminohexanoate, see Lysine
Diastereoisomers, 82, 94, 95, 98
3,7-Diazanonanedioic acid diamide, 112-120, 124
Diethylenetriamine, 248
Dihydronicotinamide adenine dinucleotide, see NADH
Dimeric complexes, see Binuclear complexes
Dioxygen, 32
as ligand, 40, 45, 64-69, 165, 169
reduction of, 169

Dipolar
  coupling (or interaction), 7, 11, 21, 36, 89, 141, 159-162, 166, 169, 171, 185, 196
  splitting, 89
Dismutases
  superoxide, see Superoxide dismutase
Dissociation constants, see Stability constants
Distribution curves, see Species distribution curves
Disulfide (and groups, see also individual names), 71
DPN, see NAD
DPNH, see NADH
Dysprosium(III), 167, 186

E

EDTA, see Ethylenediamine-N,N,N',N'-tetraacetate
Eigenvalues, 111, 119
Eigenvector
  analysis, 110, 112, 120
  decomposition, 111
  representation, 113, 114, 122
Electron-electron double resonance, 144
Electronic interactions, 212
  Zeeman, see Zeeman
Electron-nuclear coupling (or interaction), 159-162, 212, 213, 215, 234
Electron nuclear double resonance, see ENDOR
Electron paramagnetic resonance, see EPR
Electron spin, 88, 89, 214
  effective, 4
Electron spin echo
  applications, 165, 207-258
  envelope modulation, see Electron spin echo envelope modulation
  laser, 257
  signal decay, 212
  spectrometer, 211
  three pulse method, 210, 211
  two pulse method, 210, 211
  vector model of signal, 209
Electron spin echo envelope modulation
  anisotropic hyperfine interactions, 227-231
  arbitrary hyperfine interactions, 231-234
  autoregressive model, 224
  biological applications, 240-257
  ^{13}C modulation, 247
  frequency transformation, 221-224
  ^{1}H and ^{2}D modulations, 240-247
  in disordered systems, 227-240
  in single crystals, 224-227
  interaction of several nuclei, 219, 220
  ^{14}N and ^{15}N modulations, 248-256
  nuclear quadrupole interactions, 234-240
  ^{31}P modulation, 256, 257
  partial excitation, 220, 221
  primary, 215-217
  stimulated, 218, 219
  theory, 213-221
Electron spin resonance, see EPR
Electron transfer (or transport), 169, 242
  eukaryotic, 20
Elongation factor Tu, 90-93
ENDOR, 1-75, 169, 208, 217, 226, 227, 234, 251, 255, 256
  circularly polarized, 19
  cryogenic requirements, 15
  DOUBLE, 12, 18, 19
  -induced EPR, 12, 18
  instrumentation, 13-19
  theory, 4-13
  transitions, 5-7, 9
  TRIPLE, 12, 18, 19
Enzymes, see individual names
Enzyme-metal-substrate complexes, see Higher order complexes and Ternary complexes
EPR, 21, 26, 28, 29, 31-39, 42, 44-46, 49, 54, 61, 68, 69, 71-73, 226
  application of saturation methods, 129-198
  calculated spectra, 116, 123

SUBJECT INDEX

[EPR]
 continuous wave saturation technique, 144-154
 ENDOR-induced, 12, 18
 flow system, 108
 for studying solution equilibria, 105-125
 inhomogenous line broadening, 212
 line-broadening effects of ^{17}O, 81-100
 line intensity analysis, 107-111
 low frequency, 35
 overlapping spectra, 106, 107, 112, 124
 pulse saturation and recovering method, 151, 154-159, 167
 room temperature spectra, 115
 simulated absorption spectra, 147
 single crystal, 40, 71, 150, 154
 spectra, 47, 153, 164, 177, 180, 181
 spectral diffusion, 140-144
 titration, 107-109, 112-124
 transitions, 5, 6, 8, 9
Equilibria
 EPR studies, 105-125
 pH-dependent, 108, 109
 solution, 105-125
Equilibrium constants, see Acidity and Stability constants
Escherichia coli, 90
ESR, see EPR
Ethanol, 183, 184
 mercapto-, 249
Ethylenediamine, 27
Ethylenediamine-N,N,N',N'-tetraacetate, 186
Euclidian space, 193
Eukaryotes, electron transport, 20
EXAFS, 30, 35, 36
Extended absorption fine structure spectroscopy, see EXAFS

F

Factor analysis, 110, 111
 evolving, 111, 118-120, 124
Ferredoxins, 190, 244
 2Fe, 72
 2Fe-2S, 167, 168, 196, 197, 245
 4Fe-4S, 167, 168, 245
Ferricytochrome c, 191
 horse, 152, 153
 single crystal, 190
Flavin semiquinone radical, 166
Flavoproteins, 166
Fluoride, as ligand, 69, 145, 150, 171
Fluorine, ^{19}F, 234
Formate (or formic acid), 69, 94
Formation constants, see Stability constants
Fungi (see also individual names), laccase, 31-34

G

Galactose oxidase, 249
Gaussian distribution, 148
5'-GDP, 86, 87, 90, 92
 ^{17}O-labeled, 90
 thio analogs, 91
Geometries of coordination spheres, see Active sites and Coordination spheres
g-factor, 48, 52, 57, 63, 189
 nuclear, 4
Glutamine synthetase, 84
Glycera dibranchiata, 47, 49, 50, 52, 55, 57, 59, 67, 69
Glycine (and residues), 27
Glycyl-L-histidyl-L-lysine, 247
5'-GTP, 86, 87, 90, 91
 thio analogs, 91
Guanosine 5'-diphosphate, see 5'-GDP
Guanosine 5'-triphosphate, see 5'-GTP
Gyromagnetic ratio, 15

H

Halides, see individual names
Halogens, see individual names
Hamiltonian, 4, 5, 8, 23, 84, 85, 88, 114, 141, 185, 186, 188
Heart
  bovine, 34, 35
  horse, 152, 153
Heme(s), 10, 21, 22, 34, 36, 37, 44, 49, 167, 188, 243, 244, 249
  a, 169-171, 187, 189
  a_3, 46, 54, 61-64, 169, 187, 189
Heme proteins (see also individual names), 2, 40, 43-72, 134, 151, 167, 171-173, 186, 191, 194-198, 250
  iron(IV), 71, 72
  non-, 72, 73
Hemocyanin, 19
Hemoglobin(s), 3, 40-72, 250
  cobalt(II)-substituted, 40, 65
  deoxy, 44
  high-spin, 10
  human, 49
  hybrid, 45, 61
  low-spin, 10
  met-, 10, 158, 159, 243
  mutant, 45
  nitrosyl-, 44, 47, 49-61
  oxy-, 44, 45, 64-69
Hexacyanoferrate(III), 187
Higher order complexes (see also Ternary complexes), 90-99
Histidine (and residues), 21, 31, 34, 36, 37, 41, 43, 48, 49, 51, 52, 54, 57-59, 61-63, 65-67, 69-71, 184, 247, 249
Horse
  ferricytochrome c, 152, 153
  heart, 152, 153
  liver, 184
  muscle, 98
Horseradish peroxidase
  compound I, 71, 72, 172, 173
  compound II, 172, 173

Human
  blood, 38
  hemoglobins, 49
  transferrin, 25, 38-40
Hydrogen, ^2H, see Deuterium
Hydrogenases, 72, 73, 250
Hydrogen bonds, 41, 57-59, 65, 67, 184, 226, 249, 250
Hydrogen carbonate (see also Carbonate), 39, 247
Hydroperoxide, 68, 71
Hydrogen peroxide, 71, 72
Hydroxides, as ligands, 10, 69, 191
Hyperfine
  anisotropy, 7, 10, 11, 34, 217, 227-232, 234, 236, 240, 250, 255
  broadening, 143
  coupling (or interaction), 2-4, 6-9, 12, 18, 20, 22-25, 27, 29, 31-34, 41, 46, 48, 51, 52, 54-57, 61-63, 84, 115, 124, 132, 160, 165, 215, 224-234, 243, 247-251, 255, 256
  structure, 141, 143, 144, 164
  super-, see Superhyperfine
  tensor, 4, 8, 10, 19, 22, 23, 28, 31, 39, 41, 53, 62, 70
  transitions, 10

I

Imidazolate bridge, 21, 37
Imidazole (and moieties, see also Histidine), 21, 24, 26-28, 30-32, 37, 46-49, 51, 54, 55, 178, 243, 248, 249
  benz-, 40, 41
  nitroxide, 247
Inositol hexaphosphate, 49
Instrumentation
  electron spin echo spectrometer, 211
  ENDOR, 13-19
  for EPR saturation methods, 155, 156
Invertebrates, 43

SUBJECT INDEX

Iodide, 106
Ion pair formation, 106
Iron (oxidation state not defined), 21, 34-37, 131, 243
^{57}Fe, 45, 64, 73
-sulfur proteins (see also Ferredoxins), 72, 73, 166-168, 186, 195-198, 244, 245, 249
tetraphenylporphyrin, 22, 55, 64
Iron(II), 44-49, 61-64, 187, 196
tetraphenylporphyrin, 46-48
Iron(III), 21, 44, 69-71, 73, 133, 151, 153, 169, 173, 187, 190, 247, 249, 250
hexacyanoferrate(III), 187
high-spin, 10, 69, 152, 163, 171, 172
low-spin, 10, 34, 69, 134, 152, 163, 171, 190, 191, 194, 195, 249
tetraphenylporphyrin, 44
Iron(IV), 21, 172, 173
heme, 71, 72
Iron(V), 172

K

Kinases
adenylate, 96, 97
creatine, 93-96
3-phosphoglycerate, 98, 99
Kirromycin, 90
Kramer's
doublets, 135, 171-184
ions, 131, 133, 190-192

L

Laccase, 20, 25, 32-34, 37, 165, 242, 249
fungal, 31-34
tree, 31, 32
Lac tree (see also individual names), 20, 32, 33
Lanthanide ions, see individual elements

Larmor precession, 213
Least squares
minimization procedure, 110
nonlinear program, 109
procedure, 111, 120, 158, 222
Leucine (and residues), 49, 59, 67
Ligand field (see also Crystal field), 115, 122
Ligands, see individual names
Linear regression (see also Least squares), 109
Lipoxygenase, soybean, 152, 153
Liver
alcohol dehydrogenase, 163, 164, 174, 176, 178, 182-184
horse, 184
Lorentzian
distribution, 148
lineshape, 141
Lysine (and residues), 247, 249

M

Magnesium ion, 82, 83, 90, 96-99
Magnetic resonance
electron para-, see EPR
nuclear, see NMR
optically detected, 226
Magnetic susceptibility measurements, 171, 173, 178, 179
Mammalian
cytochrome oxidases, 169, 187-189
muscle, 96
Manganese(II), 106, 124, 144, 187
high spin, 84
^{55}Mn, 84
^{17}O superhyperfine coupling, 81-100
Marquardt algorithms, 111, 158
Membranes, resonance x-ray diffraction studies, 188
Mercaptoethanol, 249
Mercapto groups and ligands, see Thiols and individual names
Metalloproteins (see also Active sites and Proteins), 19-75,

[Metalloproteins], 134, 136, 151, 158, 162, 166
Methanobacterium thermoautotrophicum, 250
Methanol, 246
Methionine (and residues), 29, 30, 32
Methyl viologen, 250
Methemoglobins, see Hemoglobins
Metmyoglobins, see Myoglobins
Micelles, 246
Microorganisms (see also individual names), 20, 90
Mitochondria, respiratory chain, 169
Mixed ligand complexes, see Higher order complexes and Ternary complexes
Mollusks, 19
Molybdenum (oxidation state not defined), 73
^{95}Mo, 73
Molybdenum(V), 166, 167
Mössbauer spectroscopy, 42, 73, 172
Muscles
  contractile filaments, 98
  horse, 98
  mammalian, 96
  porcine, 96
Myoglobins, 3, 40, 42-72, 198, 249
  cobalt(II)-substituted, 40
  met-, 70, 136, 137, 145, 150, 163, 171, 172, 190, 191, 196, 243
  nitrosyl-, 44, 46, 47, 49-61, 196
  oxy-, 45, 64-69
  single crystals, see Single crystals
  sperm whale, 49
Myosin subfragment 1, 98

N

NAD$^+$, 182-184
NADH, 178, 183, 184
Neodymium(III), 133, 241, 256
Neutron diffraction, 65

Newton-Gauss-Marquardt algorithms, 111
Newton-Raphson method, 109
Nickel, hydrogenases, 250
Nickel(II), 166, 187
Nickel(III), 106
Nicotinamide adenine dinucleotide, see NAD$^+$
Nicotinamide adenine dinucleotide reduced, see NADH
Nitrate, 94
Nitrogen
  ^{14}N, 4, 8, 10, 11, 22-31, 33, 35-38, 40, 46, 48, 50-53, 56, 61-63, 68, 70, 72, 225, 226, 237, 240, 243, 248-256
  ^{15}N, 4, 36, 38, 61, 63, 70, 234, 240, 248-256
Nitrogenases, 73
Nitroxide(s)
  as ligand, 44, 46-64, 169, 187, 196
  radicals, 44, 46, 239, 245-247
  spin labels, 187
NMR, 33, 42, 82, 83, 96, 154, 160, 256
^{31}P, 94, 96
Nuclear
  electron coupling, see Electron nuclear coupling
  g-factor, 4
  interaction, 3, 4, 8, 23-25, 84, 88, 114, 159, 215, 217, 228, 234, 237-240
  magnetic moments, 18
  quadrupolar coupling, 31, 32, 53, 215, 217, 224-226, 234-242, 250, 256
  quadrupole resonance, 226, 248, 250, 251
  quantum transition induction, 19
  relaxation time, 144
  spin, 88, 89, 114, 214, 215
  spin decoupling, 19
  Zeeman splitting, 29, 35
Nuclear magnetic resonance, see NMR
Nucleoside triphosphates (see also individual names), 81-100

Nucleotides (see also individual names), metal-protein complexes, 81-100

O

Octaethylporphyrin, 46, 48, 49
Oligonuclear complexes, see Cluster
Optically detected magnetic resonance, 226
Optical spectra, see Absorption bands and spectra
Orbach process, 132, 133, 135-137, 151, 168, 171, 173, 176, 190, 196, 197
Orthophosphate, see Phosphate
Oxalate (or oxalic acid), 247
Oxidases, 19, 242
  aldehyde, 166
  ascorbate, 242, 249
  blue, 32
  cytochrome, 169, 170, 187-189, 196
  cytochrome c, see Cytochrome c oxidase
  galactose, 249
  mixed function, 171
  xanthine, 167
Oxidoreductase, succinate-ubiquinone, 249
Oxygen (see also Dioxygen)
  ^{16}O, 22, 23, 88, 89, 164
  ^{17}O, 41, 45, 64, 72, 89, 90, 96, 97, 165, 166, 181, 182
  ^{17}O-enriched water, 91, 92, 98, 163, 164, 184, 243
  ^{17}O-Mn(II) superhyperfine coupling, 81-100
  ^{18}O, 88
  ^{18}O water, 163, 164
  radical, 165, 166
Oxygenase, lip-, see Lipoxygenase
Oxygen donor ligands (see also individual names), identification of, 81-100

P

P. putida, 171, 190
Pancreatic trypsin inhibitor, 198
Paramagnetic metal ions, see individual names
Peptidases, see Carboxypeptidase A
Peptides (see also individual names), 106, 107
  poly-, 90, 134, 195
Peroxidases, 42
  cytochrome c, 71, 72
  horseradish, 71, 72, 172, 173
Peroxides (see also Hydrogen peroxide and Hydroperoxides), 71
Phenolates (or phenols, and phenolic groups), 39
Phenylalanine (and residues), 43, 67, 68
Phosphate (including hydrogen- and dihydrogenphosphate), 92, 98, 162
  creatine, 93-96
  inositol hexa-, 49
Phosphates (and groups, see also individual names), 81-100
  thio analogs, 82, 91, 97
3-Phosphoglycerate kinase, 98, 99
Phosphorus (oxidation state not defined)
  ^{31}P, 234, 256, 257
  ^{31}P-NMR, see NMR
Photometry, see Absorption bands and spectra
Photosynthesis, system I, 166, 251
Phthalocyanine, 2
2-Picoline-N-oxide, 178
Piperidine, 246, 247
Plants (see also individual names), photosystem, 251
Plastocyanin(s), 24, 28-32, 37, 242
PMR, see NMR
Point dipoles, 11, 12
Polynuclear complexes, see Cluster

Polyporus versicolor, 32
Polypeptides (*see also* Peptides
  and Proteins), 90, 134, 195
Porcine muscle, 96
Porphyrins (*see also* Hemes and
  individual names), 11, 40,
  41, 243, 249
  π-cation radical, 172, 173
  octaethyl, 46, 48, 49
  tetraphenyl-, 22, 24, 44, 46-
    49, 55, 64
Potentiometric measurements,
  107, 108, 112, 114, 120, 124
Propionic acid (or residue), 70
Proteins (*see also* individual
  names), 129-198
  cobalt substituted, 40-42, 65,
    163, 164, 173-184
  copper(II), 19-40, 241-243
  copper(II) substituted, 19-40
  flavo-, 166
  heme, *see* Heme proteins
  iron-sulfur (*see also* Ferre-
    doxins), 72, 73, 166-168,
    186, 195-198, 244, 245, 249
  metal ion clusters, 166-171
  metallo-, *see* Metalloproteins
  metal-nucleotide complexes,
    81-100
  non-heme, 72, 73
  structure, 190-198
  three dimensional structure,
    192
Protonation constants, *see*
  Acidity constants
Proton magnetic resonance, *see*
  NMR
Pyridine, 27, 46, 249
  N-oxide, 179
Pyrrole (and residues), 41, 46,
  49
  nitrogen, 50, 54, 60, 69, 70,
    251
  proton, 23
  ring-nonplanarity, 226

Q

Quadrupolar
  coupling (or interaction), 3,
    8, 9, 23-25, 27-29, 31, 32,
    34, 41, 49, 51-53, 63, 88,
    215, 217, 224-226, 234-242,
    250, 256
  resolution, 28
  resonance, 226, 248, 250, 251
  tensor, 8, 53, 70, 224, 226

R

Radical(s), 212, 225, 239, 240,
  246, 247, 256
  anion, 106
  cation, 250, 251
  cysteine, 169
  flavin semiquinone, 166
  free, 71, 72, 162
  nitroxide, 44, 46, 239, 245-247
  organic, 17, 64, 106, 107, 162,
    186
  oxygen, 165, 166
  peroxy, 71
  piperidine, 246, 247
  porphyrin π-cation, 172, 173
  spin label, 187, 245, 246
  thioether, 71
  thiyl, 35, 36
  tryptophanyl, 71
Radiation, ionizing, 45
Raman
  process, 132-136, 151, 166-168,
    170, 176
  spectroscopy, 42
  spin-lattice relaxation rates,
    190-198
Rare earth ions, *see* individual
  names
Reductase, succinate-ubiquinone
  oxido-, 249
Relaxation
  cross-, 136-141, 143
  mechanisms, 86, 90, 159

SUBJECT INDEX

[Relaxation]
rates, 136, 166, 167, 172, 190-198
spin-lattice, 131-136, 142-160, 162, 166-171, 173, 178, 179, 185-187, 190-198, 210
spin-spin, 136-140, 210
theory, 130-144
Relaxation times, 133, 148, 150, 151, 154, 160, 185, 187, 211
nuclear, 144
spin-lattice, 15, 155, 156, 165
spin-spin, 15
Rhus vernicifera, 28, 32-34, 249
Ribonucleic acid, see RNA
Ribosomes, 90
RNA, aminoacyl-tRNA, 90

S

Serum, blood, 38
Single crystals, 6, 8, 44, 45, 51, 61, 162, 223, 258
cytochrome c, 190
electron spin echo in, 224-227
ENDOR spectroscopy, 12
EPR, 40, 71, 150, 154
host, 22
-like, 10
metmyoglobin, 70, 137, 150
nitrosyl myoglobin, 49, 53, 55
oxymyoglobin, 64
Soybean lipoxygenase, 152, 153
Species distribution curves, 111, 114, 119-121, 124
Spectrophotometry (see also Absorption bands and spectra)
titrations, 109, 112, 120, 124
Sperm whale myoglobin, 49
Spin
density, 12, 45, 58, 62, 64, 70, 118, 124, 169, 247
dynamics, 29
electron, 4, 88, 89, 214
labels, 187, 245, 246

[Spin]
-lattice relaxation, 15, 131-136, 142-160, 162, 165-171, 173, 178, 179, 185-187, 190-198, 210
nuclear, 88, 89, 114, 214, 215
-orbit coupling, 132, 174, 175
-spin relaxation, 15, 136-140, 210
Spinach, 166
plastocyanin, 25, 31
Spirulina maxima, 167, 168, 196, 197
Stability constants, 106, 111, 113-115, 117, 120, 124
mathematical methods, 109-111
Stellacyanin, 20, 24, 28, 29, 31, 32, 242, 249
Stokes-Einstein relation, 85
Succinate
dehydrogenase, 167
ubiquinone oxidoreductase, 249
Sulfhydryl groups, see Thiols
Sulfur ligands (see also Thiols and individual names), 29, 30, 35-37, 72, 82, 91, 97, 166, 167
Superhyperfine
$Mn(II)$-^{17}O coupling, 81-100
splitting, 117, 169
structure, 26, 39, 72
Superoxide dismutase(s), 20, 25, 249
bovine, 21, 37, 38
Surfactants, deuterated headgroup, 246
Susceptibility, see Magnetic susceptibility measurements
Synthetase, glutamine, 84

T

Tensor
anisotropic g, 19
complete set, 22
coupling, 89, 99

[Tensor]
  g, 20, 22, 33, 36, 45-47, 54, 72, 115, 162
  hyperfine, 4, 8, 10, 19, 22, 23, 28, 31, 39, 41, 53, 62, 70
  quadrupole, 8, 53, 70, 224, 226
  symmetry, 32
  zero field splitting-, 132
Ternary complexes (see also Higher order complexes), 91-93, 96, 97, 99, 107, 182-184
1,2',2''-Terpyridine, 106
Tetraphenylporphyrin, 22, 24, 44, 46-49, 55, 64
Thiocyanate, 94
Thioethers (see also Sulfur ligands), 71
Thiols (and thiolate groups, see also individual names), 30, 169
Thiophosphates, 82, 91, 97
Titration
  concentration profiles, 111
  curves, 113, 115, 117, 119, 121, 122
  EPR, 107-109, 112-124
  model-free species distribution, 111
  pH-, see Potentiometric measurements
  redox, 187
  spectrophotometric, 109, 112, 120, 124
  spectroscopic, 111, 118
Transferrins, 247
  human, 25, 38-40
4,7,10-Triazatridecane-1,13-diamine, 120-124
Trypsin inhibitor, bovine pancreatic, 198
Tryptophan (and residues), 71
Tungsten, ^{183}W, 226
Tyrosine (and residues), 39

V

Valine (and residues), 43
Vanadium(IV), 238

Vertebrates, 43
Visible spectra, see Absorption bands and spectra
Vitamin B_{12}, 40, 41

W

Wolfram, see Tungsten

X

Xanthine oxidase, 167
X-ray
  absorption edge spectroscopy, 35
  crystal structures, 11, 21, 23, 26, 29, 31, 37, 43, 60, 64, 66, 67, 69, 72, 96, 98, 176, 182, 183, 243, 245
  extended absorption fine structure spectroscopy, see EXAFS
  resonance x-ray diffraction studies, 188

Y

Yeast, 35, 36, 70, 98

Z

Zeeman
  electronic interaction, 84
  frequency, 229, 230, 236, 243, 245, 255
  nuclear interaction, 84, 88, 159, 215, 217, 228, 234, 237-240
  nuclear splitting, 29, 35
  temperature, 142
  terms, 4, 28, 51
Zero field splitting, 84-86, 132, 135-137, 171, 173, 174, 176, 177, 183, 184
Zinc(II), 21, 29, 37, 144, 173, 176, 177, 249
  tetraphenylporphyrin, 22

RAYMOND H. FOGLER LIBRARY
DATE DUE